METHODS IN MOLECULAR BIOLOGY™

AF769077

Series Editor
John M. Walker
School of Life Sciences
University of Hertfordshire
Hatfield, Hertfordshire, AL10 9AB, UK

For further volumes:
http://www.springer.com/series/7651

Peroxisome Proliferator-Activated Receptors (PPARs)

Methods and Protocols

Edited by

Mostafa Z. Badr

School of Pharmacy, University of Missouri-Kansas City, Kansas City, Missouri, USA

Jihan Youssef

University of Missouri-Kansas City, Kansas City, Missouri, USA

Humana Press

Editors
Mostafa Z. Badr
School of Pharmacy
University of Missouri-Kansas City
Kansas City, Missouri, USA

Jihan Youssef
University of Missouri-Kansas City
Kansas City, Missouri
USA

ISSN 1064-3745 ISSN 1940-6029 (electronic)
ISBN 978-1-4939-6257-0 ISBN 978-1-62703-155-4 (eBook)
DOI 10.1007/978-1-62703-155-4
Springer New York Heidelberg Dordrecht London

© Springer Science+Business Media New York 2013
Softcover re-print of the Hardcover 1st edition 2013
This work is subject to copyright. All rights are reserved by the Publisher, whether the whole or part of the material is concerned, specifically the rights of translation, reprinting, reuse of illustrations, recitation, broadcasting, reproduction on microfilms or in any other physical way, and transmission or information storage and retrieval, electronic adaptation, computer software, or by similar or dissimilar methodology now known or hereafter developed. Exempted from this legal reservation are brief excerpts in connection with reviews or scholarly analysis or material supplied specifically for the purpose of being entered and executed on a computer system, for exclusive use by the purchaser of the work. Duplication of this publication or parts thereof is permitted only under the provisions of the Copyright Law of the Publisher's location, in its current version, and permission for use must always be obtained from Springer. Permissions for use may be obtained through RightsLink at the Copyright Clearance Center. Violations are liable to prosecution under the respective Copyright Law.
The use of general descriptive names, registered names, trademarks, service marks, etc. in this publication does not imply, even in the absence of a specific statement, that such names are exempt from the relevant protective laws and regulations and therefore free for general use.
While the advice and information in this book are believed to be true and accurate at the date of publication, neither the authors nor the editors nor the publisher can accept any legal responsibility for any errors or omissions that may be made. The publisher makes no warranty, express or implied, with respect to the material contained herein.

Printed on acid-free paper

Humana Press is a brand of Springer
Springer is part of Springer Science+Business Media (www.springer.com)

Preface

All three peroxisome proliferator-activated receptor (PPAR) subtypes—PPARα, PPARβ/δ, and PPARγ—share a high degree of structural homology yet they differ in function, tissue distribution, and ligand specificity. Since these receptors play critical roles as regulators of numerous physiological as well as pathophysiological pathways, significant efforts are currently underway to fully characterize their functioning and to develop safer and more effective PPAR modulators as therapeutic agents to treat a myriad of diseases and conditions.

This volume of *Methods in Molecular Biology* contains details of experimental protocols essential to the task of investigating these receptors. This provides researchers in the PPAR arena with a wide array of advanced techniques which, no doubt, will prove to be valuable to their efforts of advancing knowledge in the field. Chapters contributed by renowned experts contain details of methods ranging from the cloning of receptors to their knockdown. In addition, experimental protocols to explore posttranslational modifications of PPARs and coactivators as well as receptor subcellular localization are included. Furthermore, specific steps on how to screen a chemical moiety for its ability to modulate PPAR subtype activities and the quantification of ligand metabolites in biological samples are presented. In addition, methods to evaluate an emerging facet in PPAR research, namely the involvement of these receptors in behavior functions, are also detailed in this volume.

We are grateful to Dr. John Walker, the *Methods in Molecular Biology* series editor, for his kind invitation to organize this volume devoted to PPARs. We are also most indebted to all colleagues who have shared their expertise and vast experience with the world-wide PPAR research community through the chapters contributed to this volume.

Mostafa Z. Badr
Jihan Youssef
MO, USA

Contents

Contributors

TILL ADHIKARY • *Institute of Molecular Biology and Tumor Research (IMT), Philipps University, Marburg, Germany*
GERARD ARAGONÈS • *Unitat de Recerca Biomèdica (URB-CRB), Institut d'Investigació Sanitària Pere Virgili, Universitat Rovira i Virgili, Reus, Spain*
MOSTAFA Z. BADR • *University of Missouri-Kansas City, Kansas City, MO, USA*
MYUNG AE BAE • *Drug Discovery Platform Technology Team, Medicinal Science Division, Korea Research Institute of Chemical Technology, Daejon, South Korea*
RAÚL BELTRÁN-DEBÓN • *Unitat de Recerca Biomèdica (URB-CRB), Institut d'Investigació Sanitària Pere Virgili, Universitat Rovira i Virgili, Reus, Spain*
ROBERT G. BENNETT • *VA Nebraska and Western Iowa Health Care System, Omaha, NE, USA; Departments of Internal Medicine and Biochemistry and Molecular Biology, University of Nebraska Medical Center, Omaha, NE, USA*
SIMON BEUCK • *Institute of Biochemistry, Center for Preventive Doping Research, German Sport University Cologne, Cologne, Germany*
JORDI CAMPS • *Unitat de Recerca Biomèdica (URB-CRB), Institut d'Investigació Sanitària Pere Virgili, Universitat Rovira i Virgili, Reus, Spain*
JI SUK CHANG • *Laboratory of Nutrient Sensing and Adipocyte Signaling, Pennington Biomedical Research Center, Baton Rouge, LA, USA*
CATHERINE COIRAULT • *INSERM UMR-S 974, UM76 Univ. Paris 6, UMR7215, Institut de Myologie, Assistance Publique-Hôpitaux de Paris, Paris, France*
MAURIZIO CRESTANI • *Laboratorio "Giovanni Galli" di Biochimica e Biologia Molecolare del Metabolismo-Spettrometria di Massa, Dipartimento di Scienze Farmacologiche e Biomolecolari, Università degli Studi di Milano, Milan, Italy*
OSAMU EZAKI • *Nutritional Science Program, National Institute of Health and Nutrition, Tokyo, Japan*
THOMAS W. GETTYS • *Laboratory of Nutrient Sensing and Adipocyte Signaling, Pennington Biomedical Research Center, Baton Rouge, LA, USA*
FEDERICA GILARDI • *Center for Integrative Genomics, University of Lausanne, Lausanne, Switzerland*
MARCO GIUDICI • *Laboratorio "Giovanni Galli" di Biochimica e Biologia Molecolare del Metabolismo-Spettrometria di Massa, Dipartimento di Scienze Farmacologiche e Biomolecolari, Università degli Studi di Milano, Milan, Italy*
CORINE GLINEUR • *INSERM U1011, UDSL, Institut Pasteur de Lille, University Lille Nord de France, Lille, France; INSERM U1019-CNRS UMR 8204, Institut Pasteur de Lille, Center for Infection and Immunity of Lille, University Lille Nord de France, Lille, France*
CRISTINA GODIO • *Department of Chemical Physiology, The Scripps Research Institute, La Jolla, CA, USA*
DANA GREENE-SCHLOESSER • *Departments of Radiation Oncology and Cancer Biology, Wake Forest School of Medicine, Winston-Salem, NC, USA*

Aziz Guellich • *Department of Cell Biology and Molecular Medicine, Cardiovascular Research Institute, New Jersey Medical School, University of Medicine and Dentistry of New Jersey, Newark, NJ, USA; INSERM UMR-S 769, LabEx LERMIT, Université Paris-Sud 11, Châtenay-Malabry, France*
Viera Habalová • *Department of Medical Biology, Medical Faculty of P.J. Šafárik University, Trieda SNP 1Košice, Slovak Republic*
Weimin He • *Center for Cardiovascular Diseases, College of Pharmacy and Health Sciences, Texas Southern University, Houston, TX, USA*
Christian H. Ihling • *Department of Pharmaceutical Chemistry and Bioanalytics, Institute of Pharmacy, Martin-Luther-Universität Halle-Wittenberg, Halle (Saale), Germany*
Jorge Joven • *Unitat de Recerca Biomèdica(URB-CRB), Institut d'Investigació Sanitària Pere Virgili,Universitat Rovira i Virgili, Reus, Spain*
Min-Sun Kim • *Drug Discovery Platform Technology Team, Medicinal Science Division, Korea Research Institute of Chemical Technology, Daejon, South Korea*
Nobutaka Kiyokawa • *Department of Pediatric Hematology and Oncology Research, National Research Institute for Child Health and Development, Tokyo, Japan*
Carolyn M. Komar • *Department of Biological Sciences, West Virginia School of Osteopathic Medicine, Lewisburg, WV, USA*
Fabio Lannutti • *Dipartimento di "Scienze del Farmaco", Università "G. d'Annunzio" di Chieti-Pescara, Chieti, Italy*
Jian Liu • *Department of Nutrition Sciences, University of Alabama at Birmingham, Birmingham, AL, USA*
Mary Jane Long • *Veterinary Medicine, Microbiology Department, College of Veterinary Medicine, Iowa State University, Ames, IA, USA*
Alessandro Marrone • *Dipartimento di "Scienze del Farmaco", Università "G. d'Annunzio" di Chieti-Pescara, Chieti, Italy*
Ross A. McKinnon • *Flinders university, Bedford Park, SA, Australia*
Nico Mitro • *Laboratorio "Giovanni Galli" di Biochimica e Biologia Molecolare del Metabolismo-Spettrometria di Massa, Dipartimento di Scienze Farmacologiche e Biomolecolari, Università degli Studi di Milano, Milan, Italy*
Yoshitaka Miyagawa • *Department of Pediatric Hematology and Oncology Research, National Research Institute for Child Health and Development, Tokyo, Japan*
Ines Möller • *Institute of Biochemistry, Center for Preventive Doping Research, German Sport University Cologne, Cologne, Germany*
Mathias Q. Müller • *Department of Pharmaceutical Chemistry and Bioanalytics, Institute of Pharmacy, Martin-Luther-Universität Halle-Wittenberg, Halle (Saale), Germany*
Suong N.T. Ngo • *The University of Adelaide, Roseworthy, SA, Australia*
Noor Akmal Syed Husain Sharifah • *Department of Pathology, Universiti Kebangsaan Malaysia, Jalan Yaacob Latif, Bandar Tun RazakKuala Lumpur, Malaysia*
Hajime Okita • *Department of Pediatric Hematology and Oncology Research, National Research Institute for Child Health and Development, Tokyo, Japan*
Dolores Pérez-Sala • *Centro de Investigaciones Biológicas, CSIC, Madrid, Spain*
Benoit Pourcet • *Center for Clinical Pharmacology, Division of Medicine, University College of London, London, UK; INSERM U1011, UDSL, Institut Pasteur de Lille, University Lille Nord de France, Lille, France*
Nazzareno Re • *Dipartimento di "Scienze del Farmaco", Università "G. d'Annunzio" di Chieti-Pescara, Chieti, Italy*
Mike E. Robbins • *Department of Radiation Oncology, Wake Forest School of Medicine, Winston-Salem, NC, USA*

ESTHER RODRÍGUEZ-GALLEGO • *Unitat de Recerca Biomèdica (URB-CRB), Institut d'Investigació Sanitària Pere Virgili, Universitat Rovira i Virgili, Reus, Spain*
ANNA RULL • *Unitat de Recerca Biomèdica (URB-CRB), Institut d'Investigació Sanitària Pere Virgili, Universitat Rovira i Virgili, Reus, Spain*
WILHELM SCHÄNZER • *Institute of Biochemistry, Center for Preventive Doping Research, German Sport University Cologne, Cologne, Germany*
CAROLINE I. SCHNEGG • *Department of Cancer Biology and Brain Tumor Center of Excellence, Wake Forest School of Medicine, Winston-Salem, NC, USA*
ELENA SCOTTI • *Howard Hughes Medical Institute, University of California at Los Angeles, Los Angeles, CA, USA*
SUDHIR SINGH • *VA Nebraska and Western Iowa Health Care System, Omaha, NE, USA; Departments of Internal Medicine and Biochemistry and Molecular Biology, University of Nebraska Medical Center, Omaha, NE, USA*
ANDREA SINZ • *Department of Pharmaceutical Chemistry and Bioanalytics, Institute of Pharmacy, Martin-Luther-Universität Halle-Wittenberg, Halle (Saale), Germany*
BART STAELS • *INSERM U1011, UDSL, Institut Pasteur de Lille, University Lille Nord de France, Lille, France*
XIAO-FENG SUN • *Department of Oncology, Institute of Clinical and Experimental Medicine, University of Linköping, Linköping, Sweden*
TEAYOUN KIM *Department of Nutrition Sciences, University of Alabama at Birmingham, Birmingham, AL, USA*
MARIO THEVIS • *Institute of Biochemistry, Center for Preventive Doping Research, German Sport University Cologne, Cologne, Germany*
WAI KIT CHIA • *Department of Diagnostic Laboratory Services, Universiti Kebangsaan Malaysia Medical Centre (UKMMC), Jalan Yaacob Latif, Bandar Tun Razak Kuala Lumpur, Malaysia*
TOMOMI YAMAZAKI • *Nutritional Science Program, National Institute of Health and Nutrition, Tokyo, Japan*
QINGLIN YANG • *Department of Nutrition Sciences, University of Alabama at Birmingham, Birmingham, AL, USA*
LIE YANG • *Institute of Digestive Surgery and State Key Laboratory of Biotherapy, West China Hospital, Sichuan University, Chengdu, Sichuan Province, China*
JIHAN YOUSSEF • *University of Missouri-Kansas City, Kansas City, MO, USA*
ZUBAIDAH ZAKARIA • *Division of Hematology, Institute for Medical Research, Kuala Lumpur, Malaysia*
OLGA ZHELYABOVSKA • *Department of Nutrition Sciences, University of Alabama at Birmingham, Birmingham, AL, USA*
ZONG-GUANG ZHOU • *Institute of Digestive Surgery and State Key Laboratory of Biotherapy, West China Hospital, Sichuan University, Chengdu, Sichuan Province, China*
JOZEF ŽIDZIK • *Department of Medical Biology, Medical Faculty of P.J. Šafárik University, Trieda SNP 1Košice, Slovak Republic*
SILVIA ZORRILLA • *Instituto de Química-Física "Rocasolano", CSIC, Madrid, Spain*

Chapter 1

PPARs: History and Advances

Jihan Youssef and Mostafa Z. Badr

Abstract

Peroxisome proliferator-activated receptors (PPARs) are members of the steroid hormone receptor superfamily, discovered in 1990. To date, three PPAR subtypes have been identified; PPARα, PPAR β/δ, and PPARγ. These receptors share a high degree of homology but differ in tissue distribution and ligand specificity. PPARs have been implicated in the etiology as well as treatment of several important diseases and pathological conditions such as diabetes, inflammation, senescence-related diseases, regulation of fertility, and various types of cancer. Consequently, significant efforts to discover novel PPAR roles and delineate molecular mechanisms involved in their activation and repression as well as develop safer and more effective PPAR modulators, as therapeutic agents to treat a myriad of diseases and conditions, are underway. This volume of *Methods in Molecular Biology* contains details of experimental protocols used in researching these receptors.

Key words: Peroxisome proliferators, PPARs, RXRs, PPRE, TZDs

1. Discovery of Peroxisome Proliferator-Activated Receptors

Existence of a specific mediator for the effects of chemicals which are known to cause peroxisome proliferation (peroxisome proliferators) was suggested by the tissue and cell specificity of the pleiotropic effects of these chemicals. In attempting to identify such a molecular target, a cytosolic protein displaying reversible stereospecific binding to nafenopin was detected in rat liver and a receptor-mediated mechanism for peroxisome proliferation was consequently postulated (1). A peroxisome proliferator-binding protein was later purified from rat liver cytosol and was identified as a dimer protein with a molecular weight of 140,000–160,000 KDa. This protein was capable of binding to peroxisome proliferators structurally related to clofibrate (2). Further analysis of the isolated protein revealed that it is homologous with the heat

Mostafa Z. Badr and Jihan Youssef (eds.), *Peroxisome Proliferator-Activated Receptors (PPARs): Methods and Protocols*, Methods in Molecular Biology, vol. 952, DOI 10.1007/978-1-62703-155-4_1, © Springer Science+Business Media New York 2013

shock protein HSP70 but its role in the process of peroxisome proliferation remained unclear at that time (3). Thus, efforts continued in order to identify the putative mediator of these observed prominent effects.

The ability of peroxisome proliferators to modulate specific gene transcription suggested that they could act via a mechanism similar to that of steroid hormones. This assumption paved the way to a significant discovery when a novel member of the steroid hormone receptor superfamily of ligand-activated transcription factors was isolated by screening a mouse cDNA library (4). The cloned receptor was found to be structurally related to steroid hormone receptors and was activated by a wide range of molecules including fatty acids and fibrates. Since the pattern of expression of the receptor mRNA mirrored the tissue-specific effects of peroxisome proliferators and because the identified receptor was thought to mediate the peroxisome proliferative response, it was named peroxisome proliferator-activated receptor (PPAR).

Following the initial discovery of mouse PPAR, the receptor was identified in other species including rat (5) and human (6). In addition, three related *xenopus* receptors belonging to nuclear hormone receptor superfamily were cloned and named PPARα, PPARβ, and PPARγ proving the existence of more than one form of PPAR in a given species (7). PPARδ was initially identified in human as an additional form of PPAR (6) but was found later to be closely related to PPARβ described in *xenopus*.

2. Molecular Aspects of PPAR Functions

Subsequent studies indicated that the three PPAR subtypes, PPARα, PPARβ/δ, and PPARγ, share a high degree of homology but differ in tissue distribution and ligand specificity and each subtype is encoded by a different gene (8). Human PPARα gene is located on chromosome 22 slightly telomeric to a linkage group of six genes and genetic markers existing in the general region 22q12–q13.1 (9), while PPARβ/δ gene is located on chromosome 6 at position 6p21.1–p21.2 (10) and PPARγ gene is mapped to chromosome 3 at position 3p25 (11). Alternate transcription start sites and alternate splicing generate four types of PPARγ mRNA γ1, γ2, γ3, and γ4. It is believed that PPARγ1, γ3, and γ4 mRNA translate to an identical protein. Thus only two protein isoforms, PPARγ1 and PPARγ2 (which has an additional 30 amino acids at the N-terminus) exist (12).

PPARs form heterodimers with retinoic acid receptors (RXRs) which also belong to the nuclear hormones family of receptors (13, 14). Upon ligand binding, PPAR/RXR heterodimers recognize and bind to a specific DNA sequence known as PPAR response element (PPRE) consisting of a direct repeat of six nucleotides separated by a single spacer nucleotide (15). The 5′ flanking nucleotides of the core PPRE may play an important role in PPAR subtype specificity (16). PPRE is found in a large number of genes encoding proteins involved in a variety of cellular functions including lipid and carbohydrate metabolism. The three PPARs activate both overlapping and distinct sets of target genes. Both PPARα and PPARβ/δ activate genes involved in lipid oxidation (17, 18), while PPARγ activates lipogenic genes (19).

Intrinsic properties of each PPAR subtype, including post-translational modifications, are key determinant of the genes that can be activated by a certain PPAR subtype. PPAR subtype specificity may also be partly imposed by differential affinity of the receptors towards cofactors where distinct conformations could be induced by the individual coactivator resulting in differential transcriptional activity or specificity (20). With regards to receptor activators, each PPAR ligand induces a specific change in receptor conformation, resulting in the differential recruitment of cofactors and gene-specific transcriptional regulation. Thus, in addition to a panel of common genes regulated in a similar manner by all PPAR agonists, each agonist regulates its unique profile of genes, resulting in specific biological effects. This concept inspired the creation of new compounds with differential gene regulating properties as novel therapeutic agents without significant adverse effects.

2.1. Recent Advances in PPAR Research

PPARs have been implicated in several important diseases and pathological conditions. Recent discoveries suggest that PPAR subtypes present a valuable target for the treatment of several significant diseases and pathological conditions in various organs. In the central nervous system, PPARs mediate neuroprotective effects where PPAR agonists have been proven useful in animal models of several CNS diseases (21). These receptors are expressed in the heart and PPAR agonists have shown promising effects in preventing progression of atherosclerosis in experimental models as well as in clinical trials (22, 23). In addition, the discovery of anti-inflammatory and immunomodulatory roles of PPARs has prompted the investigation of these receptors as potential targets for treatment of asthma and other inflammatory lung diseases (24), as well as for several gastrointestinal diseases (25–27).

PPARγ ligands, thiazolidinediones (TZDs), are effective oral antidiabetic drugs. These agonists improve glycemic control in part by increasing insulin action in adipose tissue, skeletal muscles, and liver. TZDs also increase insulin biosynthesis and release as well as glucose transport in β-cells by up-regulating expression of genes involved in these processes (28, 29). In addition, studies revealed that the ability of TZDs to reduce β-cell apoptosis and enhance their proliferation may be responsible for the increase in β-cell mass with consequent improvement in glucose metabolism (30, 31). Ligands of PPARβ/δ have also been proposed as insulin sensitizers (32).

In the musculoskeletal system, PPARs are reported to play an interesting role. It appears that PPARα and PPARγ activation have opposite regulatory effects in bone formation (33). The role of PPARβ/δ, however, is not yet defined although studies suggest that it may contribute to bone anabolism (34). In the skeletal muscle, PPARβ/δ is the most abundant PPAR isotype with a higher expression level in oxidative type I muscle fibers compared to glycolytic type II muscle fibers (35, 36). Evidence suggests that PPARβ/δ plays an important role in regulation of skeletal muscle metabolism particularly lipid oxidation by acting as an activator of fat burning with subsequent beneficial effects in metabolic disease (37). Activation of PPARδ has been shown to increase physical performance and improve endurance performance (38). PPARδ agonists are, therefore, characterized as *exercise mimetics* (38), and are claimed to be, therefore, abused by athletes (39).

Among the most consequential involvements of PPARs is their role in cell differentiation and cancer. The literature is replete with contradictory evidence implicating PPARs in the promotion and development of cancer, as well as for a protective role against cancer. While numerous studies report that the expression level of these receptors and/or their activation correlates with a positive outcome against cancer, this does not appear to be a universal phenomenon (40).

Based on the above mentioned PPAR roles, in addition to numerous others, significant efforts to discover and develop safer and more effective PPAR modulators, as therapeutic agents to treat a myriad of diseases and conditions, are underway. The following chapters present details of state-of-the-art experimental protocols essential to the successful conduction of research in this expanding field.

References

1. Lalwani ND et al (1983) Detection of a nafenopin binding protein in rat liver cytosol associated with induction of peroxisome proliferation by hypolipidemic compounds. Biochem Biophys Res Commun 116:388–393
2. Lalwani ND et al (1987) Peroxisome proliferator-binding protein: Identification and partial characterization of nafenopin-, clofibric acid-, and ciprofibrate-binding proteins from rat liver. Proc Natl Acad Sci USA 84: 5242–5246
3. Alvares K et al (1990) Identification of cytosolic peroxisome proliferator binding protein as a member of the heat shock protein HSP70 family. Proc Natl Acad Sci USA 87:5293–5297
4. Isseman I, Green S (1990) Activation of a member of the steroid hormone receptor superfamily by peroxisome proliferators. Nature 347:645–650
5. Gottlicher M et al (1992) Fatty acids activate chimera of the clofibric acid-activated receptor and the glucocorticoid receptor. Proc Natl Acad Sci USA 89:4653–4657
6. Schmidt A et al (1992) Identification of a new member of the steroid hormone receptor superfamily that is activated by a peroxisome proliferator and fatty acids. Mol Endocrinol 6:1634–1641
7. Dreyer C et al (1992) Control of the peroxisomal beta-oxidation pathway by a novel family

of nuclear hormone receptors. Cell 68: 879–887
8. Berger J, Moller DF (2002) The mechanisms of action of PPARs. Annu Rev Med 53: 409–435
9. Sher T et al (1993) cDNA cloning, chromosomal mapping and functional characterization of human peroxisome proliferator activated receptor. Biochemistry 32:5598–5604
10. Yoshikawa T et al (1996) Assignment of the human nuclear hormone receptor, NUC1 (PPAR delta), to chromosome 6p21.1–p21.2. Genomics 35:637–638
11. Greene ME et al (1995) Isolation of the human peroxisome proliferator activated receptor gamma cDNA: expression in hematopoietic cells and chromosomal mapping. Gene Expr 4:281–299
12. Bugge A, Mandrup S (2010) Molecular mechanisms and genome-wide aspects of PPAR subtype transactivation. PPAR Res 2010:169506
13. Kliewer SA et al (1992) Convergence of 9-Cis retinoic acid and peroxisome proliferator signaling pathways through heterodimer formation of their receptors. Nature 358:771–774
14. Gearing KL et al (1993) Interaction of peroxisome-proliferator-activated receptor and retinoid x-receptor. Proc Natl Acad Sci USA 90:1440–1444
15. Ijpenberg A et al (1997) Polarity and specific sequence requirements of peroxisome proliferator-activated receptor heterodimer binding to DNA. A functional analysis of the malic enzyme PPAR response element. J Biol Chem 272: 20108–20117
16. Nielsen R et al (2006) Peroxisome proliferator-activated receptor subtype-and cell-type-specific activation of genomic target genes upon adenoviral transgene delivery. Mol Cell Biol 26:5698–5714
17. Gulick T et al (1994) The peroxisome proliferator-activated receptor regulates mitochondrial fatty acid oxidative enzyme gene expression. Proc Natl Acad Sci USA 91:11012–11016
18. Wang YX et al (2003) Peroxisome-proliferator-activated receptor delta activates fat metabolism to prevent obesity. Cell 113:159–170
19. Mueller E et al (2002) Genetic analysis of adipogenesis through peroxisome proliferator-activated receptor gamma isoforms. J Biol Chem 277:41925–41930
20. Oberkofler H et al (2002) Peroxisome proliferator activated receptor (PPAR) γ coactivator-1 recruitment regulates PPAR subtype specificity. J Biol Chem 277:16750–16757
21. Racke M, Drew P (2008) PPARs in neuroinflammation. PPAR Res 2008:638356
22. Calkin AC, Thomas MC (2008) PPAR agonists and cardiovascular disease in diabetes. PPAR Res 2008:245410
23. Duval C et al (2002) The role of PPARs in atherosclerosis. Trends Mol Med 8:422–430
24. Standiford T, Roman J (2007) PPARs in lung biology and disease. PPAR Res 2007:28765
25. Matthiessen MW et al (2005) Peroxisome proliferator-activated receptor expression and activation in normal human colonic epithelial cells and tubular adenomas. Scand J Gastroenterol 40:198–205
26. Pathak R et al (2007) Effect of peroxisome proliferator-activated receptor-alpha agonist (bezafibrate) on gastric secretion and gastric cytoprotection in rats. Fundam Clin Pharmacol 21:291–296
27. Peters JM et al (2008) Role of peroxisome-proliferator-activated receptor beta/delta (PPARbeta/delta) in gastrointestinal tract function and disease. Clin Sci (Lond) 115: 107–127
28. Kim HS et al (2008) Rosiglitazone stimulates the release and synthesis of insulin by enhancing GLUT-2, glucokinase and BETA2/ NeuroD expression. Biochem Biophys Res Commun 367:623–629
29. Masuda K et al (1995) Effects of Troglitazone (CS-045) on insulin secretion in isolated rat pancreatic islets and HIT cells: an insulinotropic mechanism distinct from glibenclamide. Diabetologia 38:24–30
30. Han SJ et al (2008) Rosiglitazone inhibits early stage of glucolipotoxicity-induced beta-cell apoptosis. Horm Res 70:165–173
31. Holloway AC et al (2008) Rosiglitazone prevents diabetes by increasing beta-cell mass in an animal model of type 2 diabetes characterized by reduced beta-cell mass at birth. Diabetes Obes Metab 10:763–771
32. Tanaka T et al (2003) Activation of peroxisome proliferator-activated receptor delta induces fatty acid beta-oxidation in skeletal muscle and attenuates metabolic syndrome. Proc Natl Acad Sci (USA) 100:15924–15929
33. Syversen U et al (2009) Different skeletal effects of the peroxisome proliferator activated receptor (PPAR) alpha agonist fenofibrate and the PPARgamma agonist pioglitazone. BMC Endocr Disord 9:10
34. Still K et al (2008) The peroxisome proliferator activator receptor alpha/delta agonists linoleic acid and bezafibrate upregulate osteoblast differentiation and induce periosteal bone formation in vivo. Calcif Tissue Int 83:285–292
35. Braissant O et al (1996) Differential expression of peroxisome proliferator-activated receptors

(PPARs): tissue distribution of PPAR-alpha, -beta, and -gamma in the adult rat. Endocrinology 137:354–366

36. Wang YX et al (2004) Regulation of muscle fiber type and running endurance by PPARdelta. PLoS Biol 2:e294
37. de Lange P et al (2008) Peroxisome proliferator-activated receptor delta: a conserved director of lipid homeostasis through regulation of the oxidative capacity of muscle. PPAR Res 2008:172676
38. Narkar VA et al (2008) AMPK and PPARdelta agonists are exercise mimetics. Cell 134: 405–415
39. Thevis M et al (2010) Characterization of two major urinary metabolites of the PPARdelta-agonist GW1516 and implementation of the drug in routine doping controls. Anal Bioanal Chem 396:2479–2491
40. Youssef J, Badr M (2011) PPARs and cancer: challenges and opportunities. Br J Pharmacol 164:68–82

Chapter 2

PPAR-Alpha Cloning, Expression, and Characterization

Suong N.T. Ngo and Ross A. McKinnon

Abstract

Peroxisome proliferator-activated receptor α (PPARα) is a member of the nuclear/steroid receptor gene superfamily that also comprises β, δ, and γ isoforms. PPARα is a ligand-activated transcription factor that plays an important role in the regulation of many genes involved in key metabolic processes. Today, PPARα has been cloned from mammalian, marsupial, and a number of marine species and its expression has been found to be relatively tissue- and species-specific. Here, we describe the methods for cloning of PPARα genes by RT-PCR and RACE approaches and related protocols for studying the expression of cloned PPARα cDNAs in mammalian cell systems.

Key words: PPARα cloning, RT-PCR, RACE, Immunoblots, Cos-7 cell expression system

1. Introduction

The peroxisome proliferator-activated receptors (PPARs) belong to a subfamily of the nuclear receptor gene that involves in energy and metabolic processes, inflammation, proliferation, and differentiation (1). Of the PPAR subfamily, which also comprises β, δ, and γ isoforms, PPARα is an important ligand-activated transcription factor that plays an essential role in lipid metabolism and energy homeostasis (2). PPARα has been found to be abundantly expressed in liver, kidney, heart, and skeletal muscle (3) and its expression is relatively species-specific. For examples, in rat and mouse livers, PPARα mRNA is highly expressed, while in human and guinea-pig livers, PPARα mRNA is much less abundant (4, 5).

PPARα was first isolated from mouse liver (6) and subsequently cloned from *Xenopus* (7, 8), rat (9), guinea-pig (4), human (5),

Mostafa Z. Badr and Jihan Youssef (eds.), *Peroxisome Proliferator-Activated Receptors (PPARs): Methods and Protocols*, Methods in Molecular Biology, vol. 952, DOI 10.1007/978-1-62703-155-4_2, © Springer Science+Business Media New York 2013

koala (10), chicken (11), and some marine species (12–15). In species studied thus far, PPARα can be activated by a class of structurally diverse compounds collectively classified as peroxisome proliferators (PPs). Fatty acids, fatty acid derivatives, and non-metabolizable fatty acids have also been identified as potent activators of PPARα (6, 9). *Xenopus* PPARα has also been suggested to be activated by naturally occurring fatty acids (9).

PPARα regulates the effects of PPs via transcriptional activation of PPARα target genes. PPARα proteins from different species have been found to have similar function when studied in vitro (4). PPARα binds as a heterodimer with the retinoid X receptor (RXR) to a peroxisome proliferator responsive element (PPRE) located in the promotor region of PPARα target genes (8, 16). PPRE has been identified in genes encoding peroxisomal β-oxidation enzymes and cytochrome P450 CYP4A subfamily, in particular human, rat, and *Xenopus* acyl CoA oxidase (AOX), rabbit CYP4A6, rat CYP4A1, as well as several other peroxisome proliferator-inducible genes (8, 16–33). To date PPRE has been found in 88 target genes as described by Lemay and colleagues (34). Evidence for the direct binding of PPs to PPARα has also been reported (35, 36).

In certain species such as mouse and rat, PPs cause peroxisome proliferation, which is characterized by a profound increase in number and size of peroxisomes in liver parenchymal cells and an increase in liver weight. In addition, exposure to PPs induces expression of peroxisomal β-oxidation enzymes, CYP4As, and other susceptible genes (4, 34, 37, 38). A transgenic mouse line with a PPARα gene disruption exhibits no peroxisome proliferation, unchanged constitutive expression of fatty acid-metabolizing enzymes, and no increase in liver size after PP treatment (39, 40).

As PPARα controls the regulation of many genes which encode enzymes involved in key metabolic processes, it is essential to accomplish adequate primary information on PPARα. The molecular data on PPARα genes across species will enable further studies to investigate the mechanism underlining the regulatory roles of PPARα and fully establish the effects of PPs on cellular function and processes via modulation of PPARα. Thus, technique for successful generation of such data will be invaluable. In this book chapter, we describe the methodologies that have been applied successfully for the isolation and cloning of PPARα by our laboratory, including step-by-step procedures for primer designs, RT-PCR approaches and optimization, gene cloning, and gene sequencing and analyses. Related methods for studying the expression of cloned PPARα cDNAs in mammalian cells and for identification of cDNA-expressed PPARα proteins and PPARα from cellular fractions will be briefly discussed.

2. Materials

2.1. Cloning of PPARα cDNAs by RT-PCR Approaches

2.1.1. Isolation of RNA by Guanidinium Thiocyanate-Phenol-Chloroform Extraction (41, 42)

1. Denaturing solution: 4 M guanidium thiocyanate, 25 mM sodium citrate, pH 4.0, 0.5% (w/v) *N*-laurylsarcosine (sarkosyl), 0.1 M 2-mercaptoethanol (2-ME) (see Note 1). Weigh 94.56 g guanidium thiocyanate and transfer to a 200 mL graduated cylinder or a glass beaker. Add diethyl pyrocarbonate (DEPC)-treated water to a volume of about 170 mL and mix to dissolve. Then add 10 mL of 0.5 M sodium citrate and 5 mL of 20% (w/v) sarkosyl. Make up to 200 mL with water. Store at 25°C (room temperature), in a bottle wrapped with aluminum foil. Just before use, add 1.4 μL of 14.26 M 2-ME to 200 mL of the solution. Store at room temperature (see Note 2).
2. 2 M Sodium acetate, pH 4.0. Dissolve 27.2 g sodium acetate. $3H_2O$ in 80 mL of DEPC-treated water. Adjust to the pH of 4.0 with glacial acetic acid. Make up to 100 mL with DEPC-treated water. Store at room temperature (see Note 3).
3. Phenol (DEPC-treated water saturated). Dissolve 50 g phenol crystal in about 1 L distilled water at 65°C. Remove as much as possible the upper phase water by aspiration. Store at 4°C (see Note 4).
4. Chloroform:isoamyl alcohol (49:1) (v/v). Prepare just before use. Mix 98 mL of chloroform with 2 mL of isoamyl alcohol (see Note 5).
5. Isopropanol. Use straight from the manufacturer container.
6. 80%v/v Ethanol. Add 80 mL absolute ethanol to 20 mL DEPC-treated water.

2.1.2. Purification of mRNA and cDNA

Purification of mRNA from total RNAs and synthesis of first-strand cDNA (Amersham Pharmacia Biotech, Buckinghamshire, UK) (see Note 6):

1. Purification of mRNA. The kit (Amersham) provides sufficient columns and reagents for 2 or 4 mRNA purifications including:
   - (i) Oligo(dT)-cellulose columns containing oligo(dT)-cellulose suspended in 0.15% Kathon™ CG/1CP Biocide storage buffer.
   - (ii) High salt buffer containing 10 mM Tris–HCl, pH 7.4, 1 mM ethylenediaminetetraacetate (EDTA), and 0.5 M NaCl.
   - (iii) Sample buffer containing 10 mM Tris–HCl, pH 7.4, 1 mM EDTA, and 3 M NaCl.
   - (iv) Low salt buffer containing 10 mM Tris–HCl, pH 7.4, 1 mM EDTA, and 0.1 M NaCl.

(v) Elution buffer containing 10 mM Tris–HCl, pH 7.4, and 1 mM EDTA.

(vi) 5–10 mg/mL glycogen solution.

Additional reagents to be prepared:

(vii) 0.1%v/v DEPC-treated water: Add 1 mL of DEPC to 1 L Milli-Q grade water. Shake vigorously to get the DEPC into solution, allowed to stand overnight at room temperature, then autoclave for 15 min at 121°C (see Note 7).

(viii) TE (Tris–EDTA) buffer: 10 mM Tris–HCl, pH 7.4, 1 mM EDTA in DEPC-treated water. This buffer can be prepared by adding stock solutions of 1 mL of 1 M Tris–HCl, pH 7.4, and 200 μL of 0.5 M EDTA, pH 8.0 to 90 mL DEPC-treated water. Make up to 100 mL with DEPC-treated water, then autoclave for 15 min at 121°C. Store at room temperature. To prepare 1 M Tris–HCl, pH 7.4 stock solution, dissolve 121.1 g Tris base in 800 mL DEPC-treated water. Adjust pH to 7.4 by adding 70 mL HCl. Make up to 1 L with DEPC-treated water, then autoclave for 15 min at 121°C. For preparation of 0.5 M EDTA, pH 8.0, dissolve 186.1 g EDTA.2H_2O in 800 mL DEPC-treated water. Adjust pH to 8.0 with approximately 20 g NaOH pellets. Add DEPC-treated water to 1 L, then autoclave. Store the stock solutions at room temperature.

(ix) 5.5 M Guanidinium thiocyanate: Weigh 118.2 g guanidinium thiocyanate and transfer to a 200 mL graduated cylinder or a glass beaker. Add DEPC-treated water to a volume of about 190 mL and mix to dissolve, then make up to 200 mL with water. Store at room temperature.

(x) Glycogen (1%w/v).

2. Synthesis of first-strand cDNA. The kit (Amersham) provides all the reagents to generate full-length first-strand cDNA from an RNA template, utilizing a number of primers. First-strand cDNA synthesized from this kit could be used as a template for DNA amplification by most polymerase chain reaction (PCR) applications. The kit contains sufficient reagents for up to 55 first-strand cDNA syntheses including:

   (i) Bulk First-Strand cDNA Reaction Mixes containing cloned FPL Cpure™ Murine reverse transcriptase, RNAguard™ (porcine), RNase/DNase-free Bovine serum albumin (BSA), deoxyadenosine triphosphate (dATP), dCTP (cytidine), dGTP (guanosine), and dTTP (thymidine) in aqueous solution.

   (ii) 200 mM dithiothreitol (DTT) aqueous solution.

(iii) pd(N6) primer containing 0.2 μg/mL random hexadeoxynucleotides in aqueous solution.

(iv) Not I $d(T)_{18}$ bifunctional primer containing 5 μg/mL 5′ d[AACTGGAAGAATTCGCGGCCGCAGGAA T_{18}]-3′ in aqueous solution.

(v) DEPC-treated RNase-free water.

Additional DEPC-treated water is required and can be prepared as in previous step (Subheading 2.1.2 step 1 and vii).

2.1.3. RT-PCR and RACE

1. Oligonucleotide primers: Custom-made (Sigma Life Science, St Louis, MO, USA) (see Note 8).
2. Rapid amplification of cDNA ends (5′-; 3′-RACE) amplification kits (Clontech Laboratories Inc, Palo Alto, CA; Ambion International, Austin, TX; GibcoBRL Life Technologies, Rockville, MD) (USA).
3. *Taq* DNA polymerases (Qiagen Pty Ltd., VIC, Australia), Clontech Advantage 2 polymerase mix, Ultra-pure deoxyribonucleoside triphosphate (dNTPs) (100 mM, Amersham Pharmacia Biotech, Buckinghamshire, UK), Clontech 50× dNTP mix, 5× Loading buffer (Promega Corporation, Madison, WI, USA; Amersham; Sigma).
4. Mineral oil, 0.5 mL sterile PCR reaction tubes, PCR pipettors and pipette tips, DNA markers (Promega), thermal cycler (Perkin-Elmer GeneAmp 480, Waltham, MA, USA), Air-cooled thermal cycler PC-960 (Corbett Research, Australia).

2.1.4. PPARα Cloning, Sequencing, and Analyses

1. DNA electrophoresis low-melt agarose gel: 1% w/v low-melt agarose, 1× Tris–acetate (TAE) buffer. Prepare the gel by melting 1 g agarose in 80 mL 1× TAE in a 100 mL glass bottle, allow the solution to cool down, make up to 100 mL with TAE buffer. Concentrated stock solution 50× TAE can be prepared by dissolving 242 g Tris base in about 0.5 L sterile or Milli-Q water, add 57.1 mL glacial acetic acid, 100 mL 0.5 M EDTA (pH 8.0), make up to 1 L with Milli-Q water. Prepare 0.5 M EDTA, pH 8.0 as in previous step (Subheading 2.1.2 step 1 and viii).
2. DNA SubCell™ electrophoresis system (Bio-Rad Laboratories, Hercules, CA, USA).
3. DNA gel extraction kits: QIAEX II gel extraction kit (Qiagen); 5× stop/loading buffer, 0.1 μg/mL ethidium bromide (EtBr) (Promega).
4. Shimadzu UV-1601UV-visible spectrophotometer (Shimadzu Corporation, Tokyo, Japan).
5. Sub-cloning vectors and competent cells for DNA transformation: pGEM®-T vector system, pCMV-Script®, pCI-neo,

Epicurian Coli® XL10-Gold™ ultracompetent cells (Promega; Stratagene, La Jolla, CA, USA), isopropyl-β-D-thio-galactoside (IPTG; 0.4 M), X-gal (50 mg/mL; Progen Industries Ltd, Brisbane, QLD, Australia). Alternately, prepare competent cells in-house using DH5α or XL1-blue *Escherichia coli* (Stratagene).

6. Restriction enzymes, DNA markers, T_4 DNA ligase (Invitrogen Life Science, VIC, Australia), Plasmid DNA purification kits (QIAGEN, QIAfilter plasmid Midi/Maxi; Qiagen). Alternately, plasmid DNAs can be purified in-house.
7. Plasmid DNA purification components. Resuspending buffer (50 mM glucose, 25 mM, Tris–HCl, pH 8.0, 10 mM EDTA), freshly prepared lysis solution 2.0 M NaOH, 1% w/v sodium dodecyl sulphate (SDS), neutralizing solution (3 M potassium acetate, pH 4.8), TE equilibrated phenol (pH 8)/chloroform (1:1). Prepare the solutions as in previous step.
8. M13 universal, M13 reverse, T3, T7, $T7_{PCI}$ sequencing primers (Geneworks, Adelaide, Australia), ABI Prism™ Big-Dye reaction mix (Perkin Elmer).
9. Perkin Elmer ABI Prism 377 sequencer.
10. Cell culture and cDNA expression components. Luria Bertani broth (LB; 1% w/v trypton, 0.5% w/v yest extract, 1% w/v NaCl, pH 7), antibiotics, Dulbecco's Modified Eagle Medium (DMEM; supplemented with 10% fetal bovine serum (FBS), 50 U/ML penicillin/streptomycin, 4 mM l-glutamine), 50 mM $CaCl_2$, sterile glycerol (Sigma), Cos-7 cells (Afican green monkey kidney cells), Lipofectamine reagent (Gibco-BRL), 0.1 M phosphate buffer (pH 7.4).
11. Orbital mixer incubator 0M10 (Ratek Instrument Pty, Boronia, VIC, Australia).

2.2. Characterization of PPARα from Cytosolic, Nuclear Fractions, and cDNA-Expressed PPARα Proteins

Cytosolic and nuclear extracts are used to detect PPAR-alpha protein from tissues. The cell fractions should be prepared from fresh tissues.

2.2.1. Cell Fraction Extraction Buffers

1. Nuclear extract homogenization buffer: 20 mM Tris–HCl (pH 7.4), 1 mM EDTA, 25 mM KCl, 1 mM phenylmethanesulfonyl fluoride (PMFS), 1 mM DTT, 10% (v/v) glycerol. This buffer can be prepared by adding stock solutions of 2 mL of 1 M Tris–HCl, pH 7.4, 200 μL of 0.5 M EDTA, pH 8, 2.5 mL of 1 M KCl, 1 mL of 0.1 M PMFS, 200 μL of 0.5 DTT, and 10 mL glycerol to 90 mL milli-Q water. Make up to 100 mL with water. Store at room temperature. Prepare 1 M Tris–HCl, pH 7.4 and 0.5 M EDTA stock solutions as in previous steps.

For preparation of 1 M KCl stock solution, dissolve 7.46 g KCl in 100 mL Milli-Q water. To prepare 0.1 M PMFS, dissolve 1.74 g in isopropanol. To make 0.5 M DTT stock solution, dissolve 7.71 g DTT in 100 mL Milli-Q water. Store DTT stock solution at −20°C, other stock solutions at room temperature (25°C).

2. Homogenization buffer with 0.4 M NaCl: Freshly prepared by dissolving 2.33 g NaCl in 100 mL nuclear extract homogenization buffer.
3. Potassium phosphate 0.1 M buffer (pH 7.4) with 1.15% w/v KCl. Prepare two stock solutions of 1 M K_2HPO_4 and 1 M KH_2PO_4. To obtain phosphate 0.1 M buffer (pH 7.4), mix 80.2 mL 1 M K_2HPO_4 to 19.8 mL 1 M KH_2PO_4, then dissolving 1.15 g KCl in 100 mL of this phosphate buffer. To prepare the 1 M K_2HPO_4 and 1 M KH_2PO_4 stock solutions, dissolve 45.6 g K_2HPO_4 in 200 mL Milli-Q water and 34.0 g KH_2PO_4 in 250 mL water, respectively.
4. Teflon or glass Potter Elvehjem homogenizer (Omni International, Kennesaw, GA, USA).
5. Reagents for measurement of protein concentration: Freshly prepared protein measurement solution (0.28 M Na_2CO_3, 0.1 M NaOH, 0.8 mM CuSO4, 1.4 mM potassium sodium tartrate), Folin and Ciocalteus reagent (1 in 3 dilution in water), BSA (1 mg/mL).

2.2.2. SDS Polyacrylamide Gel Components

Prepare as stock solutions:

1. Resolving gel buffers: 1.5 M Tris–HCl, 0.4% (w/v) SDS, pH 8.8. Weigh 181.7 g Tris and 4 g SDS and transfer to a 1 L graduated cylinder or a glass beaker containing about 100 mL distilled or Milli-Q water (see Note 9). Add water to 900 mL, mix until Tris and SDS dissolved (see Note 10), and adjust pH to 8.8 with concentrated HCl (see Note 11). Make up to 1 L with water. Store at 4°C (see Note 12).
2. Stacking gel buffer: 0.5 M Tris–HCl, 0.4% (w/v) SDS, pH 6.8. Weigh 60.6 g Tris and 4 g SDS. Prepare a 1 L solution as in previous step (Subheading 2.2.2 step 1) for the resolving gel buffer. Store at 4°C.
3. Acrylaimide:bis-acrylamide solution (29.2:0.8 acrylamide:bis-acrylamide): Dissolve 29.2 g acrylamide and 0.8 g bis-acrylamide in about 40 mL water in a 100 mL graduated cylinder. Make up to 100 mL with water and filter through a 0.45 μm filter (see Note 13). Store at 4°C, in a bottle wrapped with aluminum foil (see Note 14).
4. SDS polyacrylamide gel electrophoresis (PAGE) running buffer: 0.25 M Tris–HCl, pH 8.3, 0.192 M glycine, 0.1% (w/v)

SDS. Weigh 3 g Tris, 14.4 g glycine, and 1 g SDS, mix, and make it to 1 L with water.

5. SDS lysis buffer: 0.2 M Tris–HCl, 10.2% (w/v) SDS, 17% (v/v) 2-ME, 0.03% (w/v) bromophenol blue (BPB), 33% (v/v) glycerol. Dissolve 3 g SDS and 10 mg BPB in 12.5 mL stacking gel buffer, add 5 mL 2-ME and 10 mL glycerol, and mix well. Leave one aliquot at 4°C for current use and store remaining aliquots at –20°C (see Note 12).
6. *N,N,N,N'*-tetramethyl-ethylenediamine (TEMED; 10% v/v). Dilute 200 μL TEMED in 2 mL water.
7. 2%, w/v SDS: Dissolve 2 g SDS in 100 mL water.

2.2.3. Immunoblotting Components

Prepare as stock solutions:

1. Western blot transfer buffer: 0.025 M Tris, 0.192 M glycine, 20% (v/v) methanol. Weigh 7.7 g Tris–HCl, 9.2 g Tris, and 72 g glycine, mix, and dissolve in about 3.5 L water. Add 1 L methanol and make up to 5 L with water.
2. Tris buffered saline (5× TBS): 2.5 NaCl, 0.1 M Tris–HCl, pH 7.5. Weigh 10.2 g Tris–HCl, 2.4 g Tris, and 146 g NaCl, mix and make to 1 L with water. Adjust pH to 7.5 with HCl.
3. Tween-20 (10% v/v). Measure 10 mL Tween-20 and make to 100 mL with1× TBS (see Note 15). Freshly prepared before use:
4. Nitrocellulose membranes (NitroBind, Micron Separations Inc, MA, USA).
5. Ammonium persulphate (APS; 10% w/v). Dissolve 1 g APS in 10 mL water.
6. Resolving gel (7.5%). Add 10 mL resolving gel buffer then 10 mL acrylamide–bis to 20 mL water, mix, and add 120 μL APS and 100 μL TEMED (see Note 16).
7. Stacking gel (5%): Prepare as per resolving gel by mixing 2.5 mL stacking gel buffer, 1.5 mL acrylamide–bis, 6 mL water, 30 μL APS, and 100 μL TEMED (see Note 17).
8. First overlay. Add in order 250 μL SDS (2% w/v), 35 μL APS, and 150 μL TEMED to 4.25 mL water.
9. Second overlay. Mix 3 mL resolving gel buffer with 2.7 mL water.
10. Wash buffer: 5 mL stacking gel buffer, 0.5 μL APS, 1 mL TEMED, 3 mL water.
11. TBS containing tween-20 (TTBS; 0.05% v/v): Measure 10 mL 10% (v/v) Tween-20 and make to 2 L with 1× TBS.
12. Blocking solution: 5% (w/v) skim milk in TBS.
13. Antibody buffer: 0.02% (w/v) skim milk in TBBS.

14. Medium binder clips, plastic container.
15. Hoefer (Inc, San Francisco, CA, USA) TE 42 transphor electro-transfer unit.
16. Imidazole buffer (20 mM, pH 7.4). Dissolve 136 mg imidazole in 100 mL water and adjust pH to 7.4 with HCl.
17. Horseradish peroxidase (HRP) development solution: 0.05% (w/v) 3,3′-diaminobenzidine (DAB) and 0.05% (v/v) H_2O_2 in 20 mM imidazole buffer (see Note 18).

2.2.4. Antigens and Conjugates

1. Rabbit anti-human PPARα polyclonal primary antibody (1:200 dilution; Santa Cruz Biotechnology, Inc, Heidelberg, Germany).
2. Mouse anti-rabbit IgG HRP-conjugated (1:2,000 dilution; Santa Cruz).
3. Pre-stained protein ladder (GibcoBRL).

3. Methods

3.1. Isolation of RNA and cDNA

1. For optimal RNA extraction, fresh tissue is preferable. Alternately snap-frozen tissue can be used. Remove quickly the tissue from the animal at the time of death, immediately freeze in liquid nitrogen, and store at −80°C until required.
2. For RNA extraction and purification (see Note 19), bake all general laboratory glassware at 180°C overnight prior to use. Soak plastic ware and other equipment in DEPC-treated water (0.1% v/v) overnight, then autoclave the glassware and equipment for 15 min at 121°C (see Note 20).
3. Prepare all solutions for the preparation of RNA using Milli-Q grade water to which add 0.1% w/v DEPC, leave overnight at room temperature, then autoclave the solutions for 15 min at 121°C. Use molecular grad reagents and chemicals free of RNAase where possible.
4. Sterilize all solutions used for DNA preparation.

3.1.1. Total RNA Extraction

Prepare according to Chomcynski and Sacchi (42) with modifications (see Note 21).

Carry out centrifugation at 11,000 × *g* at 4°C.

1. Homogenize 1 g tissue with 10 mL denaturing solution in a 20 mL Teflon glass homogenizer. Add 1 mL sodium acetate, 10 mL phenol and 2 mL chloroform:isoamyl, mix vigorously and stored on ice for 15 min.

2. Transfer the contents to a 30 mL centrifuge tube (see Note 22), centrifuge for 20 min, collect the supernatant in a fresh 30 mL centrifuge tube.
3. Add 10 mL denaturing solution and repeat the procedures in steps 1–2.
4. Add 10 mL isopropanol and store at −20°C for at least 1 h to precipitate the RNA.
5. Centrifuge the tube for 20 min, collect the pellet, briefly dry, and then resuspend in 300 μL denaturing solution. Add 300 μL isopropanol and allow the RNA to precipitate at −20°C overnight.
6. Centrifuge for 15 min, recollect the pellet, rinse with ethanol, dry briefly, and then resuspend in 200 μL DEPC-treated water.
7. Store RNA at −80°C in small aliquots as ethanol precipitates (see Note 23) or in DEPC-treated water with RNaseOUT Ribonuclease inhibitor (0.2 I.U./L; Invitrogen).

3.1.2. Purification of mRNA (Amersham mRNA Purification Kit)

Prepare in accordance with the manufacturer's protocol. Carry out centrifugation at 350×*g* for 2 min room temperature otherwise stated.

1. Prepare the sin column by washing the oligo(dT)-cellulose twice with 1 mL high salt buffer.
2. Dissolve 1.3 mg total RNA in 1 mL elution buffer, heat the sample at 65°C for 5 min then place on ice. To the tube, add 200 μL sample buffer, mix then pour into the prepared column and allow the column to drain by gravity. Centrifuge and wash the column twice with 250 μL high salt buffer and three times with 250 μL low salt buffer. Centrifuge the column between each wash.
3. Apply 4× 250 μL elution buffer (preheated at 65°C), centrifuge and collect the eluted RNA solution. To the mRNA solution, add 100 μL sample buffer, 10 μL glycogen, and 2 mL ethanol then stored at −20°C overnight. Collect the mRNA pellet by centrifugation at 12,000×*g* for 20 min at 4°C, and then resuspend in DEPC-treated water. Store mRNA in small aliquots as in previous step (Subheading 3.1.1 step 7).

3.1.3. Synthesis of cDNA (Amersham First-Strand cDNA Synthesis Kit)

Prepare in accordance with the manufacturer's protocol. Briefly, heat mRNA (10 ng) in 8 μL RNase-free water at 65°C for 10 min, chill on ice, add 5 μL bulk first-strand cDNA reaction, 1 μL DTT, 1 μL Not 1-d(T)18 primer, mix, and incubate at 37°C for 1 h. Stored cDNAs at −20°C.

3.2. Reverse Transcription Polymerase Chain Reactions (RT-PCR) and Rapid Amplification of cDNA Ends (RACE)

3.2.1. Primer Design

1. RT-PCR degenerative primers: Design degenerative primers based on PPARα cDNA sequences already been cloned from other species. These sequences are often available in the GenBank database:

PPARα cDNA	Accession number	Authors
(i) Mouse	NM_011144	Issemann and Green (1990) (6)
(ii) Rat	NM_013196	Göttlicher et al. (1992) (9)
(iii) Guinea-pig	AJ006218	Tugwood et al. (1998) (4)
(iv) Human	NM_005036	Qi et al. (2001) (5)
(v) Koala	AF_463455	Ngo et al. (2007) (10)
(vi) Chicken	NM_001001464	Diot and Douaire (1999) (11)
(vii) *Xenopus*	NM_001089813	Dreyer et al. (1992); Krey et al. (1993) (7, 8)
(viii) PPARα cDNA has also been cloned in a number of marine species (12–15).		

2. Gene-specific primers (GSPs): For RACE reactions to obtain 3′- and 5′- cDNA ends in order to clone full-length PPARα cDNA, design GSPs from already know partial sequences of the full-length cDNA. For a complete RACE protocol, at least two GSPs are required: an antisense for the 5′-RACE PCR and a sense primer for the 3′-RACE PCR. If only 5′- or 3′- RACE is carried out, only one GSP is needed. Design the primers to create overlapping 5′- and 3′-RACE products, which, if a suitable restriction site is located in the region of overlap, can subsequently be joined by restriction digestion and ligation to create the full-length cDNA. The relationship of the primers used in RT-PCR and RACE reactions to the template and resulting PCR, RACE products is shown in Fig. 1.
3. Primer sequence: It is best and where possible degenerative and GSPs should be:
   (i) 23–28 nt (see Note 24).
   (ii) $T_m \geq 65°C$; best results are achieved if $T_m \geq 70°C$ (43) (enable the use of touchdown PCR if needed) (see Note 25).
   (iii) 50–70% GC.
   (iv) The 3′-terminal ends of each primer should not be complementary to each other and should contain a low GC content.
4. Location of primer sequences within the gene: If possible, choose the location of primer sequences within the gene so that the 5′- and 3′-RACE products will be 2 kb or less

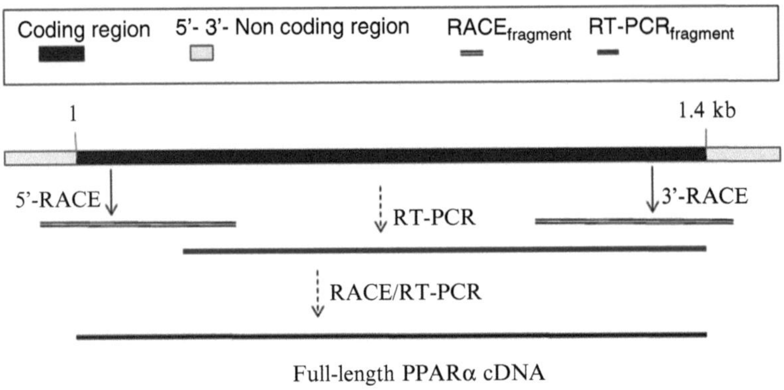

Fig. 1. Schematic diagram of the cloning strategy to obtain full-length PPARα cDNA by PCR approaches-RT-PCT and RACE. The solid box represents the coding region of PPARα and the lighter boxes depict the 5′- and 3′- non coding regions. Single and double lines represent PPARα cDNAs: 5′-cDNA amplified fragment obtained by RACE is double-line; partial PPARα cDNA obtained by RT-PCR is light solid; full-length PPARα cDNA is dark solid.

(see Note 26). Design GSPs that produce overlapping 5′- and 3′-RACE products and it is best when the overlap between primers is at least 100–200 bases (see Note 27).

3.2.2. Template Quality

It is important to use high quality and high molecular weight DNA, particularly when amplifying large sequences and/or the highest possible sensitivity is required (see Note 28).

1. Use first-strand cDNA synthesis kit (Amersham) to obtain pure and high quality cDNA template.
2. The longer or more complex the sequence, the more important the template quality is. This is because the number of un-nicked, full-length targets decreases as the target length increases, poor- quality DNA will have very few large un-nicked targets.
3. Poor cDNA template due to incomplete reverse transcription can lead to an absence of product, truncated products, or a mix of truncated and full-length products, resulting in a smeared band on a gel. This problem can be minimized when highest quality template is used.
4. During thermal cycling, some depurination occurs when DNA is denatured which can result in truncated products. This can be minimized by using high quality cDNA template.

3.2.3. PCR Master Mix

The use of PCR master mix can greatly reduce tube-to-tube variation and save time.

1. Assemble PCR master mix by the appropriate volumes of all reagents required for multiple PCR reactions. Vortex gently to

mix (without bubbling) PCR master mix, or to mix by gentle pipetting.

2. To test multiple primer sets with one cDNA template, include the template in PCR master mix. To test the same primers with multiple templates, include the primers in the master mix. For setting up several sets of parallel samples, prepare multiple PCR master mixes, i.e., each with a different set of primers or templates.

3.2.4. Touchdown PCR

Touchdown PCR approach can significantly improve the specificity of RT-PCR and RACE reactions (44, 45). It involves the use of an annealing/extension temperature that is several degrees higher than the T_m of the primer (>3–10°C) during the initial PCR cycles (5–10 cycles). The annealing/extension temperature is then reduced to the primer T_m for the remaining PCR cycles (see Note 29).

3.2.5. Recommended Cycling Parameters

1. Optimal RT-PCR conditions (see Note 30):
   (i) Denaturing at 94°C for 3 min.
   (ii) 30 cycles of 94°C for 45 s, 55°C for 45 s, and 72°C for 1 min.
   (iii) Final extension at 72°C for 10 min.
2. Touchdown PCR conditions for 5′- and 3′-RACE reactions (see Note 31):
   (i) 5 cycles of 94°C for 5 s and 72°C for 2 min.
   (ii) 5 cycles of 94°C for 5 s, 70°C for 10 s, and 72°C for 2 min.
   (iii) 32 cycles of 94°C for 5 s, 68°C for 10 s, and 72°C for 2 min.

3.2.6. RT-PCR Procedure

1. Before preparing the RT-PCR master mix: Place all components on ice and allow the reagents to thaw completely. Mix each component thoroughly by vortexing on low, gentle speed, or by gentle pipetting.
2. PCR reactions: 1× PCR reaction buffer, 200 μM of each dNTP, 0.1 μg of each forward and reverse primer, 1.5 I.U. Tag DNA polymerase, magnesium (Mg^{++}; 0.5–2.0 mM) in a final volume of 20 μL with PCR water.
3. Master mix: Prepare enough PCR master mix required for all PCR reactions plus one additional tube (see Note 32). Mix well, spin briefly using a microcentrifuge, and transfer the master mix into each PCR reaction tube.
4. Add each of the remaining reagents, mix gently, then overlay the contents of each tube with a few drops of mineral oil, and place cap on firmly.

5. It is best to include a negative and/or positive control PCR reaction (see Note 33).
6. Commence thermal cycling using a standard or touchdown PCR program.

3.2.7. RACE Procedure (Clontech Smart RACE cDNA Amplification Kit)

1. RACE reactions: Prepare PCR master mix as in previous step (Subheading 3.2.6 step 3), however using a 50 μL RACE reaction (see Note 34). The same master mix can be used for both 5′- and 3′- RACE reactions.
2. To prepare 5′- and 3′-RACE reactions, add the components in order in a 0.5 mL PCR tube:

PCR reactions	5′-RACE	3′-RACE
(i) Test cDNA template	5 μL	5 μL
(ii) RACE AP (10 μM)	1 μL	1 μL
(iii) 5′-GSP (antisense, 10 μM)	1 μL	–
(iv) 3′-GSP (sense, 10 μM)	–	1 μL
(v) Master mix	43 μL	43 μL
Final volume	50 μL	50 μL

3. Mix, overlay the contents with a few drops of mineral oil, and commence thermal cycling using a touchdown PCR program.

3.2.8. Isolation of Full-length PPARα by PCR

1. Design 5′- GSP and 3′- GSP based on the sequence obtained from 5′- and/or 3′-RACE products. These primers are derived from the actual 5′- and/or 3′- end of the cloned PPARα cDNA sequences (Subheading 3.2.1 steps 2, 3 and 4).
2. Assemble PCR master mix as in previous step (Subheading 3.2.7 step 1).
3. To prepare PCR reactions, add the components in order in a 0.5 mL PCR tube (can include 5′- and/or 3′-RACE reaction as optional controls):

PCR reactions	Full-length	5′-RACE	3′-RACE
(i) Test cDNA template	5 μL	5 μL	5 μL
(ii) 5′-RACE GSP (antisense, 10 μM)	1 μL	1 μL	–
(iii) 3′-RACE GSP (sense, 10 μM)	1 μL	–	1 μL
(iv) 5′-GSP (antisense, 10 μM)		1 μL	–
(v) 3′-GSP (sense, 10 μM)		–	1 μL
(vi) Master mix	43 μL	43 μL	43 μL
Final volume	50 μL	50 μL	50 μL

4. Mix, overlay the contents with a few drops of mineral oil, and commence thermal cycling.
5. Optimal PCR conditions for full-length PPARα amplification:
   (i) 30 cycles of 94°C for 30 s, 55°C for 30 s, and 72°C for 2 min.
   (ii) Final extension at 72°C for 10 min (see Note 35).

3.3. Cloning and Sequencing of PPARα cDNA RT-PCR and RACE Products

3.3.1. DNA Electrophoresis and Extraction of DNA from Agarose Gel

1. When cycling is completed, transfer 5 μL of the PCR reaction to a fresh tube (see Note 36), and add 1 μL 5× loading or stop buffer. Analyze PCR sample(s) along with suitable DNA size markers (see Note 37) on a 1% (w/v) agarose gel containing EtBr. A 0.5% (w/v) gel can be used for cDNA fragments size ≥1 kb. For resolution of amplified PCR products, low-melt agarose gel is used. View the gel under UV illumination.
2. Excise portions of the agarose gel containing the desired insert using a sterile scalpel, subdivide into 250 μg portions, and place in sterile eppendorf tubes. Purify the PCR products (QIAEX II protocol) and store DNA at 4°C.

3.3.2. Sub-cloning of PCR Products into a Plasmid Vector

Clone the purified PCR product directly into a T/A-type PCR cloning vector. Alternately, clone the product into a conventional vector using restriction enzymes and introduce restriction enzymes into the GSP (see Note 38).

1. Ligation reactions (see Note 39):
   (i) Ligation into pGEM-T® or pCI-neo: 150 ng cDNA insert, 1 μL T4 DNA ligase 10× buffer, 1 μL T4 DNA ligase (3 I.U./μL), 1 μL pGEM-T®, or pCI-neo (10 ng/μL) in a final volume of 10 μL with sterile water.
   (ii) Ligation into pCMV-Script XR®: 150 ng cDNA insert, 0.5 μL T4 DNA ligase 10× buffer, 0.5 μL rATP, 0.5 μL T4 DNA ligase (4 I.U./μL), 1 μL pCMV-Script XR® (10 ng/μL) in a final volume of 5 μL with sterile water.
   (iii) Set up ligation reactions, mix gently by pipetting, and incubate at 4°C overnight.
2. Bacterial cell culture preparation:
   (i) Cell culture and growth. Sterile all media for bacterial cell culture by autoclaving at 121°C for 15 min. Grow cells in LB containing an appropriate antibiotic (see Note 40) in a mixer incubator at 37°C overnight with vigorous shaking.
   (ii) LB plates. Add 1% (w/v) agar to the LB broth, autoclave, and cool the media to 50°C then add an appropriate antibiotic (steriled by filtration through a 0.22 μm filter) and pour into 90 mm Petri dishes.

3. Competent cell preparation (see Note 41):
   (i) Inoculate 0.5 mL of an overnight culture in 30 mL LB broth containing an appropriate antibiotic and incubate at 37°C for about 3 h (see Note 42).
   (ii) Centrifuge the suspension at 3,000 × *g* at 4°C for 15 min, collect the cell pellet, and suspend in 15 mL cold $CaCl_2$. Centrifuge the suspension, recollect the cells, then resuspend in 3 mL cold $CaCl_2$, and place on ice for 45 min. Store cells in 25% (v/v) sterile glycerol at –80°C.
4. Transformation:
   (i) Mix 2 μL (up to 5 μL) ligation reaction with 100 μL competent cells and place on ice for 45 min. Heat shock the cells at 42°C for 90 s, place on ice for 2 min, then suspend in 1 mL LB broth at 37°C for 90 min with gentle agitation to recover the cells. Centrifuge briefly, resuspend in 100 μL LB, then plate the cells onto LB plates containing a suitable antibiotic, and incubate at 37°C overnight.
   (ii) For screening of cDNA insert cloned in pGEM-T® or pCMV-Script XR®, add 20 μL IPTG and 20 μL X-gal to the cell suspension prior to plating. This enables color selection of the plasmid. Select the bacterial cells harboring recombinant plasmid DNA which will appear as white colonies.

3.3.3. Plasmid DNA Preparation

1. Small scale purification (alkaline lysis method):
   (i) Inoculate a single isolated bacterial colony containing the recombinant plasmid DNA in 5 mL LB containing an appropriate antibiotic and incubate at 37°C overnight with vigorous shaking. Pour 1.5 mL of the culture into an eppendorf tube, centrifuge at 1,000 × *g* for 1 min, remove the medium by aspiration, and leave the bacterial pellet as dry as possible. Resuspend (by vortexing) the pellet in 100 μL ice-cold re-suspending buffer and store at room temperature for 5 min (with the cap of the tube opened).
   (ii) Add 200 μL cell lysis solution, close the cap, and invert the tube rapidly up and down three times to mix the contents (do not vortex). Store the sample mixture on ice for 5 min. Add 150 μL ice-cold neutralizing solution (see Note 43), mix gently, and place on ice for 5 min. Centrifuge for 5 min, transfer the supernatant to a new tube, add equal volume of calibrated phenol/chloroform, mix, and centrifuge for 2 min. To the supernatant, add two volume of ethanol, mix, incubate at room temperature for 2 min, then centrifuge for 10 min to collect the DNA pellet. Allow the DNA to dry.

(iii) Wash the DNA pellet with 1 ml 70% (v/v) ethanol, vortex briefly, recover the pellet by centrifugation for 10 min, dry briefly, and resuspend in 20 μL sterile water. Store DNA at 4°C.

2. Qiaquick spin mini-prep (Qiagen):
   (i) Carry out at room temperature essentially as per manufacturer's protocol.
   (ii) It is best to use Qiaquick protocol to obtain high quality plasmid DNA for sequencing.
3. Large scale purification (Qiagen maxi plasmid purification kit):
   (i) Perform essentially as per manufacturer's protocol.
   (ii) Once the pellet DNA has been obtained, wash DNA with ice-cold 70% (v/v) ethanol, dry briefly, then resuspend in appropriate amount of sterile water. Stored DNA at 4°C.
4. Purification of plasmid DNA for transfection (Qiagen QIAfilter midiprep spin kit): For transfection of cloned cDNA into a mammalian expression cell system (e.g., Cos-7 cells), it is best to use high quality plasmid DNA, i.e., DNA prepared by this kit.
5. Quantification of DNA:
   (i) Prepare a 1 in 500 dilution of DNA solution in Milli-Q or sterile water.
   (ii) Measure the absorbance of the solution at 260–280 nm (using a UV spectrophotometer) to determine the DNA concentration.

3.4. Sequencing

1. Sequencing primers (see Note 44):

Name	Nucleotide sequence (5′-3′)
M13 universal sequencing primer	GTA AAA CGA CGG CCA GT
M13 reverse sequencing primer	CAC ACA GGA AAC AGC TAT GAC CAT G
T3 primer	AAT TAA CCC TCA CTA AAG GG
T7 primer	GTA ATA CGA CTC ACT ATA GGG C
$T7_{PCI}$ primer	TTA ATA CGA CTC ACT ATA GGC

2. Sequencing reactions: 400 ng DNA (1–2 μL), 8 μL ABI PrismTM Big-Dye reaction mix, and 3.2 pmole primer (1–2 μL each primer) in a final volume of 20 μL with water.
3. Thermal cycling conditions: 25 cycles of 96°C for 10 s, 50°C for 5 s, and 60°C for 4 min.

4. To each reaction, add 80 μL 75% (v/v) isopropanol, mix, incubate for 15 min, and centrifuge at 10,000 × *g* at room temperature for 25 min to obtain the DNA pellet. Before subjected to analysis by electrophoresis in a sequencer, wash the DNA pellet with 200 μL 75% isopropanol and dry briefly (see Note 45).
5. Use a computer program, i.e., MacVector™ (Oxford Molecular Lt.) to analyze the data obtained from the sequencing reactions. To identify the cloned cDNAs, accomplish ClustalW alignment of cDNA cloned sequences against sequences from other species (GenBank database, http://www.ncbi.nlm.nih.gov/blast).

3.5. Characterization of PPARα in Cytosolic and Nuclear Extracts and cDNA-Expressed PPARα Protein

3.5.1. Preparation of Cytosolic and Nuclear Extracts

Use a Teflon/glass Potter homogenizer.

1. Cytosolic extract: Carry out centrifugation at 4°C. Homogenize 1 g tissue in 10 mL potassium phosphate buffer, transfer the contents into a 30 mL ultracentrifuge tube (see Note 22), and centrifuge at 20,000 × *g* for 20 min. Transfer the supernatant into a new tube, centrifuge at 280,000 × *g* for 1 h, and collect the supernatant. Store the cytosol at 4°C.
2. Nuclear extract (46): Carry out centrifugation at 1,000 × *g*. Homogenize 1 g tissue in three volumes of homogenization buffer, centrifuge the contents in a 30 mL tube for 5 min, collect the pellet, and suspend in three volumes of homogenization buffer. Centrifuge the suspension for 5 min (see Note 46), collect the pellet, and resuspend in two volumes of homogenization buffer containing 0.4 M NaCl. Mix the suspension gently at 4°C for 20 min (by stirring with a magnetic stir bar), centrifuge for 60 min, and collect the supernatant. Store the nuclear extract at −80°C.

3.5.2. Expression of PPARα in Cos-7 Cells (see Note 47)

1. Grown Cos-7 cells in a monolayer in supplemented DMEM at 37°C and 5% (v/v) CO_2. Seed the cells at 6×10^5 cells per well, i.e., using a 6-well 33-mm tissue culture plate, then allow the cells to grow in DMEM containing 10% (v/v) FBS until 90% confluent. Immediately prior to transfection, wash the cells gently with DMEM with no serum, and remove the media.
2. Use 10 μL LipofectAMINE reagent (GibcoBRL) to transfect full-length PPARα cDNA (cloned in an appropriate mammalian expression vector) into Cos-7 cells as per manufacturer's protocol. Before overlaying the mixture sample on the cells, incubate the sample for 30 min at room temperature to allow the LipofectAMINE-DNA complexes to form. Incubate for 16 h, replace with fresh supplemented DMEM, and harvest the cells 3 days following transfection. Rinse cells briefly in 0.1 M phosphate buffer (pH 7.4), homogenize using a

hand-held eppendorf homogeniser, and use this crude cell lysate directly for further analyses.

3.5.3. Measurement of Protein Concentrations (47)

1. Dilute the protein sample (cytosol, nuclear extract, or cell lysate) 1 in 20 to a final volume of 300 μL with water in a 10 mL glass tube. Similarly, prepare a range of BSA standard samples, i.e., from 0 to 1 mg/mL in 300 μ final volume in each 10 mL glass tube, for constructing the standard absorbance curve (see Note 48).
2. To each sample and standard tube, add 5 mL protein measurement solution, incubate at room temperature for 15 min. Add 500 μL Folin reagent, mix briefly by vortexing, and incubate for 15 min.
3. Measure absorbance at 660 nm and calculate the protein concentration based on the standard absorbance curve constructed from a range of known BSA concentrations.

3.5.4. Immunoblot Analysis of PPARα from Cytosolic, Nuclear Fractions, and cDNA-Expressed PPARα Proteins (48)

1. 12.5% SDS polyacrylamide gel electrophoresis (see Note 49):
   (i) Prepare the resolving gel by mixing the reagents (Subheading 2.2.3 step 6) in a 50 mL conical flask and pour into the apparatus. Add the first overlay and allow the gel to set. Remove the first overlay (by pouring off), add the second overlay, and leave overnight (covered with aluminum foil) (see Note 50).
   (ii) Remove the second overlay, add the wash, then remove the wash (use blotting paper to remove excess wash if needed).
   (iii) Prepare the stacking gel by mixing the reagents as in previous step (Subheading 3.5.4 step 1 and i) and insert the gel combs (10-well) immediately without introducing air bubbles (by sliding in slowly and pushing down hard). Allow the gel to set for 1 h.
   (iv) Solubilize the protein (50 μg/lane, in 10 μL) with an equal volume of solubilising buffer and heat the samples to 100°C for 10 min, then centrifuge the heated sample at 3,000×*g* for 30 s to bring down condensate (see Note 51).
   (v) Load the samples starting from the second lane and add the pre-stained protein ladder in the first lane. Do not touch the bottom of well or sides when loading samples. Stack and run the gel (see Note 52).
   (vi) Immediately following SDS-PAGE, when the dye front reaches the end of the gel, turn off the power supply. Separate the gel plates with the help of a spatula or a similar tool, with the gel to remain on one of the glass plates. Remove the stacking gel. Rinse the gel with water and transfer carefully to a container with transblotting buffer.

2. Immunoblot:
   - (i) Cut a nitrocellulose membrane to the shape of the gel. Rinse twice in distilled water and immerse in transblotting buffer. In another dish, immerse the botting paper and the sponge pads.
   - (ii) After rinsing the gel (still supported by the bottom glass plate) in deionized water for 10 min (to remove traces of SDS-PAGE running buffer), excise the gel and leave to air-dry for 5 min. Gentle lay the nitrocellulose membrane on top of the gel.
   - (iii) Gently lift the gel-membrane sandwich from the glass plate and place the sandwich on two sheets of Whiteman no. 3 filter paper (place membrane side directly on the filter paper and the exposed gel side on top), cut to the size of the gel. Place another two sheets of Whiteman no. 3 filter paper on top of the gel, cut to the shape of the gel.
   - (iv) Place the nitrocellulose-gel-filter paper sandwich between the two sponges, then place the sandwich-sponge between the two white racks and secure with clamps.
   - (v) Add in transblotting buffer to the mesh tank (3/4 of the way up), place this assembly in (by sliding in), and arrange so that the nitrocellulose faces the positive anode (red).
   - (vi) Allow the gel to electrophorese at 50 mA overnight.
   - (vii) Remove the membrane for immunoblotting.

     Use a shaking water bath at 37°C from now on:
   - (viii) Wash the membranes in TBS for 5 min to remove the blotting buffer. Block the membranes with blocking solution for 1 h.
   - (ix) Wash twice in TTBS, 5 min each wash.
   - (x) Add the primary antibody and incubate for 3 h.
   - (xi) Wash as in step ix of this section.
   - (xii) Transfer the membrane to a glass plate (facing upwards) and add 15 mL secondary antibody. Use a plastic dispensing pipette to keep topping up the secondary antibody to ensure the nitrocellulose membrane is not left dry.
   - (xiii) Wash as in step ix of this section, then wash twice in TBS, 5 min each time.
   - (xiv) Add 20 mL HRP development solution and let bands to develop.
   - (xv) Rinse two to three times with deionized water.

4. Notes

1. Guanidium thiocyanate is toxic; minimize handling as much as possible. Handle the 2-mercaptoethanol under a fume hood.
2. Make up a stock solution minus the 2-ME and store up to 3 months at room temperature, wrapped in foil. The working Denaturing solution should be prepared just before use by adding in the 2-ME and can be stored up to 1 month at room temperature.
3. The solution can be stored up to 1 year at room temperature. Handle the glacial acetic acid in a fume hood.
4. Use nucleic acid grade phenol. The saturated phenol can be stored up to 1 month at 4°C. Handle the phenol under a fume hood.
5. Prepare the solution under the fume hood.
6. The Amersham mRNA Purification kit utilizes oligo(dT)-cellulose pre-packed spun columns and suitable for mRNA purification from total RNA extracted from 25 mg up to 1 g tissue.
7. Autoclave the solution to inactivate DEPC. Handle the DEPC under a fume hood.
8. For optimal DNA amplification, we custom-make oligonucleotide primers (both degenerate- and gene-specific) so that they are within 20–30 base pairs length.
9. Having the cylinder prefilled with 100 mL water helps to dissolve Tris and SDS more easily, allowing the magnetic bar to go to work immediately. If using a glass beaker, Tris can be dissolved faster given that water is warmed to about 37°C. However, the solution should be brought to room temperature before adjusting pH.
10. Care should be taken to mix and dissolve SDS since it makes bubbles.
11. Can use 12 M HCl to adjust pH. Care should be taken to avoid sudden drop in pH below the required pH. Alternately, use 12 M HCl at first to narrow the gap from the starting pH to the required pH, and from then use a series of lower concentrations (e.g., 6 and 1 M) to get to pH 8.8.
12. SDS precipitates at 4°C. Thus, the buffer needs to be warmed to dissolve SDS prior to use.
13. Acrylamide is highly toxic. It is a neurotoxin, thus care should be exercised to avoid skin contact and exposing the chemical to yourself and coworkers as much as possible. Handle all weighing, transfer, mixing of acrylamide in a fume hood. Wear a mask when weighing.

14. The acrylamide solution can be stored at 4°C up to 1 month. It is best to prepare this fresh each month when the gels are prepared.
15. Dilute 400 mL 5× TBS to 2 L with water to obtain 1× TBS.
16. Add TEMED last.
17. Once the resolving gel has set and been ready, prepare the stacking gel.
18. Freshly prepare the development solution just before use. Dissolve DAB in the buffer then add to the plastic container. Add H_2O_2 last.
19. Should take maximum care not to contaminate RNA samples with RNase. Appropriate treatment of glass/plastic wares and other equipment by DEPC and autoclaving is critically important as this will help to minimize the activity of RNase deliberated during cell lysis and to avoid the accidental introduction of trace amount of RNase from other potential sources in the laboratory.
20. After soaking, it is important to autoclave to remove traces of DEPC that might modify purine residues in RNA by carboxymethylation.
21. Our laboratory uses this method, however we repeat the phenol-chloroform extraction step twice in order to obtain cleaner RNA.
22. We use sterile and DEPC-treated compatible plastic centrifuge tubes for RNA preparation and keep these tubes only for RNA work to avoid contamination with RNase. For preparation of cellular fractions, we use plastic centrifuge tubes which are compatible for high speed centrifugation.
23. After the RNA pellet is suspended in 200 μL DEPC-treated water, add three volumes of ethanol, mix, pipette 200 μL into each sterile DEPC-treated eppendorf tube, and store at −80°C. When the RNA is required, centrifuge the tube for 5 min to recover the pellet, wash briefly with 80% isopropanol and resuspend in DEPC-treated water.
24. Generally there is no advantage to using primers longer than 30 nucleotides (nt). The T_m should be 70°C or higher to achieve optimal PCR results as determined by nearest neighbor analysis (43). This also allows using touchdown PCR. The T_m of primers can be calculated or determined experimentally by doing PCR at different temperatures. Do not use self-complementary primer sequences which can fold back and form intra-molecular hydrogen bands or primers that have complementarity to the provided RACE primers, in particular in the 3′- ends.
25. We find that longer primers with $T_m \geq 70$°C give more robust amplification in RT-PCR and RACE, particularly from difficult samples.

26. PPARα full-length cDNA is about 1.5 kb, we have had good success using degenerative primers to amplify partial cDNA fragments of ≥1 kb and amplifying 5′-cDNA fragments that extend up to 800 bp from the GSP sites by RT-PCR and RACE, respectively.
27. The overlap 100–200 bases between primers will allow a stretch of known sequence incorporated into the amplified RT-PCR and RACE 5′- and 3′- fragments and can be used to verify that the correct gene was amplified.
28. Although most PCR applications work well with templates of average or even low quality (as PCR amplification proceeds exponentially), we find that fewer problems are encountered and best results achieved in RT-PCR and RCAE protocols when the starting DNA material is of highest quality.
29. Often use touchdown PCR approach to achieve high specificity, particularly in 5′-, 3′- RACE PCRs. If the T_m of GSP >70°C and T_m of RACE adapted primer (AP) is 71°C, only gene-specific synthesis occurs during these initial cycles, enabling a critical amount of gene-specific product to accumulate. The annealing/extension temperature is then decreased to the AP primer T_m for remaining PCR cycles, allowing efficient, exponential amplification of the gene-specific template.
30. We find these parameters are optimal conditions for PCR amplifications to obtain PPARα cDNAs of ~1 kb. Manufacturers' protocols often recommend different parameters and a shorter denaturing time, e.g., initial denaturing at 95°C for 1 min; 25–35 cycles of 95°C for 1 min and 68 for 1 min (<1 kb) or 3 min (1–5 kb); final extension 68°C for 1 min.
31. We have had good success amplifying 5′ PPARα cDNA fragments of 500–800 bp using these touchdown PCR parameters.
32. In our laboratory, we prepare a master mix for 10 PCR reactions testing the same primers:

Master mix	10 reactions (×11)	Each 20 μL reaction (×1)
(i) PCR water:	121 μL	11 μL
(ii) 10× PCR buffer	22 μL	2 μL
(iii) 50× dNTP mix (10 mM each)	4.4 μL	0.4 μL
(iv) Forward primer	11 μL	1 μL
(v) Reverse primer	11 μL	1 μL
(vi) Tag DNA polymerase	11 μL	1 μL
Final volume	180.4 μL	16.4 μL

Add the components in order in a 0.5 mL PCR tube. Use 1 μL cDNA synthesis reaction (cDNA template) and 2.6 μL $MgCl_2$ (testing a range of concentrations).

33. For negative control, we use a PCR reaction without the addition of the cDNA template.

    For positive control, we include a cDNA template that we have previously used successfully to generate a known PCR product.

34. We prepare a master mix for three RACE reactions:

Master mix	**3 reactions (×4)**	**Each 50 μL reaction (×1)**
(i) PCR water:	144 μL	36 μL
(ii) 10× PCR buffer	20 μL	5 μL
(iii) 50× dNTP mix (10 mM each)	4 μL	1 μL
(vi) Advantage 2 Polymerase Mix	4 μL	1 μL
Final volume	172 μL	43 μL

35. We have had good success amplifying full-length PPARα cDNA using this PCR protocol. The final extension is optional; however, it may reduce background in some cases. We also find that better results are obtained by reducing the denaturation step (as exposure of DNA to high temperatures causes some depurination of single-stranded DNA during denaturation, which eventually leads to strand scission). High temperature also causes gradually loss of Tag DNA polymerases enzyme activity. In experiments with larger template, it is best and particularly important to minimize the denaturation time.

36. Store the remaining 45 μL of the reaction mixture on ice so that it can be subjected to further cycling if needed.

37. For 0.5–2 kb insert, we use 1 kb DNA ladder. For larger size DNA fragment, we use *Hind III*. Carry out electrophoresis at 10 V/cm of the gel and use lower voltage, e.g., 5 V/cm for low-melt agarose gel to minimize the amount of heat generated in the tank.

38. For directional cloning of full-length PPARα cDNA into pCMV-Script XR® expression vector, we include restriction sites *EcoRI* and *XhoI* at the 5′- and 3′- ends of the forward and reverse GSP respectively. For sequencing of PCR, RACE products, we subclone directly into pGEM-T® vector.

39. In our laboratory, we set up a series of control ligation reactions to check the efficiency of ligation into a vector (as described below). For optimization of insert:vector molar ratios, we perform ligation using 1:3, 1:1, and 3:1 ratio.

Testing	Reaction set up with
DNA ligase is active	*Hind III* restricted lambda DNA
Ligation efficiency	control insert DNA (supplied)
Number of background colonies	cut plasmid
Transformation efficiency	un-cut plasmid

40. We add an appropriate antibiotic (as describe below) when carrying out transformation of plasmid DNA into cells and competent cells preparation.

(i) Preparation of competent cells		
	Media	Working concentration
DH5α	LB with no antibiotic	–
XL1-blue *E. coli*	LB tetracycline	50 μg/mL (in 50% ethanol)
(ii) Transformation	LB agar plate	
pGEM-T®, pCI-neo	Ampicillin	100 μg/mL (in water)
pCMV-Script X®	Kanamycin	50 μg/mL (water)

w

41. We have had good success using in-house prepared XL1-blue *E. coli* and DH5α competent cells for transformation of plasmid DNA containing PPARα cDNA.
42. After incubation of the cells for about 3 h, we start to take a small aliquot of the culture and measure the OD (optical density) at 600 nm (using a spectrophotometer). We stop incubating the cells when the OD is between 0.3 and 0.4 (log phase).
43. By adding potassium acetate solution, it neutralizes the solution and precipitates cell debris and chromosomal DNA.
44. In our laboratory, we sequence PPARα cDNA cloned in pGEM-T®, pCMV-Script XR®, and pCI-neo vectors using the M13, T3, T7, $T7_{PCI}$ obtained desalted (Geneworks) and internal primers derived from known PPARα cDNA sequences. We modify the T7 primer sequence to obtain $T7_{PCI}$ in order to sequence cDNA cloned in pCI-neo vector.
45. We dry the DNA pellet on a heating block for about 1–2 min at 55°C.
46. This step helps to remove the remainder of cytosol.
47. In our laboratory, we use Cos-7 cells for expression of cloned PPARα cDNAs. Cos-7 cells are able to amplify and over express a transfected cDNA under the control of the SV40 promoter such as cDNA clone in pCMV-Script XR® or pCI-neo vector.

48. We prepare the cytosolic/nuclear extract and BSA standard samples by adding:

Tube	Conc (mg/mL)	Cytosol/ nuclear	BSA (1 mg/mL, μL)	Water (μL)
Sample	–	10 μL	–	290
1 (BSA)	0.0		–	300
2	0.16		50	250
3	0.332		100	200
4	0.500		150	150
5	0.667		200	100
6	0.833		250	50
7	1.0		300	–

49. Must wash the gel glass plates separately to avoid scratching, dry with a Chux cloth, and wipe over with methanol. Apply the Cello-seal (supplied as part of the Hoefer electro-transfer unit) onto the plastic lining and put the apparatus together. The clamps and the lower glass plate should face inwards.
50. Make the resolving gel and the first overlay together. The first overlay must be added immediately after casting gel within the gel cassette. Allow space for stacking the gel.
51. While the gel is setting, we solubilize and heat the protein samples. Fill the gel tank with gel running buffer 1/3 of the way up. Once the gel is set, place the gel apparatus in the tank. Remove the comb slowly and quickly fill the tank with gel running buffer. Remove air bubbles along the bottom of the apparatus.
52. We stack the gel at 100 V for 1 h and run at 200 V for 1 h 45 min. It is stacked when it reaches the interphase, i.e., top of the lower gel.

References

1. Escher P, Wahli W (2000) Peroxisome proliferator-activated receptors: insight into multiple cellular functions. Mutat Res 448:121–138
2. Desvergne B, Wahli W (1999) Peroxisome proliferator-activated receptors: nuclear control of metabolism. Endocr Rev 20:649–688
3. Mandard S, Muller M, Kersten S (2004) Peroxisome proliferator receptor a targetgenes. Cell Mol Life Sci 61:393–416
4. Tugwood JD, Holden PR, James NH, Prince RA, Roberts RA (1998) A peroxisome proliferator-activated receptor-alpha (PPAR-α) cDNA cloned from guinea-pig liver encodes a protein with similar properties to the mouse PPAR-α: implication for species differences in responses to peroxisome proliferators. Mol Toxicol 72:169–177
5. Qi C, Zhu Y, Pan J, Usuda N, Maeda N, Yeldandi AV, Rao MS, Reddy JK, Hashimoto T (2001) Peroxisomal β-oxidation and peroxisome proliferator-activated receptor alpha: an adaptive metabolic system. Annu Rev Nutr 21:193–230

6. Issemann I, Green S (1990) Activation of a member of the steroid hormone receptor superfamily by peroxisome proliferators. Nature 347:645–650
7. Dreyer C, Krey G, Keller H, Givel F, Helftenbein G, Wahli W (1992) Control of peroxisome β-oxidation pathway by a novel family of nuclear hormone receptors. Cell 68:879–887
8. Krey G, Keller H, Mahfoudi A, Medin J, Ozato K, Dreyer C, Wahli W (1993) Xenopus peroxisome proliferator activated receptors: genomic organization, response element recognition, heterodimer formation with retinoid X receptor and activation by fatty acids. J Steroid Biochem Mol Biol 47:65–73
9. Göttlicher M, Widmark E, Li Q, Gustafson JÅ (1992) Fatty acids activate a chimera of the clofibric acid-activated receptor and the glucocorticoid receptor. Proc Natl Acad Sci USA 89:4653–4657
10. Ngo SNT, Mckinnon RA, Stupans I (2007) Hepatic nuclear receptor PPARalpha in the koala (Phascolarctos cinereus): cloning and molecular characterization. Comp Biochem Physiol C Toxicol Pharmacol 146:375–382
11. Diot C, Douaire M (1999) Characterization of a cDNA sequence encoding the peroxisome proliferator activated receptor alpha in the chicken. Poult Sci 78:1198–1202
12. Batista-Pinto C, Rodrigues P, Rocha E, Lobo-da-Cunha A (2005) Identification and organ expression of peroxisome proliferator activated receptors in brown trout (Salmo trutta f. fario). Biochim Biophys Acta 1731:88–94
13. Tsai ML, Chen HY, Tseng MC, Chang RC (2008) Cloning of peroxisome proliferators activated receptors in the cobia (Rachycentron canadum) and their expression at different life-cycle stages under cage aquaculture. Gene 425:69–78
14. Raingeard D, Cancio I, Cajaraville MP (2006) Cloning and expression pattern of peroxisome proliferator-activated receptor alpha in the thicklip grey mullet Chelon labrosus. Mar Environ Res 62:S113–S117
15. Raingeard D, Cancio I, Cajaraville MP (2009) Cloning and expression pattern of peroxisome proliferator-activated receptors, estrogen receptor alpha and retinoid X receptor alpha in the thicklip grey mullet Chelon labrosus. Comp Biochem Physiol C Toxicol Pharmacol 149:26–35
16. Hsu M, Palmer CNA, Griffin KJ, Johnson EF (1995) A single amino acid change in the mouse peroxisome proliferator activated receptor alpha alters transcriptional response to peroxisome proliferators. Mol Pharmacol 48: 559–567
17. Johnson EF, Palmer CN, Hsu MH (1996) The peroxisome proliferator-activated receptor: transcriptional activation of the CYP4A6 gene. Ann N Y Acad Sci 804:373–386
18. Johnson EF, Palmer CN, Griffin KJ, Hsu MH (1996) Role of the peroxisome proliferator-activated receptor in cytochrome P450 4A gene regulation. FASEB J 10:1241–1248
19. Varanasi U, Chu R, Huang Q, Castellon R, Yeldandi AV (1996) Identification of a peroxisome proliferator-responsive element upstream of the human peroxisomal fatty acid acyl-coenzyme A oxidase gene. J Biol Chem 271: 2147–2155
20. Fang HL, Strom SC, Cai H, Falany CN, Kocarek TA, Runge-Morris M (2005) Regulation of human hepatic hydroxysteroid sulfotransferase gene expression by the peroxisome proliferator-activated receptor alpha transcriptionfactor. Mol Pharmacol 67:1257–1267
21. Targett-Adams P, McElwee MJ, Ehrenborg E, Gustafsson MC, Palmer CN, McLauchlan J (2005) A PPAR response element regulates transcription of the gene for human adipose differentiation-related protein. Biochim Biophys Acta 1728:95–104
22. Seree E, Villard PH, Pascussi JM, Pineau T, Maurel P, Nguyen QB, Fallone F, Martin PM, Champion S, Lacarelle B, Savouret JF, Barra Y (2004) Evidence for a new human CYP1A1 regulation pathway involving PPARalpha and 2 PPRE sites. Gastroenterology 127:1436–1445
23. Lee GY, Kim NH, Zhao ZS, Cha BS, Kim YS (2004) Peroxisomalproliferator-activated receptor alpha activates transcription of the rat hepatic malonyl-CoA decarboxylase gene: a key regulation of malonyl-CoA level. Biochem J 378:983–990
24. Noon LA, Clark AJ, King PJ (2004) A peroxisome proliferatorresponse element in the murine mc2-r promoter regulates its transcriptional activation during differentiation of 3T3-L1 adipocytes. J Biol Chem 279: 22803–22808
25. Girnun GD, Domann FE, Moore SA, Robbins ME (2002) Identification of a functional peroxisome proliferator-activated receptor response element in the rat catalase promoter. Mol Endocrinol 16:2793–2801
26. Lopez D, Irby RB, McLean MP (2003) Peroxisome proliferator-activated receptor alpha induces rat sterol carrier protein x promoter activity through two peroxisome proliferator-response elements. Mol Cell Endocrinol 205:169–184

27. Barrero MJ, Camarero N, Marrero PF, Haro D (2003) Control of human carnitine palmitoyltransferase II gene transcription by peroxisome proliferatoractivated receptor through a partially conserved peroxisome proliferatorresponsive element. Biochem J 369:721–729
28. Boulanger A, McLemore P, Copeland NG, Gilbert DJ, Jenkins NA, Yu SS, Gentleman S, Redmond TM (2003) Identification of beta-carotene 15, 15′-monooxygenase as a peroxisome proliferator-activated receptor target gene. FASEB J 17:1304–1306
29. Teboul L, Febbraio M, Gaillard D, Amri EZ, Silverstein R, Grimaldi PA (2001) Structural and functional characterization of the mouse fatty acid translocase promoter: activation during adipose differentiation. Biochem J 360:305–312
30. Vu-Dac N, Gervois P, Jakel H, Nowak M, Bauge E, Dehondt H, Staels B, Pennacchio LA, Rubin EM, Fruchart-Najib J, Fruchart JC (2003) Apolipoprotein A5, a crucial determinant of plasma triglyceride levels, is highly responsive to peroxisome proliferator-activated receptor alpha activators. J Biol Chem 278:17982–17985
31. Staels B (2003) Peroxisome proliferator-activated receptor alpha induces hepatic expression of the human bile acid glucuronidating UDP-glucuronosyltransferase 2B4 enzyme. J Biol Chem 278:32852–32860
32. Hunt MC, Yang YZ, Eggertsen G, Carneheim CM, Gafvels M, Einarsson C, Alexson SE (2000) The peroxisome proliferator-activated receptor alpha (PPARalpha) regulates bile acid biosynthesis. J Biol Chem 275:28947–28953
33. Louet JF, Chatelain F, Decaux JF, Park EA, Kohl C, Pineau T, Girard J, Pegorier JP (2001) Long-chain fatty acids regulate liver carnitine palmitoyltransferase I gene (L-CPT I) expression through a peroxisomeproliferator-activated receptor alpha (PPARalpha)-independent pathway. Biochem J 354:189–197
34. Lemay DG, Hwang DH (2006) Genome-wide identification of peroxisome proliferator response elements using integrated computational genomics. J Lipid Res 47:1583–1587
35. Devchand PR, Keller H, Peters JM, Vazquez M, Gonzalez FJ, Wahli W (1996) The PPARα-leukotriene B4 pathway to inflammation control. Nature 384:39–43
36. Dowell P, Peterson VJ, Zabriskie TM, Leid M (1997) Ligand-induced peroxisome proliferator-activated receptor alpha conformational change. J Biol Chem 272:2013–2020
37. Osumi T, Wen JK, Hashimoto T (1991) Two cis-acting regulatory sequences in the peroxisome proliferator-responsive enhancer region of rat acyl-CoA oxidase gene. Biochem Biophys Res Commun 175:866–871
38. Sher T, Yi HF, Mcbridge OW, Gonzalez FJ (1993) cDNA cloning, chromosome mapping, and functional characterisation of the human peroxisome proliferator activated receptor. Biochemistry 32:5598–5604
39. Lee SST, Pineau T, Drago J, Lee EJ, Owens JW, Kroetz DL, Fernandez-Salguero PM, Westphal H, Gonzalez FJ (1995) Targeted Disruption of the alpha isoform of the peroxisome proliferator-activated receptor gene in mice results in abolishment of the pleiotropic effects of peroxisome proliferators. Mol Cell Biol 15:3012–3022
40. Aoyama T, Peters JM, Iritani M, Nakashima T, Furihata K, Hashimoto T, Gonzalez FJ (1998) Altered constitutive expression of fatty acid-metabolizing enzymes in mice lacking the peroxisome proliferator-activated receptor alpha (PPAR-α). J Biol Chem 273:5678–5684
41. Chomczynski P, Sacchi N (2006) The single-step method of RNA isolation by acid guanidium thiocyanate-phenol-chloroform extraction: twenty-something years on. Nat Protoc 1: 581–585
42. Chomczynski P, Sacchi N (1998) Single step method for RNA isolation by acid guanidinium thiocyanate-phenol-chloroform extraction. Anal Chem 162:156–159
43. Freier SM, Kierzek R, Jaeger JA, Sugimoto N, Caruthers MH, Neilson T, Tumer DH (1986) Improved free-energy parameters for predictions of RNA duplex stability. Proc Natl Acad Sci USA 83:9373–9377
44. Don RH, Cox PT, Wainwright BJ, Baker K, Mattick JS (1991) 'Touchdown' PCR to circumvent spurious priming during gene amplification. Nucleic Acids Res 19:4008
45. Roux KH (1995) Optimisation and trouble shooting in PCR. PCR Methods Appl 4: 5185–5194
46. Geber AM, Oesch F (1992) Induction of peroxisome proliferator activated receptor by fenofibrate in rat liver. FEBS Lett 309:37–40
47. Lowry OH, Rosenbrough NJ, Farr AL, Randall RJ (1951) Protein measurement with the Folin phenol reagent. J Biol Chem 193:265–275
48. Laemmli UK (1970) Cleavage of structural protein during the assembly of the head of bacteriophage T4. Nature 227:680–685

Chapter 3

PPARGC1A and PPARG Genotyping: Beginner's Guide to Genotyping with Unlabeled Probes

Viera Habalová and Jozef Židzik

Abstract

Amplicon melting and genotyping with unlabeled probes has become wide spread technique for SNP genotyping. In both methods PCR is performed with new generation of double-strand DNA binding dyes. Nucleic acid melting generates distinct melting curves that can be used to identify the presence of sequence variation within the amplicon. Unlabeled probes are inexpensive and provide the sequence specificity. This chapter provides comprehensive instructions about design of primers and probes and evaluation of their quality. Chapter also contains useful advices on how to perform DNA purification, optimize PCR conditions, how to prepare working solutions, and how to interpret the data of melting analysis.

Key words: Unlabeled probes, Genotyping, Melting analysis, PPARGC1A, PPARG, Primers and probe design

1. Introduction

In the last decade, in conjunction with real-time polymerase chain reaction (PCR) two kinds of cost effective closed-tube methods become popular for SNP (Single Nucleotide Polymorphism) genotyping: amplicon melting and genotyping with unlabeled probes (LUNA probes). Both methods use the new generation of saturating dye that fluorescence in the presence of double-stranded DNA. Genotyping by method of amplicon melting depends on the specificity of the PCR primers. It is commonly believed that the extra specificity of a probe is necessary in clinical assays at least. Unlabeled probes on the other hand are helpful also when amplicon melting alone does not provide adequate detail in highly polymorphic region. Any variation in sequence between primers will affect amplicon melting while only a variant under the probe will

Mostafa Z. Badr and Jihan Youssef (eds.), *Peroxisome Proliferator-Activated Receptors (PPARs): Methods and Protocols*, Methods in Molecular Biology, vol. 952, DOI 10.1007/978-1-62703-155-4_3, © Springer Science+Business Media New York 2013

affect probe melting (1, 2). Genotyping with unlabeled probes was first published in 2004. It is based on monitoring of the melting probe-target duplexes after asymmetric PCR, in which one primer in excess overproduces the target strand recognized by probe. Different probe melting temperatures are recorded depending upon the sequence of different alleles under the probe. Heterozygous samples are easily distinguished from homozygous samples by a double peak on derivate melting curve plots. To prevent PCR elongation of the probes, those are usually blocked at the 3′-end often with phosphate, although other blockers are more stable. The best discrimination of unlabeled probes is on high-resolution melting instruments but it can be also easily detected on standard real-time instrument. LightCycler 1.5 (Roche) or LightScanner 32 (Idaho) instruments, for example, perform amplification and melting analysis on samples in capillaries. The following procedure is for genotyping of SNP rs8192678 (Gly482Ser) in PPARGC1A and rs1801282 (Pro12Ala) in PPARG gene.

2. Materials

2.1. DNA Purification Equipment and Chemicals

Equipment

1. High speed centrifuge.
2. Tube thermoblock.
3. Nanodrop 2000 (Thermo Scientific).
4. Vortex.

Chemicals

1. Wizard genomic purification kit (Promega).
2. 96–100% Ethanol molecular biology grade.
3. Isopropanol molecular biology grade.
4. Samples: Anti-coagulated peripheral venous blood in EDTA/sodium heparin/sodium citrate. Anticoagulant blood samples may be stored at 2–8°C for up to 2 months, but DNA yield will be reduced with increasing length of storage.

2.2. Gradient PCR Components

Instruments

1. Table top mini-centrifuge.
2. Vortex.
3. Gradient Tprofessional Basic thermocycler (Biometra).
4. Electrophoresis chamber.
5. Microwave oven.

6. Power supply.
7. Parafilm "M" (American National Can).
8. UV Transilluminator.

Chemicals

1. Molecular grade water.
2. 1× TE (Tris–EDTA) buffer (10 mM Tris, 1 mM EDTA) pH 8.0, i.e., DNA rehydration solution from Wizard genomic purification kit (Promega).
3. 10× BioThermAB reaction buffer with $MgCl_2$ (160 mM $(NH_4)_2SO_4$, 670 mM Tris–HCl ph 8.8 (at 25°C), 15 mM $MgCl_2$, 0.1% Tween 20) (GeneCraft, Munster, Germany).
4. 12.5 mM dNTP mix (Jena Bioscience, Jena, Germany).
5. BioThermAB Hot Start Taq DNA polymerase (5 U/μL) (GeneCraft, Munster, Germany).
6. Working solutions of primers and probes (according to Habalová et al. (3)):

   Primers for Pro12Ala (rs1801282) polymorphism in PPARG gene

   10 μM solution of PPARG-Forward primer 5′-CCCTATT CCATGCTGTTATG-3′.

   10 μM solution of PPARG-Reverse primer 5′-AATACTG GAGTGTACACATGATAG-3′.

   10 μM solution of PPARG-Probe 5′-TTCTCCTATTGA CCCAGAAAGCGAT-P-3′.

   For Gly482Ser (rs8192678) polymorphism in PPARGC1A gene

   10 μM solution of PPARGC1A-Forward primer 5′-ATCC CAGTCAAGCTGTTTTTGA-3′.

   10 μM solution of PPARGC1A-Reverse primer 5′-ATT GTTCATTACTGAAATCACTGTCC-3′.

   10 μM solution of PPARGC1A-Probe 5′-GAAGCAGA CAAGACCAGTGAACTG-P-3′ (see Note 1).
7. 10× TBE buffer (Tris–Borate–EDTA), pH 8.0 (see Note 2).
8. Agarose for DNA electrophoresis.
9. Distilled water.
10. 6× Loading dye (see Note 3).
11. 10 mg/mL Ethidium bromide solution.
12. Quick-Load® 100 bp DNA Ladder (New England Biolabs).

2.3. Real-Time PCR Components

Instruments

1. Light Cycler 1.5 or 2. (Roche) or LightScanner32 (Idaho Technology).
2. Centrifuge.

Chemicals

1. Molecular grade water.
2. 10× BioThermAB reaction buffer with $MgCl_2$ (160 mM $(NH_4)_2SO_4$, 670 mM Tris–HCl pH = 8.8 (at 25°C), 15 mM $MgCl_2$, 0.1% Tween 20) (GeneCraft, Munster, Germany).
3. 12.5 mM dNTP (Jena Bioscience, Jena, Germany).
4. 10× LCGreenPlus (Idaho Technology Inc., Salt Lake City, USA).
5. 25 mM $MgCl_2$.
6. BioThermAB Hot Start Taq DNA polymerase (GeneCraft, Munster, Germany).
7. Working solutions of primers and probes:

   For Polymorphism rs1801282

   1 μM solution of PPARG-Forward primer 5′-CCC TATTCCATGCTGTTATG-3′.

   10 μM solution of PPARG-Reverse primer 5′-AATACTG GAGTGTACACATGATAG-3′.

   10 μM solution of PPARG-Probe 5′-TTCTCCTATT GACCCAGAAAGCGAT-P-3′.

   For polymorphism rs8192678

   1 μM solution of PPARGC1A-Forward primer 5′-ATCC CAGTCAAGCTGTTTTTGA-3′.

   10 μM solution of PPARGC1A-Reverse primer 5′-ATTG TTCATTACTGAAATCACTGTCC-3′.

   10 μM solution of PPARGC1A-Probe 5′-GAAGCA GACAAGACCAGTGAACTG-P-3′.

3. Methods

3.1. Design of Primers and Probe

While designing primers and probes it is important to locate the mutation or polymorphism of interest and determine all the known variants present in the whole region of the amplicon to minimize the chance that the primers are placed over polymorphisms or the probe will sit over another polymorphism different from polymorphism of interest. Many databases free of charge are available online, which can help to determine target sequence. For example: *Ensembl* is a joint scientific project between the European Bioinformatics Institute and the Wellcome Trust Sanger Institute. Ensembl's aim remains to provide a centralized resource for geneticists, molecular biologists, and other researchers studying the genomes of our own species and other vertebrates and model organisms. Commercially available computer software (e.g., OLIGO 6, LightScanner Primer Design) or Web-based tools (e.g., Primer3) can be used for primer design. The software may reduce the cost and time involved in experimentation by lowering the chances of failed experimentation. Free software OligoAnalyzer 3.1 (Integrated DNA Technologies) allows simple checking of primers and probes. It is a useful tool to determine primer properties like T_m, GC%, primer loops, and primer dimers formation and primer–primer compatibility.

3.1.1. Target and Primer Sequence Checking (Example for rs8192678 of PPARGC1A Gene)

Ensembl database can be used for target sequence and primer sequence checking and stepwise instructions how to use it follows:

Oligonucleotide sequences from Habalová et al. (3):

PPARGC1A-Forward primer 5′-ATCCCAGTCAAGCTGTTTTTGA-3′.

PPARGC1A-Reverse primer 5′-ATTGTTCATTACTGAAATCACTGTCC-3′.

PPARGC1A-Probe 5′-GAAGCAGACAAGACCAGTGAACTG-P.

Target Sequence Checking

1. Type "http://www.ensembl.org/" in address bar of your internet browser.
2. Scroll down the page to the part marked "Search." In drop down list select: "Human."
3. Type the name of gene in search bar (in our case type PPARGC1A) and Click "Go."
4. Click "Homo sapiens" and then gene in "Results summary."
5. Click *PPARGC1A* [Ensembl/Havana merge gene: ENSG00000109819].
6. Click "Sequence" in column "Gene-based displays" on the left.
7. Sequence of gene PPARGC1A will be displayed.
8. Click "Configure this page" in the left column.
9. Scroll down Display options to the box "Show variations" and select "Yes and show links." Close and save the configuration in this window by clicking on a tick in top right corner.

10. Press "CTRL+F" on your keyboard for browser's search bar to appear.
11. Type "rs8192678" press "Enter." Sequence around wanted polymorphism/mutation with all other polymorphisms will be found and displayed.
12. Copy out the sequence (about 200 bp around selected polymorphism) to the MicrosoftWord®. Next delete all unnecessary links. Use this sequence for the design of primers, or for primers checking, or amplicon checking (see Notes 4 and 5).

In our example the selected sequence around rs8192678 of PPARGC1A gene is:

AAAACAGCTCCAAGACCAGGAAATCCGAGCCGA
GCTGAACAAGCACTTCGGTCATCCCAGTCAAGCT
GTTTTTGA**Y**GACGAAGCAGACAAGACC**G**GTGAACTG
AGGGACAGTGATTTCAGTAATGAACAATTCTC
CAAACTACCTATGTTTATAAATTCAGGACTAG
CCATGGATGGCCTGTTTGA

(Underlined **G** is polymorphic site rs8192678; **Y**—other polymorphic site; gray shade indicates the sequence of the studied oligonucleotides—primers and probe).

Oligonucleotides: Primers and Probe—Quality Assessment

1. Open OligoAnalyzer3.1.—http://eu.idtdna.com/analyzer/Applications/OligoAnalyzer/.
2. Type the sequence of the forward primer (5′-ATCCCAGTCAAGCTGTTTTTGA-3′ in our example) in 5′ to 3′ direction into input field (see Note 6).
3. Adjust Oligo conc. to 0.05 mM (concentration of limit primer). Mg^{++} conc. to 3 mM, dNTPs conc. to 0.2 mM (see Note 7).
4. Click "Analyze." As a result you can see Length (22 bp), T_m (61.4°C), and GC content of primer (40.9%) (see Note 8).
5. Next click "Hairpin" to analyze the hairpin formation in primer.
6. Adjust "Temperature" according to predicted T_m value of primer obtained in step 4 (about 62°C for PPARGC1A forward primer in our example).
7. Check hairpin structure according to T_m. Check whether T_m of hairpin secondary structure is below T_m of primer (see Note 9).
8. Next click "Self-dimer" to analyze undesirable self-dimer formation of the primer.
9. Check self-dimer structure according to Delta G (see Note 10).
10. Click "clear sequence."
11. Type the sequence of reverse primer (5′-ATTGTTCATTACTGAAATCACTGTCC-3′ in our example) in 5′to 3′ direction into input field.

12. Adjust Oligo conc. to 0.5 mM (concentration of excess primer). Mg^{++} conc. to 3 mM, dNTPs conc. to 0.2 mM (see Note 7).
13. Repeat steps 4–10 to assess the quality of reverse primer.
14. Type the sequence of probe (GAAGCAGACAAGACCA GTGAACTG in our example) in 5′to 3′ direction into input field (see Note 11).
15. Adjust Oligo conc. to 1 mM, Mg^{++} conc. to 3 mM, dNTPs conc. to 0.2 mM (see Note 7).
16. Repeat steps 4–5 to assess the quality of probe.
17. Adjust Temperature for synthesis temperature, usually 72°C, and click "Update."
18. Check hairpin structure according to T_m (see Note 9).
19. Perform primer pair comparison, than comparison of forward primer with probe and reverse primer with probe for heterodimeric structures (see Note 12).
20. Type the sequence of first oligonucleotide (e.g., forward primer) into input field.
21. Click "Default settings."
22. Click "Hetero-Dimer."
23. Type the sequence of second oligonucleotide (e.g., reverse primer) to the field "Secondary sequence."
24. Click "Calculate."
25. Analyze forward and reverse primer heterodimers according to delta G (see Note 12).
26. Use the same procedure (steps 20–25) for analysis of forward primer (first oligonucleotide) and probe (second oligonucleotide) heterodimers.
27. Use the same procedure (steps 20–25) for analysis of reverse primer (first oligonucleotide) and probe (second oligonucleotide) heterodimers.
28. Click "Clear sequence."
29. Insert the amplicon sequence delimited by forward and reverse primer ATCCCAGTCAAGCTGTTTTGAYGACGAAGCA GACAAGACCGGTGAACTGAGGGACAGTGA TTTCAGTAATGAACAAT (see Note 13).
30. Click "Hairpin."
31. Adjust Temperature for synthesis temperature usually 72°C and click "Update."
32. Check whether T_m of amplicon secondary structures are below T_m of primer pairs (i.e., 64°C; Fig. 1).
33. Avoid cross homology. Open http://www.ncbi.nlm.nih.gov/tools/primer-blast/ (see Note 14).

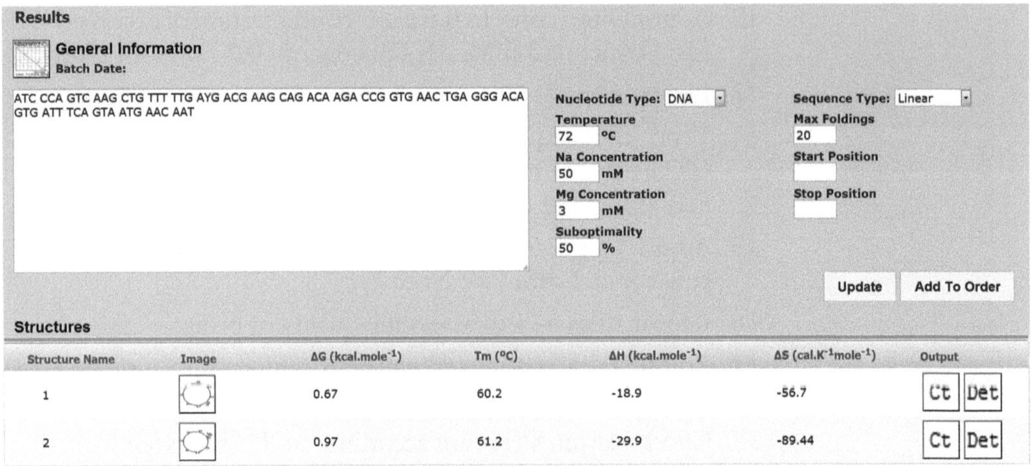

Fig. 1. Oligo analyzer settings and results for amplicon checking.

34. Type the sequence of forward and reverse primer in 5′ to 3′ direction into input field.
35. Adjust min. PCR product size (according to expected PCR product) and max. product size (e.g., 1,000).
36. Adjust primer melting temperatures according to predicted T_m from OligoAnalyzer 3.1 software.
37. Use default settings for exon/intron selection box.
38. In box "Primer Pair Specificity Checking Parameters" select organism (Homo sapiens), database (Genome reference assembly from selected organisms), Misprimed product size deviation (1,000).
39. Click "Get Primers."
40. Avoid of primer which have at least two mismatches within the last 5 bp at the 3′-end or which have more than 35% of total mismatches.

3.2. DNA Purification

DNA extraction and purification should be standardized so that all samples are prepared in the same buffers to minimize ionic strength differences that affect melting curves. Additionally, spectrophotometric quantification and dilution of sample DNA to a standard concentration for input into PCR is recommended (4). An example of simplified instructions (protocol used in our laboratory) of how to extract, purify, and prepare DNA samples for subsequent real-time PCR follows:

1. Process anti-coagulated peripheral venous blood according to the Wizard Genomic Purification Kit Handbook (see Note 15).
2. Measure the concentration and purity of DNA on spectrophotometer (see Note 16).

3. Adjust the DNA concentration to ~50 ng/μL for gradient PCR and 20 ng/μL for real-time PCR using 1× TE buffer (i.e., DNA rehydration solution from Wizard genomic purification kit—Promega) in new tube. Undiluted and diluted DNA can be stored for short-term period at 4°C or long-term period at −20°C.

3.3. PCR Optimization, Gradient PCR

Problems with the PCR amplification of specific DNA fragment occur very often. Unspecific secondary amplicons may amplify during the PCR reaction, which hinder an unequivocal assessment of the melting analysis results. In such cases, PCR conditions must be optimized. The selection of the annealing temperature is possibly the most critical step for optimizing the specificity of PCR reaction. Other steps of optimization include: titrating the magnesium, template, primer, dNTP, and TaqDNA polymerase concentration and adding detergents.

It is recommended to test annealing temperature empirically by gradient PCR. Starting gradient PCR temperature is normally 5°C below the calculated temperature of the primer melting point (T_m). However, the possible formation of unspecific secondary amplicons shows that the optimum temperature is often much higher than the calculated temperature. Currently, there are many Gradient Thermal Cyclers on the market (e.g., from Biometra, Bio-rad, Eppendorf, etc.) which enable rapid testing of the optimum temperature conditions on one block and in one experiment. During the PCR with these devices a temperature gradient is built up across the thermoblock.

An annealing temperature is selected for all primer pairs to produce specific PCR products. Optimization of parameters beyond the annealing temperature, including primer and Mg^{2+} concentrations can be performed, although it is often easier to replace problematic primer sets.

3.3.1. PCR Mix Preparation for Gradient PCR

PCR mix contains all necessary reaction components for PCR including DNA template, primers, and probe. In gradient PCR we perform symmetric PCR usually.

Dilution of primers and probes to proper concentration. Preparation of stock and working solutions

1. Before opening a tube containing lyophilized primer or probe, spin the tube briefly to collect all material at the bottom of the tube.
2. Prepare concentrated stocks of primers and probes. Add the required volume of 1× TE buffer, mix, and leave for 30 min to allow the oligonucleotides to completely dissolve (see Note 17). Concentrated stocks of primers and probes

should be stored in small aliquots at −20°C to prevent the degradation of oligonucleotides.

3. Prepare working solutions of oligonucleotides. Vortex and spin concentrated stock and then dilute tenfold stock solution with molecular grade water (example: mix 90 μL of water and 10 μL of stock solution into sterile tube). These working solutions should be stored at 4°C for a period no longer than 1 month.

Gradient PCR. Preparation of PCR mix for gradient PCR (see Note 18)

1. Briefly vortex and spin all chemicals. Do not vortex BioThermAB Hot Start Taq DNA polymerase, spin it in a centrifuge only. Prepare master mix with DNA according to Table 1 in sterile 1.5–2 mL tube.
2. Vortex, mix, and aliquot 20 μL of PCR mix to sterile 0.2 mL PCR tubes.
3. Spin tubes in the centrifuge.
4. Put tubes in one row of Gradient Tprofessional Basic thermocycler or one column of thermal block, depending on the instrument used and amplify them according to PCR reaction conditions (Table 2; see Note 19):

Table 1
Master mix for gradient PCR

	1 sample (μL)	10 samples (μL)
Water for PCR	11.8	118
10× BioThermAB reaction buffer with $MgCl_2$	2	20
12.5 mM dNTP mix	0.4	4
25 mM $MgCl_2$	0.6	6
10 μM forward primer	1	10
10 μM reverse primer	1	10
10 μM probe	1	10
BioThermAB Hot Start Taq DNA polymerase (5 U/μL)	0.2	2
DNA (50 ng/μL)	2	18
Total reaction volume	20	180

Table 2
Cycling conditions for gradient PCR

	Temperature (°C)	Time	
Initial denaturation	95	5 min	
Denaturation	95	30s	40×
Annealing	50–70	30s	
Synthesis	72	30s	
Cooling	16	Hold	

3.3.2. Preparation of Agarose Gels and Electrophoretic Separation of DNA

Agarose gel electrophoresis is the easiest and most common way of separating and analyzing DNA fragments. The DNA is visualized in the gel by ethidium bromide or environmentally safe GelRed nucleic acid stain (Biotium). These chemicals bind to DNA and are fluorescent, meaning that they absorb invisible UV light and transmit the energy as visible orange light.

Preparation of the 3% agarose gel (size of gel casting tray: 13×14 cm in this example) (see Note 20)

1. Prepare 1× TBE buffer (mix one part of 10× TBE with nine parts of distilled water). Mix thoroughly.
2. Add 91 mL 1× TBE and 10 mL distilled water (for evaporation) into a 250 mL conical flask. Weigh 2.73 g of agarose and transfer it to the flask, swirl to mix.
3. Heat the flask on medium power in microwave oven for 2 min to dissolve the agarose. Boil the mixture for two to three times.
4. Then mix thoroughly and cast gel slowly into the plastic tray with comb. Push any bubbles away to the side using a disposable pipette tip (see Note 21).
5. Allow the molten agarose to cool and gel at room temperature for 30 min.
6. Carefully remove the comb and insert tray horizontally into the electrophoresis chamber and pour the electrophoresis buffer (1× TBE) into chamber. The buffer level must be 2–5 mm over the gel.
7. Put strip of parafilm on the laboratory desk and make small drops (3 μL) of 6× loading dye (see Note 3).
8. Load the first well with 100 bp ladder (size marker).
9. Mix samples (15 μL) with loading dye on parafilm and pipette it into the sample wells.
10. Close the gel tank with lid, switch on the power-source, and run the gel at 112 V (It is recommended not to exceed 8 V/cm of gel). Mind the polarity of electrodes!

11. Monitor the progress of electrophoresis by reference to the marker dye, till the dye (bromophenol blue) reaches the 2/3 of the gel length (see Note 22).
12. Mix 50 μL of ethidium bromide stock solution and 500 mL 1× TBE buffer into nontransculent container.
13. After the electrophoresis is over, transfer the gel to a container with ethidium bromide solution (see Note 23). Stain the gel approximately for 10–15 min.
14. Rinse the gel in the container with deionized water for 10 min.
15. Place the gel on the UV transilluminator (see Note 24) in a darkroom and perform visual analysis. You can make a picture of the gel, to store it for later analysis and documentation. Figure 2 shows typical pattern for gradient PCR results.

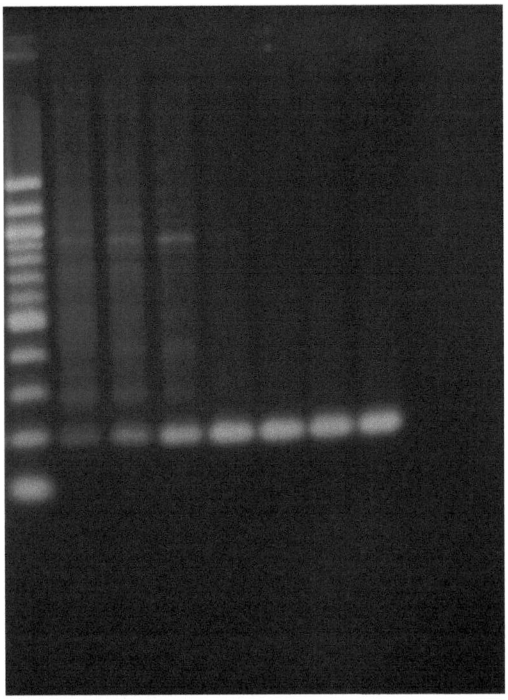

Fig. 2. Typical pattern for gradient PCR results. Agarose gel (3%) stained with ethidium bromide used for experimental determination of optimal annealing temperature. Gradient PCR temperature is from 54 to 68°C. The calculated annealing temperature of primer pair was 59°C. *Lane 1*: 1 kb DNA ladder, *Lanes 2–7*: (200 bp PCR products amplified at different temperature in gradient PCR), *Lane 2*: 54.4°C, *Lane 3*: 55.1°C, *Lane 4*: 57.2°C, *Lane 5*: 59.9°C, *Lane 6*: 63.3°C, *Lane 7*: 66°C. At temperatures lower than 63.3°C (*Lane 1–4*), a pattern of nonspecific bands result from nonspecific product amplification, where the intensity of that product was inversely proportional with the intensity of the specific product. The best temperatures for obtaining target amplicon products of outstanding quality were in the range of 63.3–68.0°C. This range of temperatures resulted in excellent amplification, as evidenced by the robust, unique, and clean fragment obtained, without any secondary products such as primer dimers. So the optimal annealing temperature should fit within this temperature range. We chose the temperature of 65°C for annealing.

3.4. Real-Time PCR Genotyping of rs1801282 in PPARG and rs8192678 in PPARGC1 with Unlabeled Probe

3.4.1. Preparation of Working Solution of Oligonucleotides for Asymmetric PCR

1. Vortex and spin concentrated stocks of primers and probes prepared according to instructions in Subheading 3.1.
2. Tenfold dilute the stock solution with molecular grade water for excess primers (in our case both reverse primers) and probe and 100-fold dilute the stock solution for limit primer (in our case both forward primers) (example: Mix 90 μL of molecular grade water and 10 μL of stock solution of reverse primer/probe into sterile tube for tenfold dilution. Add 99 μL of water and 1 μL of stock solution of forward primer for 100-fold dilution). These working solutions should be stored at 4°C for a period no longer than 1 month.

3.4.2. Preparation of PCR Mix for Asymmetric PCR

The usual total volume reaction is 10 μL however it can be scaled down to 5 or up to 15 μL if necessary depending on the instrument used for real-time PCR. Asymmetric PCR (1:5 to 1:10) produces double-stranded amplicon in earlier phase of PCR and the appropriate single-stranded DNA for probe binding in the later phase of PCR (5).

1. Prepare master mix according to Table 3, separately for each polymorphism without DNA.
2. Aliquot 9 μL of the above master mix into each capillary.
3. Add 1 μL of DNA to each capillary.
4. Spin all samples into the capillaries on a centrifuge.

Table 3
Master mix for asymmetric PCR

	1 sample (μL)	10 samples (μL)
Water for PCR	4.8	48
10× BioThermAB reaction buffer with $MgCl_2$	1	10
12.5 mM dNTP mix	0.4	4
25 mM $MgCl_2$	0.6	6
1 μM forward primer	0.5	5
10 μM reverse primer	0.5	5
10 μM probe	1	10
BioThermAB Hot Start Taq DNA polymerase (5 U/μL)	0.2	2
DNA sample (20 ng/μL)	1	–
Total volume	10	100

5. Amplify on a LightCycler1.5 according the instruction. For C>G PPARG polymorphism the PCR conditions were as follows: Initial denaturation at 95°C for 5 min, 65 cycles at 95°C for 10 s, 55°C for 10 s, and 72°C for 20 s and for PPARGC1 G>A polymorphism the PCR conditions were: initial denaturation at 95°C for 5 min, 65 cycles at 95°C for 15 s, 61°C for 15 s, and 72°C for 15 s. Amplicon sizes is 85 base pairs for PPARG, 78 base pairs for PARGC1A (see Note 25).
6. Perform melting on the LightCycler 1.5 with a denaturation at 95°C for 15 s and renaturation at 50°C for 1 min. Acquired the fluorescence data over the 50–90°C range at the thermal transition rate of 0.1°C/s. Identify genotypes by the different melting temperatures indicated by peaks on the derivate plots using instrument software (Figs. 3 and 4).

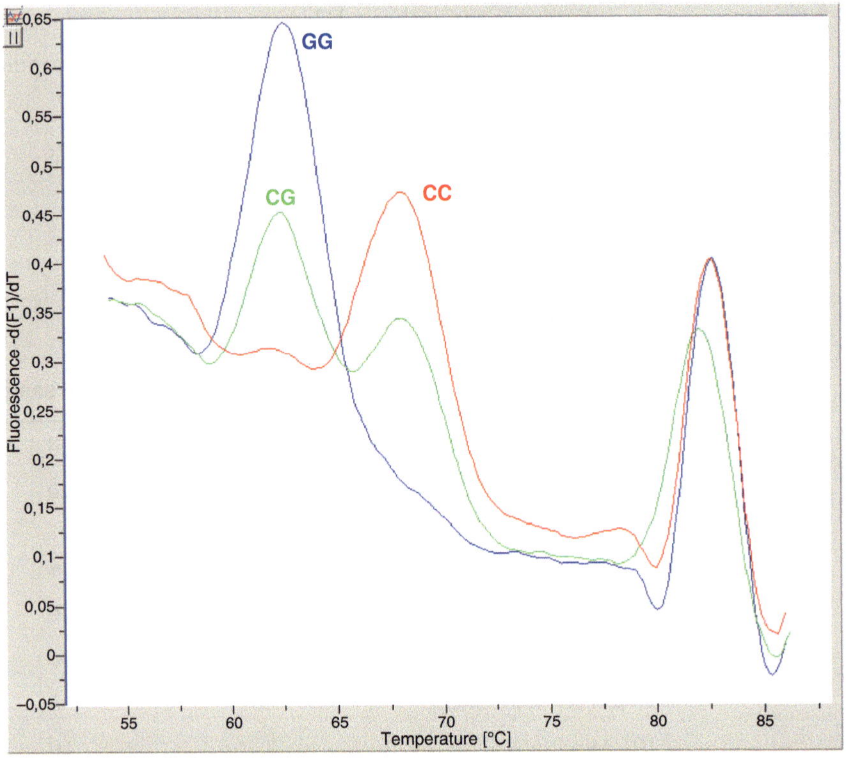

Fig. 3. Genotyping by unlabeled probe melting analysis for PPARG gene. Data are displayed as melting curves. Melting curves show of both probe/product (*left side of curve*—probe peaks) and product/product melting (*right side of curve*—amplicon peaks). Homozygous samples are represented by a single probe melting peak, whereas heterozygous samples have two peaks corresponding to both alleles. Probe-target melting is observed between 58 and 74°C. A probe was designed to ancestral allele C. A perfectly matched probe-target hybrid has a characteristic melting temperature that is higher than a mismatched hybrid.

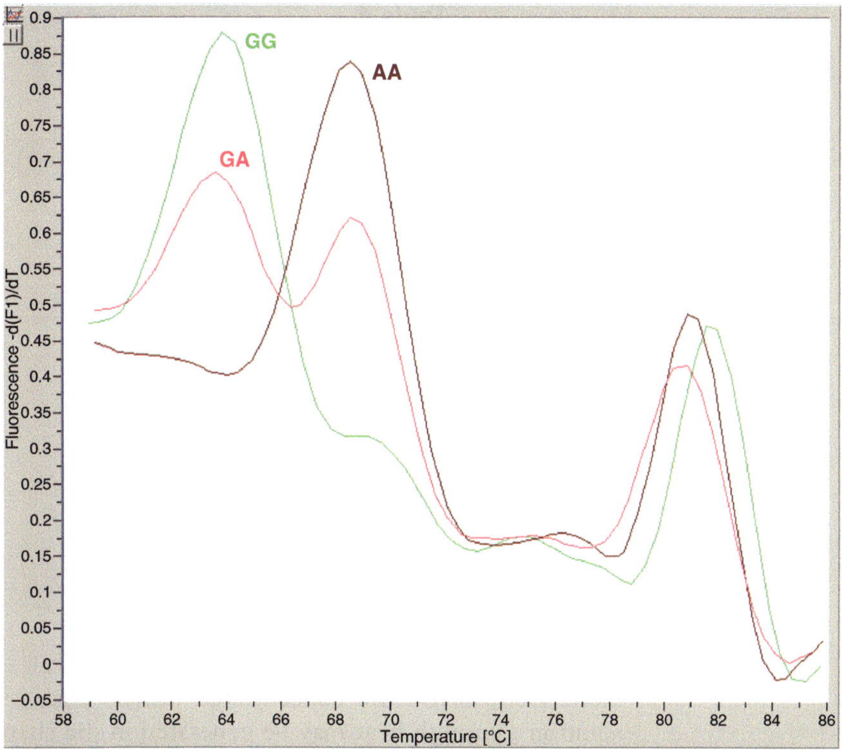

Fig. 4. Genotyping by unlabeled probe melting analysis for PPARGCA1 gene. Data are displayed as melting curves. Probe-target melting is observed between 55 and 75°C. A probe was designed for ancestral allele A. Samples with the A allele had a derivative melting peak at 69°C, whereas samples with the G allele showed a melting peak at 64°C.

4. Notes

1. Probes had to be blocked on 3′-end to prevent from elongation. Incompletely blocked or degraded probe produces extra peaks on derivate plots between the expected probe and product melting temperatures. The cheapest blocker is phosphate but the amino-modified C6, inverted dT, and C3 spacer provide the better blocking efficiencies than phosphate (6).
2. Several different buffers have been recommended for electrophoresis of DNA. The most commonly used for duplex DNA are TAE (Tris–Acetate–EDTA) and TBE (Tris–Borate–EDTA). DNA fragments will migrate at somewhat different rates in these two buffers due to differences in their ionic strength. You must use the same buffer at this stage as you used to prepare the gel, i.e., if you used 1× TBE in the gel, then use 1× TBE as a running buffer. If you mistakenly use water instead of buffer, there will be essentially no migration of DNA in the gel.

Table 4
IUPAC single-letter code for nucleotide sequences

R	A or G
W	A or T
S	C or G
Y	C or T
K	G or T
V	A or C or G
H	A or C or T
D	A or G or T
B	C or G or T
X/N	A or C or G or T

Conversely, if you use concentrated buffer (e.g., a 5× stock solution), enough heat may be generated in the chamber that will eventually melt the gel.

3. *Loading buffer*, usually contains something dense (e.g., glycerol) to allow the sample to land into the sample wells, and one or two tracking dyes (bromophenol blue, xylene cyanol), which migrate in the gel and allow visual monitoring or how far the electrophoresis has proceeded.
4. Polymorphisms are usually marked by IUPAC (International Union of Pure and Applied Chemistry) single-letter codes (Table 4).
5. When you want to check sequence of reverse primer you have to create reverse complementary chain. Reverse complementary chain for reverse primer 5′-ATTGTTCATTACTGAAA TCACTGTCC-3′ is 5′-GGACAGTGATTTCAGTAATGAAC AAT-3′.
6. A set of rules for primer sequence design is as follows (adapted from Premier Biosoft International, Palo Alto, CA) (7): Primers should be 17–30 bases in length. *Avoid runs of 4 or more Gs or Cs at the 3′* end namely at G or C-rich sequences. Runs of three or more Cs or Gs at the 3′-ends of primers promote mispriming (because of stability of annealing), and should be avoided. *Avoid a T on 3′-end.* Primers with a T at the 3′-end have a greater tolerance of mismatch. *Repeats*: A repeat is a di-nucleotide occurring many times consecutively and should be avoided because they can misprime. For example: ATATATAT. A maximum

number of di-nucleotide repeats acceptable in an oligo is 4 di-nucleotides. *3′-end Stability*: It is the maximum deltaG value of the five bases from the 3′-end. An unstable 3′-end (less negative deltaG) will result in less false priming. The presence of G or C bases within the last five bases from the 3′-end of primers (GC clamp) helps promote specific binding at the 3′-end due to the stronger bonding of G and C bases. More than 3G's or C's should be avoided in the last five bases at the 3′-end of the primer. Irrelevant polymorphism under primer can be masked by incorporating universal bases or mismatches, but mismatches destabilize oligonucleotides and lower T_m (8).

7. Typically, real-time PCR requires higher concentrations of $MgCl_2$ for optimal results. We usually start with 3 mM concentration of $MgCl_2$ and 0.2 mM of dNTP mix. *Asymmetric PCR* is used to preferentially amplify one strand of the original DNA more than the other. It is probe-target strand, i.e., strand complementary to the probe which will be produced in excess. PCR is carried out as usual, but with a great excess of the primer for the chosen strand. Tenfold or fivefold primer asymmetry is usually optimal. In the asymmetric PCR we start with 0.05 mM of lower concentration primer (i.e., for rs 8192678 in PPARGC1A, it is forward primer), 0.5 mM (reverse primer), and 1 mM of probe.
8. *The GC content* (the number of G's and C's as a percentage of the total bases) of primer should be 40–60%. When designing primers a T_m of about 55–70°C is recommended. The melting temperature of both primers should be similar and should be as high as possible, within reason, in order to increase specificity but not to be higher than the reaction temperature of TaqDNApolymerase (72°C). Probes should typically be designed to have a T_m of about 5°C above the T_m of the primers. T_m between 50 and 85°C has been successfully used. Probe T_m can be lowered by using dUTP instead of dTTP or increased by inserting locked nucleic acids into probe (1, 9).
9. *Primer/probe Secondary Structures*: Presence of the primer secondary structures (dimers, hairpins) produced by intermolecular or intramolecular interactions can lead to poor or no yield of the product. They adversely affect primer template annealing and thus the amplification. They greatly reduce the availability of primers to the reaction. The stability of secondary structures is commonly represented by its deltaG value (Gibbs free energy), the energy required to break the secondary structure or T_m. The more negative value for deltaG indicates more stable, undesirable secondary structures. *Hairpins*: They are formed by intramolecular interactions within the primer/probe/amplicon and should be avoided. Presence of hairpins

at the 3′-end most adversely affects the PCR reaction. T_m of hairpin secondary structure of primers should be below of T_m of primers. T_m of hairpin secondary structure of probe should be below T_m of synthesis temperature.

10. *Self-dimer*: A primer self-dimer is formed by intermolecular interactions between the two (same sense) primers, in case that the primer is homologous to itself. Generally a large amount of primers are used in PCR compared to the amount of target gene. When primers form intermolecular dimers much more readily than hybridizing to target DNA, they reduce the product yield. Optimally a 3′-end self-dimer with a deltaG of −5 kcal/mol and an internal self-dimer with a deltaG of −6 kcal/mol is tolerated generally.
11. Amplicons less than 100 bp and unlabeled probes from 25 to 40 bp are best suitable for genotyping purposes. The length of an unlabeled probe depending on the GC content and the desired T_m. Longer probes with shorter PCR products give stronger signals (10).
12. *Heterodimers*: Avoid complementary sequences between the primer pair or primer–probe sequences. Avoid complementarity of two or more bases at the 3′-ends of oligonucleotides pairs to minimize primer-dimer or primer–probe formation. Heterodimers are formed by intermolecular interaction between complementary sequences of oligonucleotides. Optimally a 3′-end cross dimer with a delta G of −5 kcal/mol and an internal cross dimer with a delta G of −6 kcal/mol is tolerated generally.
13. *Template secondary structure*: A single-stranded nucleic acid sequences are highly unstable and fold into conformations (secondary structures). The stability of these template secondary structures depends largely on their free energy and melting temperatures(T_m). Consideration of template secondary structures is important in primer design. If primers are designed with secondary structures which are stable even above the annealing temperatures, the primers are unable to bind to the template and the yield of PCR product is significantly affected. Hence, it is important to design primers in the regions of the templates that do not form stable secondary structures during the PCR reaction.
14. *Cross homology*: To improve specificity of the primers it is necessary to avoid regions of homology. Primers designed for target sequence must not amplify other genes in the genome. Commonly, primers are designed and then BLASTed to test their specificity. The basic local alignment search tool (BLAST) finds regions of local similarity between sequences. The program compares nucleotide or protein sequences to sequence databases and calculates the statistical significance of matches.

15. Any DNA purification procedure can be used, but the same procedure is recommended for all samples.
16. The most common technique to determine DNA yield and purity is UV spectroscopy. Nucleic acids strongly absorb at 260 nm and less strongly at 280 nm while proteins do the opposite. DNA purity can be estimated from the A_{260}/A_{280} ratio and A_{260}/A_{230} ratio. Pure solutions of nucleic acid absorb approximately twice as much at 260 nm as at 280 nm. Experimentally, the A_{260}/A_{280} ratio of a pure DNA solution is between 1.8 and 2.0. As protein contamination increases, the ratio decreases. Additionally, the presence of contaminating organic solvents, such as phenol, can affect estimations of concentration and purity. EDTA, carbohydrates and phenol all have the maximum absorbance near 230 nm. The TRIzolreagent (used for DNA extraction) is a phenolic solution which absorbs in the UV both at 230 and ~270 nm. Guanidine HCL used for DNA isolations will absorb at ~230 nm. The 260/230 values for "pure" nucleic acid are often higher than the respective 260/280 values. Expected 260/230 values are commonly in the range of 2.0–2.2. If the ratio is appreciably lower than expected, it may indicate the presence of contaminants which absorb at 230 nm. Residual contaminants and proteins could influence the melting analysis. The NanoDrop 2000 is a cuvette free spectrophotometer. Small volume of just 2 μL is enough to measure DNA concentration in range from 5 to 3,000 ng/μL. Solution in which the DNA was diluted, is usually used for blank measurement (usually TE buffer).
17. Primers and probes are prepared with standard desalting and they are usually provided in lyophilized state or form. For better stability lyophilized oligos should be dissolved in a small volume of low-salt buffer, i.e., in 1× TE buffer (10 mM Tris, pH 7.5–8.0, 1 mM EDTA) to give a concentrated stock solution. Concentration of stock solution should be 100 μM, i.e., 100 pmol/μL. To resuspend an oligo at 100 μM concentration, add the volume of TE (in μL) equal to ten times the number of nanomoles of DNA present in the tube (as usually noted on the sheet provided with the oligo), which will produce a stock solution of recommended concentration (example: If you are supplied with 46 nmol of oligonucleotide it is necessary to add 460 μL of 1× TE buffer for preparation of concentrated 100 μM stock solution). Working solution (concentration 10 μM, i.e., 10 pmol/μL) is prepared by tenfold dilution of stock solution, but now molecular grade water is used for dilution instead of TE. Sterile pipette tips with filter should be used all the time to prevent contamination.
18. Aliquoting of all chemicals is recommended. It can help in the event of contaminations or freeze/thaw cycling degradation,

namely probes. If a no template control (NTC) comes up positive, you can always discard all aliquots you have and take fresh ones without contamination. Various Taq DNA polymerases can be used including FastStartTaq DNA polymerase (Roche), KlenTaqI (AB peptides). dNTP mix that includes dUTP may also be used but it is not considered necessary in closed-tube systems. Saturating dyes, LCGreen I, LCGreen Plus, LightCycler® 480 ResoLight Dye, or others can be used for unlabeled probe genotyping.

19. The predicted T_m of each primer pair is bracketed with an annealing temperature range extending 10–15°C above and 0–5°C below the T_m. For example, if one primer has predicted $T_m = 57$°C, second one $T_m = 61$°C, the arithmetic mean of predicted T_m for this pair of primers is 59°C. So annealing temperature used in gradient PCR program should be set from 54 to 70°C.
20. Gel casting trays composed of UV-transparent plastic are available in a variety of sizes. The open ends of the trays are closed with tape while the gel is being cast, then removed prior to electrophoresis. By using gels with different concentrations of agarose, one can resolve different sizes of DNA fragments. Higher concentrations of agarose facilitate separation of small DNAs, while low agarose concentrations allow resolution of longer DNAs. When you want to visualize the nonspecific fragments of various lengths (especially shorter) different from you PCR product band (~75–150 bp), use 3–4% gel.
21. When high concentrated gels (3–4%) are prepared, do not wait until the agarose solution is cooled, instead immediately pour it into a casting tray containing a sample comb.

    Typical composition of loading buffer

    - 25 mg bromphenol blue or xylene cyanol.
    - 3 mL glycerol.
    - H_2O to 10 mL.

    The exact amount of dye in buffer is not so important. Store the loading buffer at 4°C. 10 mL of loading buffer will last for years.
22. You can confirm that current is flowing by observing bubbles coming off the electrodes. DNA will migrate towards the positive electrode, which is usually colored red.
23. Ethidium bromide is an intercalating agent and a fluorescent dye used for staining of nucleic acids. *Ethidium bromide is a known mutagen and should be handled as a hazardous chemical–wear gloves while handling.*
24. *Transilluminator* (an ultraviolet lightbox) is used to visualize ethidium bromide-stained DNA in gels. *NOTE*: *to prevent*

damage to the eyes from UV light always wear protective eyewear while observing DNA on a transilluminator.

25. Due to the slow (arithmetic) amplification later in the reaction after the limiting primer has been used up, addition of extra PCR cycles is required.

Acknowledgment

This work was supported by Centre of Excellence for the Study of Atherosclerosis OPVaV—2009/2.1/02-SORO (50%) and research grant of the Ministry of Education of the Slovak Republic (Grant No: VEGA 1/0380/10) (50%).

References

1. Zhou L, Wang L, Palais R et al (2005) High-resolution DNA melting analysis for simultaneous mutation scanning and genotyping in solution. Clin Chem 51:1770–1777
2. Liew M, Seipp M, Durtschi J et al (2007) Closed-tube SNP genotyping without labeled probes/a comparison between unlabeled probe and amplicon melting. Am J Clin Pathol 127:341–348
3. Habalová V, Klimcakova L, Zidzik J et al (2009) Rapid and cost effective genotyping method for polymorphisms in PPARG, PPARGC1 and TCF7L2 genes. Mol Cell Probes 23:52–54
4. Erali M, Wittwer CT (2010) High resolution melting analysis for gene scanning. Methods 50:250–261
5. Erali M, Palais R, Wittwer C (2008) SNP genotyping by unlabeled probe melting analysis. Methods Mol Biol 429:199–206
6. Dames S, Margraf R, Pattison D et al (2007) Characterization of aberrant melting peaks in unlabeled probe assays. J Mol Diagn 9: 290–296
7. http://www.premierbiosoft.com/company/company.html
8. Margraf RL, Mao R, Wittwer CT (2006) Masking selected sequence variation by incorporating mismatches into melting analysis probes. Hum Mutat 27:269–278
9. Chou LS, Meadows C, Wittwer C et al (2005) Unlabeled oligonucleotide probes modified with locked nucleic acids for improved mismatch discrimination in genotyping by melting analysis. Biotechniques 39:644–650
10. Zhou L, Myers A, Vandenrsteen J et al (2004) Closed-tube genotyping with unlabeled oligonucleotides probes and a saturating DNA dye. Clin Chem 50:1328–1335Agnim illutem vero ea augiam, sequametum venis autet ad tem

Chapter 4

Generation of an Inducible, Cardiomyocyte-Specific Transgenic Mouse Model with PPAR β/δ Overexpression

Teayoun Kim, Olga Zhelyabovska, Jian Liu, and Qinglin Yang

Abstract

Peroxisome proliferator-activated receptors (PPARs) consist of three subtypes, each displaying distinctive tissue distribution. In general, the three PPAR subtypes exert overlapping function in transcriptional regulation of lipid metabolism. However, each PPAR subtype possesses distinctive functions in different tissues dependent on their expression abundance, endogenous ligands, and the PPAR coregulators in a specific tissue. Transgenesis is an invaluable technique in defining the in vivo function of a particular gene and its protein. Cre/LoxP-mediated gene targeting has been extensively used to explore the tissue-specific function of PPARs. While this tissue-specific loss-of-function approach is extremely useful in determining the essential role of a PPAR, the tissue-specific gain-of-function approach is another important technique used to understand the effects of PPAR activation in a particular tissue. Transgenic overexpression of PPAR in a specific tissue has been used. However, this conventional technique requires generating the transgenic models individually for each target tissue. In this chapter, we describe the methodology for a more efficient generation of transgenic mouse models with a constitutively active form of PPARβ/δ in different tissues.

Key words: Conditional transgenic, Inducible, Transgenic, PPARdelta, , Mouse model

1. Introduction

Peroxisome proliferator-activated receptors (PPARs) consist of three subtypes, each displaying distinctive tissue distribution. While it is well established that the three PPAR subtypes exert overlapping function in transcriptional regulation of lipid metabolism, it is recognized that each PPAR subtype possesses distinctive functions in different tissues, dependent on their expression abundance, endogenous ligands, the PPAR coregulators, and/or other unknown factors in a specific tissue. Therapies targeting PPARs are clinically used and PPARs remain the main drug targets for a growing number of diseases and conditions. However, our understanding

Mostafa Z. Badr and Jihan Youssef (eds.), *Peroxisome Proliferator-Activated Receptors (PPARs): Methods and Protocols*, Methods in Molecular Biology, vol. 952, DOI 10.1007/978-1-62703-155-4_4, © Springer Science+Business Media New York 2013

of the tissue-specific actions of each PPAR subtype remains poor, hindering a broader clinical use of PPAR agonists in specific groups of patients. For example, Avandia has been extensively used for the treatment of type II diabetes. However, recent reports of the potential risk of increasing heart attack incidences in patients using Avandia dramatically raise the concerns of the potentially fatal side effects of this PPARγ activator (1, 2). Therefore, preclinical studies on animal models to identify tissue-specific actions of a PPAR agonist are pivotal in designing novel PPAR agonists with minimal side effects and in the treatment of specific patients.

It is challenging to define the tissue-specific effects of PPAR activation in the intact animal. Cre/LoxP-mediated gene targeting has been extensively used to explore the tissue-specific function of PPARs. While this tissue-specific loss-of-function approach is extremely useful in determining the essential role of a PPAR subtype, tissue-specific gain-of-function approach is also important to fully understand the effects of PPAR activation in a specific tissue. Transgenic overexpression of PPAR in a specific tissue can be achieved and has been extensively used. However, this technique requires generating a transgenic model individually for each tissue to be assessed. In this chapter, we describe the methodology for a more efficient generation of transgenic mouse models with a constitutively active form of PPARβ/δ in the heart. The constitutively active PPARβ/δ was achieved by fusing the VP16 transactivation domain with the wild type PPARβ/δ. There are many reports using the same strategy on studying LXRα and PPARs with VP16 (3–5). We first generated a ubiquitous and inducible transgenic model driven by the CAG promoter (human cytomegalovirus immediate early enhancer/chicken β-actin promoter), in which the expression of the fusion protein VP16-PPARβ/δ was silenced with a stop sequence (a bacteria chloramphenicol acetyltransferase gene, CAT) before the initial codon of the VP16-PPARβ/δ. The stop sequence can then be removed by excising one of the LoxP sequences flanking it by the Cre recombinase specifically expressed in the particular tissue being assessed. This method has been successfully used (6, 7). By crossing the above transgenic line with another transgenic mouse line carrying Cre, the overexpression of VP16-PPARβ/δ can then be achieved as desired. In this chapter, we will demonstrate the successful establishment of transgenic line with cardiomyocyte-specific overexpression of the VP-16-PPARβ/δ fusion protein in cardiomyocytes.

2. Materials

1. CAG-LoxP-CAT-LoxP-X vector. The CAG-LoxP-CAT-LoxP-X vector is modified based on the CAG-CAT-lacZ vector (7).
2. DH5α competent cells (Invitrogen).

3. cDNAs for VP16 transactivation domain (pVP16 vector, Clontech).
4. Expand High Fidelity PCR System kit (Roche).
5. Relevant restriction enzymes, ligase, and buffers (New England Biolabs).
6. Antibodies from Santa Cruz Biotechnology: Polyclonal anti-PPARβ/δ antibody (sc-7197); polyclonal anti-GAPDH antibody (sc-25778); HRP-conjugated anti-rabbit IgG (sc-2004).
7. QIAEX II Gel Extraction kit (Qiagen).
8. Titanium Taq DNA polymerase (Clontech).
9. DNA elution buffer: 5 mM Tris–HCl, 0.1 mM EDTA, pH 7.4.
10. Slide-a-lyzer cassettes (Pierce).
11. Dialysis buffer: 10 mM Tris pH 7.2, 0.1 mM EDTA + 100 mM NaCl.
12. DNA extraction buffer: 50 mM Tris–HCl, 1.0 mM EDTA, pH 8.0, 0.5 % Tween-20, and 200 μg/mL proteinase K.
13. Total RNA extraction: RNeasy® Mini kit (QIAGEN).
14. Real-time PCR kit: *Power* SYBR® Green PCR master mix. cDNA preparation: Advantage® RT-for PCR kit (Clontech).
15. Nuclear protein extraction kit: CelLytic NuCLEAR Extraction Kit (Sigma).
16. Western blot kit: Membranes were visualized by using Kodak image station 4000R (Kodak) with Super Signal West Dura Extended Duration Substrate (Pierce).
17. Nitrocellulose membrane.
18. Real-Time PCR machine.
19. Thermal cycler.
20. The transgenic mouse line of tamoxifen inducible cardiac specific overexpression of Cre (α-MyHC-Mer-Cre-Mer mice) is a generous gift from Dr. Jeffrey Molkentin's group (8).
21. Tamoxifen (Sigma T-5648).
22. Sunflower seed oil (Sigma).
23. Primers for Real-time PCR: PPARβ/δ: 5′TCGGGCTTCCACTACGG3′ and 5′ACTGACACTTGTTGCGGTTCT3′; β-actin: 5′CTGTCCCTGTATGCCTCTG3′and5′ATGTCACGCACGATTTCC3′.

3. Methods

3.1. Generation of a Transgenic Mouse Model of VP-16-PPARβ/δ That is Inducible in Any Tissue by Cre Recombinase

1. Generation of the constitutively active form of PPARβ/δ gene (VP16-PPARβ/δ, or VPD): The constitutively active form of PPARβ/δ is created by fusing the activation domain of the herpes simplex virus Vp16 protein to the N-terminal of mouse PPARβ/δ (Fig. 1). This mutant (VP16-PPARβ/δ) can transactivate a PPARβ/δ reporter to a magnitude similar to what is observed with the natural receptor in the presence of ligand (3). More importantly, VP16-PPARβ/δ signals only through PPARβ/δ pathways. Overexpression of this mutant forms of PPARβ/δ should ensure the constitutive transactivation of its target genes in a specific tissue of the intact mice. The cDNA of the VP16 transactivation domain is cloned from pVP16 vector. The cDNA of the mouse PPARβ/δ is cloned from mouse cDNA extracted from the heart of C57 BL/6 mice. Overlapping PCR is performed to obtain the VP16- PPARβ/δ using the Expand High Fidelity PCR System kit (Roche).
2. Generation of the transgenic construct of CAG-LoxP-CAT-LoxP-VPD:

   The CAG-CAT-LacZ vector (7) is modified by removing the LacZ gene (see Note 1). The human growth hormone poly A (HGH-PA) is inserted at the end of the cloning site for the candidate transgene to stabilize transgenic expression. The vector is engineered to enable a candidate gene be readily subcloned into the vector using the Not I site (Fig. 1).
3. To establish transgenic lines of CAG-LoxP-CAT-LoxP-VPD:

   After proofreading and orientation confirmation of the DNA sequence, the transgenic construct of CAG-LoxP-CAT-

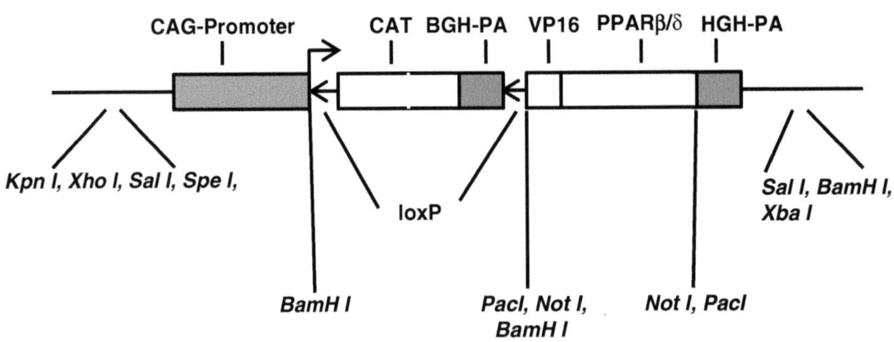

Fig. 1. Transgenic construct for the generation of the CAG-LoxP-CAT-LoxP-VPD mice. The key cloning site and the specific restricted enzymes are shown. The transgenic construct is released from the Bluescript II vector using *Sal I* restricted enzyme.

LoxP-VPD is released from the vector using the Sal I restriction enzyme, followed by gel purification (QIAEX II, Qiagen) and dialysis in Slide-a-lyzer cassettes (Pierce). For dialyzing the DNA sample, 4 L of dialysis buffer is used and is kept at 4°C overnight. The purified transgenic DNA construct is microinjected into the pronucleus of fertilized mouse oocytes derived from superovulated C57BL/6 female. Transgenic founder mice are identified using PCR genotyping as described later in Subheading 3.5.

3.2. Generation of Transgenic Mice with Inducible, Cardiomyocyte-Restricted VPD Overexpression

To generate transgenic mice with tamoxifen inducible, cardiomyocyte-restricted VPD overexpression, the two founders of CAG-LoxP-CAT-LoxP-VPD mice, designated as VPD mice are crossed with the αMyHC-Mer-Cre-Mer (TMCM) mice (8). Offsprings of the double transgenic line are confirmed by genotyping as described in Subheading 3.4. Most importantly, we confirmed that the CAG-LoxP-CAT-Loxp-VPD mice are overtly normal with ubiquitous transgene expression without a change in PPARβ/δ transcript expression in any tissue (data not shown). Two VPD lines (line 1 and line 2) with similar transgenic expression levels are established and they show normal life span with no overt phenotype. The cardiomyocyte-restricted expression of VPD is achieved by treating the mice with tamoxifen as described in Subheading 3.5 to remove the CAT and triggered VPD overexpression (Fig. 2).

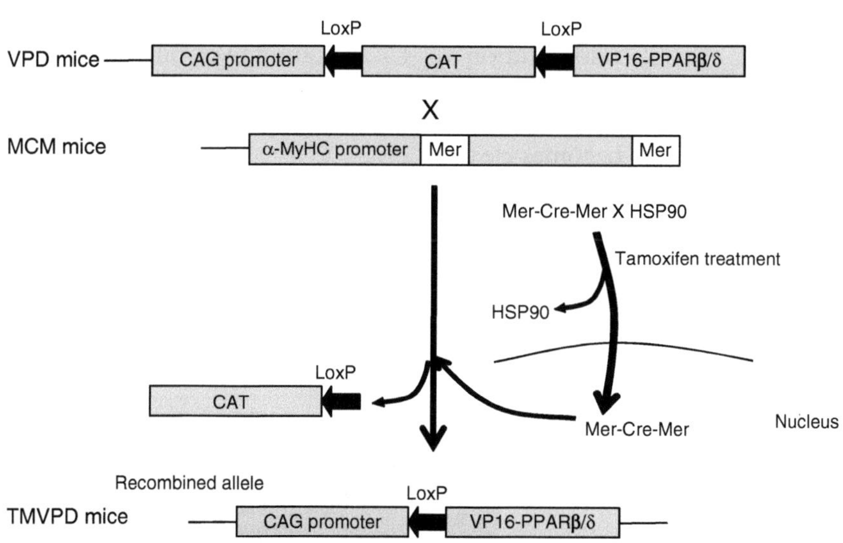

Fig. 2. Schematic illustration for the generation of transgenic mice with tamoxifen-inducible, cardiomyocyte-restricted VP16-PPARβ/δ overexpression. The transgenic mice with tamoxifen-inducible, cardiomyocyte-restricted VP16-PPARβ/δ overexpression are generated by crossing the two parent transgenic lines (CAG-LoxP-CAT-LoxP-VPD and α-MyHC-Mer-Cre-Mer). The tamoxifen-induced translocation of Cre from the cytosol to the nucleus and the Cre-mediated the recombination events are shown.

3.3. Genotyping

Genomic DNA is extracted from tail samples digested with the DNA extraction buffer. Transgenic mice are identified by PCR on the above DNA samples using the following primer pairs.

1. To identify the CAG-LoxP-CAT-LoxP-VPD transgenic mice, the following primer pair was used: 5′ TTA CAT GGT GGT AAG CTT3′ and 5′CAG TCA GTT GCT CAA TGT ACC3′ (94°C 8 min, 50°C 1 min, 72°C 13 min, 35 cycles) (see Note 2).
2. To identify the α-MyHC-Mer-Cre-Mer/CAG-LoxP-CAT-LoxP-VPD double transgenic mice, in addition to primer pair used in Subheading 3.3, step 1, a primer pair recognized α-MyHC-Mer-Cre-Mer are used: 5′GTC TGA CTA GGT GTC CTT CT3′ and 5′CGT CCT CCT GCT GGT ATA G3′ (94°C 5.5 min, 56°C 30 s, 72°C 11 min, 30 cycles).

Figure 4 shows an example of genotyping for the double transgenic line of α-MyHC-Mer-Cre-Mer/CAG- LoxP-CAT-LoxP-VPD (TMVPD) (Fig. 3).

3.4. Induction of Transgenic Expression of VP16-PPARβ/δ Specifically in Cardiomyocytes of Adult Mice

The transgenic mice with both CAG-LoxP-CAT-LoxP-VPD and α-MyHC-Mer-Cre-mer are treated with tamoxifen (Sigma) by intraperitoneal injection once a day for 2 or 5 days at a dosage of 20 mg/kg/day (see Notes 3 and 4).

3.4.1. Preparing the Tamoxifen Stock for IP Injection

1. Dissolve 50 mg tamoxifen in 300 μL 100 % ethanol. Vortex the vial well, although the solution remains cloudy.
2. Add solution to 4.7 mL sunflower oil (final concentration: 10 mg/mL tamoxifen in 0.06 % ethanol). Sonicate for 15–20 min, vortex briefly for 5–10 min until the solution becomes clear.
3. Aliquot and store the stock solution at −20°. If the stock solution undergoes more than two freeze/thaw cycles it is recommended that it should be resonicated for 5 min prior to use.

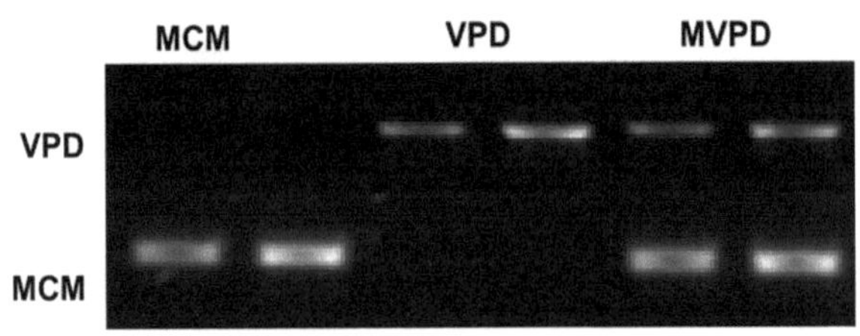

Fig. 3. TMVPD genotyping results. PCR products obtained as described in the text were analyzed on 1 % agarose gel electrophoresis.

3.4.2. Cre Excision

1. Administer 2 mg (200 μL) tamoxifen once daily per mouse by IP injection.
2. Continue administrations for 5 days. We typically wait at least additional 5 days so that tamoxifen is metabolized and removed from the body.

3.5. Determination of Tissue-Specific Transgenic Expression in Transcript and Protein Levels

3.5.1. Quantitative Real-Time RT-PCR Analyses

Quantitative real-time RT-PCR analysis is carried out using the Roche LightCycler 480 system (Roche).

We mainly use the α-MyHC-Mer-Cre-Mer mice with identical tamoxifen treatment as a control for potential toxic effect of Cre (9). A stable upregulation of VPD could be detected as late as 5 days after the end of the tamoxifen treatment. The hearts from line 1 displayed a ~2fold increase of the total PPARβ/δ mRNA level (Fig. 4a) with no significant change in PPARα and PPARγ mRNA levels compared with the control heart. The expression of total PPARβ/δ in cardiac samples of line 2 was slightly lower (~1.79-fold) than that of line 1 (data not shown).

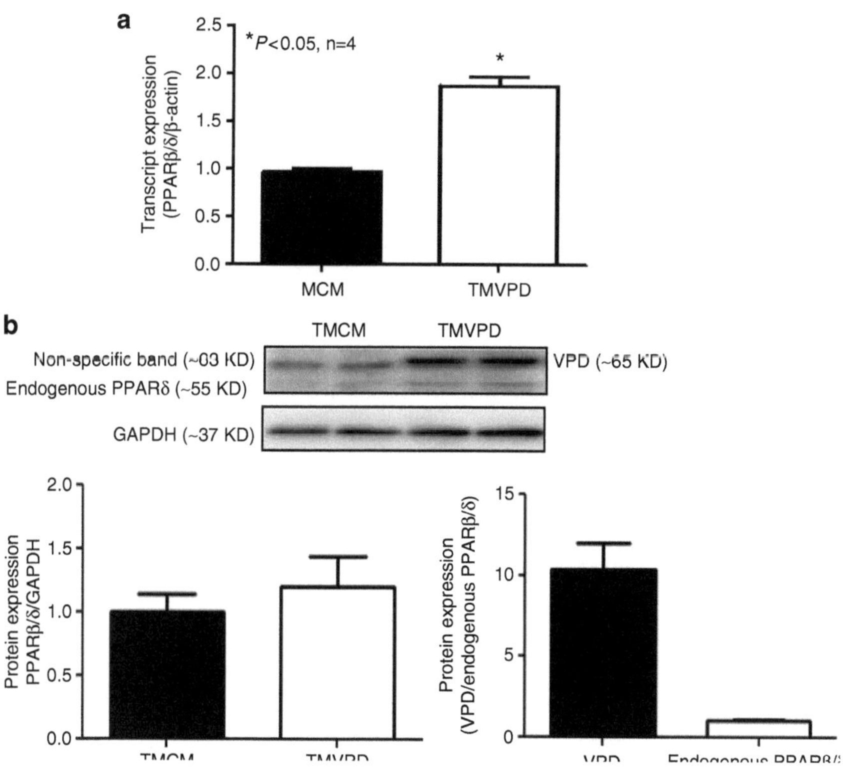

Fig. 4. Confirmation of PPARβ/δ overexpression. Hearts from TMCM, TMVPD mice, are processed for real-time PCR analysis for transcript and protein expression. (**a**) Transcript expression of PPARs. (**b**) Western blot analysis of protein expression of VPD and endogenous PPARβ/δ (see Note 6). Data are mean ± SEM. *$P < 0.05$, $n = 4$.

3.5.2. Western Blot Analysis

To detect the protein expression of the transgenic protein (VPD) and the endogenous PPARβ/δ, Western blots can be conducted using commercially available antibodies (see Note 5). The Immunoblotting images are captured using KODAK image Station 4000R (Carestream Health Inc.) by developing the membranes in Supersignal West substrates (Thermo Scientific), and analyzed with KODAK IM software (Ver 4.5.1).

In isolated cardiomyocytes from line 1, the abundance of VPD protein (see Note 6) is about ~11-fold higher relative to the endogenous PPARβ/δ protein, whereas the endogenous PPARβ/δ level in the TMVPD hearts is similar to that of control hearts (TMCM) (Fig. 4b). Alternatively, VPD protein but not the endogenous PPARβ/δ could be detected by an antibody recognized the transactivation domain of VP16 (see Note 7).

4. Notes

1. The CAG promoter construct has been proven highly effective in the ubiquitous expression of genes. However, expression is not as robust as with some other promoters, such as the cardiac-specific promoter, α-MyHC. However, the transgenic expression driven by CAG is uniform. The relatively modest transgenic expression in mice using this promoter should be enough for most physiologically related studies. However, it may not be strong enough to mimic a situation with a robust overexpression in response to pathophysiological stimulations. Other promoter constructs like CMV may be used to improve the expression level.
2. To increase the accuracy of genotyping, primer pairs that recognize the CAT sequence can be used.
3. The tamoxifen-induced Cre expression has been extensively used for temporal control of transgenic and gene targeting in mice. However, caution should be taken on the dose and duration of tamoxifen treatment. A recent report has demonstrated that tamoxifen-mediated Mer-Cre-Mer nuclear translocation can induce a severe but transient dilated cardiomyopathy, and a dramatic increase of mortality in mice with or without LoxP transgene (10). However, we have not observed similar phenotypic changes using the protocol presented here. A protocol with 2-day tamoxifen treatment with the same dose has been proposed and suggested minimal effects on cardiac gene expression (11). Even though a major cytotoxic effect of the 5-day protocol is not observed, the 2-day protocol appears to be feasible with even fewer nonspecific effects.
4. Considering the potential effects of Cre expression in the heart, it is important to use the littermate mice with Cre expression

as a control to exclude any subtle effect of Cre that could alter the phenotypic response of VPD overexpression.

5. Western blot for PPARβ/δ could be challenged. It is easier to perform the Western blot on nuclear protein samples with more enrichment of the protein.
6. The fusion protein VPD contains the VP16 transactivation domain (80 amino acids). The size for VPD is about 10 KD larger than the endogenous PPARβ/δ. Therefore, Western blot using an anti-PPARβ/δ antibody should reveal two bands (~65 and 55 KD) as shown in Fig. 4.
7. An antibody recognizing the transactivation domain of VP16 could be used to identify the VPD protein. Its size should be ~65 KD.

Acknowledgement

This work was supported by grants from National Institute of Health (1R01HL085499, 1R01HL084456 and R21 AT003734).

References

1. Nissen SE, Wolski K (2007) Effect of rosiglitazone on the risk of myocardial infarction and death from cardiovascular causes. N Engl J Med 356:2457–2471
2. Stockl KM, Le L, Zhang S, Harada AS (2009) Risk of acute myocardial infarction in patients treated with thiazolidinediones or other antidiabetic medications. Pharmacoepidemiol Drug Saf 18:166–174
3. Saez E, Rosenfeld J, Livolsi A, Olson P, Lombardo E, Nelson M, Banayo E, Cardiff RD, Izpisua-Belmonte JC, Evans RM (2004) PPAR gamma signaling exacerbates mammary gland tumor development. Genes Dev 18:528–540
4. Sugii S, Olson P, Sears DD, Saberi M, Atkins AR, Barish GD, Hong SH, Castro GL, Yin YQ, Nelson MC et al (2009) PPAR activation in adipocytes is sufficient for systemic insulin sensitization. Proc Natl Acad Sci 106:22504–22509
5. Lo Sasso G, Murzilli S, Salvatore L, D'Errico I, Petruzzelli M, Conca P, Jiang ZY, Calabresi L, Parini P, Moschetta A (2010) Intestinal specific LXR activation stimulates reverse cholesterol transport and protects from atherosclerosis. Cell Metab 12:187–193
6. Kisanuki YY, Hammer RE, Miyazaki J, Williams SC, Richardson JA, Yanagisawa M (2001) Tie2-Cre transgenic mice: a new model for endothelial cell-lineage analysis in vivo. Dev Biol 230:230–242
7. Araki K, Araki M, Miyazaki J, Vassalli P (1995) Site-specific recombination of a transgene in fertilized eggs by transient expression of Cre recombinase. Proc Natl Acad Sci USA 92:160–164
8. Sohal DS, Nghiem M, Crackower MA, Witt SA, Kimball TR, Tymitz KM, Penninger JM, Molkentin JD (2001) Temporally regulated and tissue-specific gene manipulations in the adult and embryonic heart using a tamoxifen-inducible Cre protein. Circ Res 89:20–25
9. Buerger A, Rozhitskaya O, Sherwood MC, Dorfman AL, Bisping E, Abel ED, Pu WT, Izumo S, Jay PY (2006) Dilated cardiomyopathy resulting from high-level myocardial expression of Cre-recombinase. J Card Fail 12:392–398
10. Koitabashi N, Bedja D, Zaiman AL, Pinto YM, Zhang M, Gabrielson KL, Takimoto E, Kass DA (2009) Avoidance of transient cardiomyopathy in cardiomyocyte-targeted tamoxifen-induced MerCreMer gene deletion models. Circ Res 105:12–15
11. Andersson KB, Winer LH, Mork HK, Molkentin JD, Jaisser F (2010) Tamoxifen administration routes and dosage for inducible Cre-mediated gene disruption in mouse hearts. Transgenic Res 19:715–725

Chapter 5

Specific Knockdown of PPARδ Gene in Colon Cancer Cells by Lentivirus-Mediated RNA Interfering

Lie Yang, Zong-Guang Zhou, and Xiao-Feng Sun

Abstract

Peroxisome proliferator-activated receptors (PPARs) are ligand-activated transcription factors. Of this family, PPARδ has been implicated in the pathogenesis of colorectal cancer, whereas its exact role is highly controversial as available studies yield conflicting results. Therefore, further studies using powerful and duplicable methods are needed to clarify the functions of PPARδ in colorectal cancer. RNA interference (RNAi) is such a powerful tool for specific suppression of gene expression at the posttranscriptional level. Of available RNAi methods, lentivirus has been shown to mediate RNAi most efficiently and stably based on its remarkable advantages, such as powerful transduction efficiency and a wider range of target cells, integration of expression cassette into host genomics, low immunogenicity and self-inactivated and replication-incompetent after integrating into host genomics. In this chapter, we will introduce the details of using lentivirus-mediated RNAi to specifically knockdown the expression of PPARδ in colon cancer cell lines. This method includes: designing and cloning short-hairpin RNA cassette into lentivector, constructing and titrating lentiviral particles, infecting target cells with lentivirus and assaying the knockdown efficiency.

Key words: Peroxisome proliferator-activated receptor, Transcription factor, Colorectal neoplasm, Pathogenesis, RNA interference, Lentivirus

1. Introduction

Peroxisome proliferator-activated receptors (PPARs) are ligand-activated transcription factors that are members of nuclear hormone receptor superfamily, including three distinct isoforms, α, β/δ, and γ (1). PPARs bind peroxisome proliferators and control the size and number of peroxisome produced by cells, and mediate a variety of biological processes, and may be involved in the development of several chronic diseases, including diabetes, obesity,

Mostafa Z. Badr and Jihan Youssef (eds.), *Peroxisome Proliferator-Activated Receptors (PPARs): Methods and Protocols*, Methods in Molecular Biology, vol. 952, DOI 10.1007/978-1-62703-155-4_5, © Springer Science+Business Media New York 2013

atherosclerosis, and cancer. PPARα and γ subtypes play fundamental roles in such diverse physiological processes as adipogenesis, lipid homeostasis, immune function, cell proliferation/apoptosis and carcinogenesis (2–5). Compared with PPARα and PPARγ, the identification of PPARδ function has been lagging behind. Available literatures demonstrate that PPARδ can induce reverse cholesterol transport and rectify lipoprotein profiles and triglyceride levels (6). It is a potential regulator of adipocyte differentiation and a critical mediator of embryo implantation (7), and a novel regulator of wound healing and inflammation (8, 9). PPARδ therefore represents a potential drug target for the treatment of diseases such as obesity and metabolic syndrome, which has led to the development of several synthetic drug agonists (10).

In recent decade, PPARδ is implicated in the pathogenesis of colorectal cancer as its expression is elevated in colorectal cancer cells, and this expression can be repressed by adenomatosis polyposis coli (APC), a tumor suppressor protein related to APC/β-catenin signaling pathway (11). So far, the exact role of PPARδ in colorectal cancer is unclear as the available literatures yield highly conflicting results. Some studies support a promotive role of PPARδ in colorectal carcinogenesis, while some others provide opposite evidences (12). Therefore, further studies using sophisticated techniques are needed before there is a better understanding of how PPARδ gene functions and how it influences colorectal cancer.

RNA interference (RNAi) is a powerful tool for specific suppression of gene expression at the posttranscriptional level. Introduction of 21–23 nt small interfering RNAs (siRNA), plasmids or viral vectors expressing short-hairpin RNA (shRNA) has been shown to mediate RNAi successfully in different mammalian cells (13). However, although siRNA-mediated RNAi can result in rapid reduction of mRNA levels, its effect is transient lasting only for a week or so (14). Plasmid-based RNAi may last longer with drug selection, but there still have numerous disadvantages, such as low transfection efficiency, incompetence of integrating into host DNA, and unstable expression of shRNAs (15). In contrast, the vectors of adenovirus and adeno-associated virus (AAV) have also been used to deliver siRNA successfully and tend to provide high transducing efficiency and more stable gene silencing. However, adenovirus can't integrate into the host genome and as a result are not useful for long-term gene silencing (16). AAV, in contrast, does integrate into the host genome, but its RNAi efficiency could be impaired by host immune response (17).

To overcome these limitations, a novel powerful vector, lentiviral vector (lentivector), has been established and shown to mediate RNAi most efficiently and stably (16). Lentivector can carry shRNA expression cassette into host genomics and express shRNA using RNA Pol-III *promoter*. The expressed shRNA is cut into

21 bp siRNA by an RNase-III family nuclease called dicer enzyme and subsequently guide degradation of cognate mRNA (18). Comparing with other vectors, lentivector has demonstrated remarkable advantages (19–21): (1) It can transduce both proliferating and nonproliferating cells, thus resulting in powerful transduction efficiency and a wider range of target cells; (2) As lentivector is capable of integrating shRNA expression cassette into host genomics, it can realize the stable and heritable but still cell-autonomous RNAi in mammalian cells; (3) Due to its low immunogenicity, it hardly causes immune response of host cells; and (4) It will become self-inactivated and replication-incompetent after integration into genomic DNA, avoiding the cytolysis effect of host cells.

Based on the obvious advantages of Lentivector, the RNAi technique mediated by lentivector represents a powerful tool for the research of gene function. Applying this technique, we have successfully established the colon cancer cells with stably silenced PPARδ. In this chapter, we will introduce the procedures of how to use lentivector-mediated RNAi technique to establish the colon cancer cells with stably silenced PPARδ. This method includes the following steps: (1) Designing shRNA oligos for lentivector; (2) Cloning shRNA oligos into lentivector; (3) Screening for inserts; (4) Producing lentiviral particles; (5) Titrating of the pseudotyped lentiviral particles; (6) Transducing lentiviral particles into colon cancer cells; (7) Screening stably transduced cells; and (8) Testing knockdown efficiency and maintaining the established cells.

2. Materials

2.1 Materials for Molecular Cloning

1. AgeI, EcoRI, T4 DNA ligase, DNA Miniprep Kit.
2. Buffers:
   - 10× *Annealing buffer*:

     500 mM NaCl, 100 mM Tris–HCl, 100 mM $MgCl_2$, 10 mM Dithiothreitol, pH 7.9 at 25°C.
   - 10× *Digesting buffer*:

     100 mM Bis–Tris–Propane–HCl, 100 mM $MgCl_2$, 10 mM Dithiothreitol, pH 7.0 at 25°C.
   - 10× *T4 DNA Ligase Reaction Buffer*:

     500 mM Tris–HCl, 100 mM $MgCl_2$, 100 mM Dithiothreitol, 10 mM ATP, pH 7.5 at 25°C.
3. DH5 alpha competent cells.
4. Gel extraction kit.
5. Low melting point agarose.
6. Luria Broth Agar (LB agar).

7. Terrific Broth (TB):

12 g	Tryptone
24 g	Yeast extract
2.31 g	KH_2PO_4
12.54 g	K_2HPO_4
4 mL	Glycerol

8. Ampicillin, carbenicillin.

2.2 Materials for Lentivirus Production

1. HEK-293T cell line.

   The modified human embryonic kidney cell line, 293T (American Type Culture Collection, VA), is used for lentivirus production and titration. Cells are maintained in Dulbecco's Modified Eagle's Medium (DMEM), supplemented with 10% FBS, 0.7 mM L-glutamine, and 5% Penicillin–Streptomycin in a humidified atmosphere of 5% CO_2 at 37°C.

2. Lentivectors for virus production (see Note 1).

   (a) Transfer vector—pLKO.1-TRC-Cloning Vector.

   pLKO.1 is a replication-incompetent lentivector chosen by the RNAi consortium (TRC) for cloning and expressing shRNA sequences. The original pLKO.1-TRC-cloning vector (8,901 bp) has a 1.9 kb stuffer that is released by digestion with AgeI and EcoRI (Fig. 1, Table 1). shRNA oligos are cloned into the AgeI and EcoRI sites in place of the stuffer. The AgeI site is destroyed in most cases (depending on the target sequence), while the EcoRI site is preserved. pLKO.1 can be introduced into cells via direct transfection, or can be converted into lentiviral particles for subsequent infection of a target cell line. Once introduced, the puromycin resistance marker encoded in pLKO.1 allows for convenient stable selection (see Note 2).

   (b) Packaging vector and Envelope vector (see Note 3):

   Packaging vector—pCMV-dR8.2 dvpr.

   Envelope vector—pCMV-VSV-G.

3. Transfection solutions for virus production.

   (a) 2× HEPES buffer solution (HeBS):

   50 mM SigmaUltra HEPES.

   280 mM SigmaUltra NaCl.

   1.5 mM SigmaUltra Na_2HPO_4.

   Dissolved in sterile water adjusted to PH 7.00 using NaOH.

   (b) 0.5 M SigmaUltra $CaCl_2$, dissolved in sterile water.

   (c) HEPES water: 6.25 mM HEPES solution dissolved in sterile water.

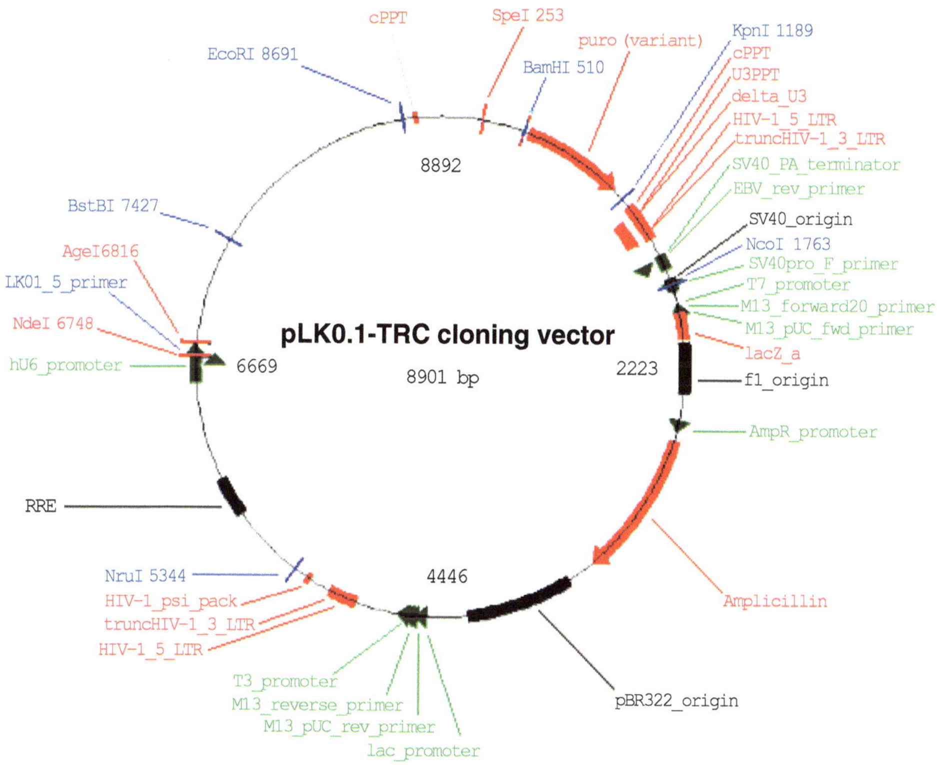

Fig. 1. Map of pLKO.1-TRC. The pLKO.1-TRC-cloning vector (8,901 bp) has a 1.9 kb stuffer, which can be replaced with shRNA oligos cloned into the AgeI and EcoRI sites.

Table 1
The functional elements of pLKO.1-TRC

Description	Vector element
U6	Human U6 promoter drives RNA Pol-III transcription for generation of shRNA transcripts
cPPT	Central polypurine tract, cPPT, improves transduction efficiency by facilitating nuclear import of the vector's preintegration complex in the transduced cells
hPGK	Human phosphoglycerate kinase promoter drives expression of puromycin
Puro R	Puromycin resistance gene for selection of pLKO.1 plasmid in mammalian cells
sin 3′LTR	3′ Self-inactivating long terminal repeat
f1 ori	f1 bacterial origin of replication
Amp R	Ampicillin resistance gene for selection of pLKO.1 plasmid in bacterial cells
pUC ori	pUC bacterial origin of replication
5′LTR	5′ long terminal repeat
RRE	Rev response element

4. Fluorescent microscope.
5. Fluorescence-activated cell sorter (FACS).
6. Filters (0.45 μm).
7. Puromycin, polybrene.
8. Centrifuge (>1,200 × *g*).

3. Methods

3.1. Designing shRNA Oligos for pLKO.1 Vector

(a) Selection of suitable 21-mer targets in PPARδ mRNA is the first step toward efficient gene silencing. Below are suggestions for target selection:

- Use an siRNA selection tool to determine a set of top-scoring targets for PPARδ gene.

  For example, the Ambion company hosts an siRNA Target Finder program that can be freely accessed (http://www.ambion.com/techlib/misc/siRNA_finder.html) (see Note 4). To use this design program, paste the mRNA sequence into the window, choose the preferred end structure (3′ TT or UU), and the program will scan the sequence for AA dinucleotide beginning with the AUG start codon of your transcript (see Note 5). A report is generated indicating the position of the AA dinucleotide, the 21 base target and the corresponding sense and antisense siRNA oligonucleotides (see Note 6).

- Select 4–6 target sequences: Choose target sites from among the sequences identified in the above step based on the following guidelines:
  - siRNAs with 30–50% GC content are more active than those with a higher G/C content.
  - Since a 4–6 nucleotide poly (T) tract acts as a termination signal for RNA Pol III, avoid stretches of >4Ts or As in the target sequence when designing sequences to be expressed from an RNA Pol-III promoter.
  - Select siRNA target sites at different positions along the length of the gene sequence.
  - Using NCBI's BLAST programme (www.ncbi.nlm.nih.gov/BLAST) to eliminate from consideration any target sequences with more than 16–17 contiguous base pairs of homology to other coding sequences.

(b) Ordering oligos compatible with pLKO.1 vector.

To generate shRNA oligos for cloning into pLKO.1, insert your sense and antisense sequences from step (a) into the oligos below (see Note 7):

Forward oligo: 5′ CCGG—21 bp sense—CTCGAG—21 bp antisense—TTTTTG 3′.

Reverse oligo: 5′ AATTCAAAAA—21 bp sense—CTCGAG—21 bp antisense 3′.

3.2. Cloning Oligos into pLKO.1

The pLKO.1-TRC-cloning vector contains a 1.9 kb stuffer that is released upon digestion with EcoRI and AgeI. The oligos from Subheading 3.1 contain the shRNA sequence flanked by sequences that are compatible with the sticky ends of EcoRI and AgeI. Forward and reverse oligos are annealed and ligated into the pLKO.1 vector, producing a final plasmid that expresses the shRNA of interest.

(a) Annealing oligos.

- Resuspend oligos in ddH_2O to a concentration of 20 μM, then mix:

5 μL	Forward oligo
5 μL	Reverse oligo
5 μL	10× Annealing buffer
35 μL	ddH_2O

- Incubate for 4 min at 95°C in a PCR machine or in a beaker of boiling water.
- If using a PCR machine, incubate the sample at 70°C for 10 min then slowly cool to room temperature over the period of several hours. If using a beaker of water, remove the beaker from the flame, and allow the water to cool to room temperature (see Note 8).

(b) Digesting pLKO.1 TRC-cloning vector (see Note 9).

- Digest pLKO.1 TRC-cloning vector with Age I. Mix:

6 μg	pLKO.1 TRC-cloning vector
5 μL	10× Digesting buffer
1 μL	AgeI
to 50 μL	ddH_2O

– Incubate at 37°C for 2 h.

- Digest with EcoR I. Mix:

30 μL	pLKO.1 TRC-cloning vector digested with AgeI
5 μL	10× Digesting buffer for EcoRI
1 μL	EcoRI
14 μL	ddH_2O

  - Incubate at 37°C for 2 h.
- Run digested DNA on 0.8% low melting point agarose gel until you can distinctly see two bands, one 7 kb and one 1.9 kb. Cut out the 7 kb band and place in a sterile micro-centrifuge tube (see Note 10).
- Purify the DNA using a gel extraction kit. Elute in 30 μL of ddH_2O.
- Measure the DNA concentration.

(c) Ligating and Transforming into Bacteria:

- Use a standard T4 DNA ligase for ligation, mix:

2 μL	annealed oligo from step (**a**)
20 ng	digested pLKO.1 TRC-cloning vector from step (**b**) (see Note 11)
2 μL	10× T4 DNA ligase buffer
1 μL	T4 DNA ligase
to 20 μL	ddH_2O

  - Incubate at 16°C for 4–20 h.
- Transform 2 μL of ligation mix into 25 μL competent DH5 alpha cells, gently shake, place on ice for 30 min.
- Place in 42°C water bath for 60–90 s, rapid cooling on ice for 2–3 min (see Note 12).
- Add 100 μL LB broth, shake for 45 min on a shaker (150 times/min) (see Note 13).
- Plate on LB agar plates containing 100 μg/mL ampicillin or carbenicillin at 37°C for 15–20 h (see Note 14).

(d) Screening for Inserts.

Once you have identified the positive clones, it is important to verify the insert by conducting a sequencing reaction.

Day 1

(1) Inoculate 3 colonies into 3 mL TB with 100 μg/mL carbenicillin, incubate for 16–20 h at a shaker (300 rpm) (see Note 15).

Day 2

(2) Spin down the cultures and use a miniprep kit to obtain DNA (see Note 16).

(3) Conduct a restriction digest with EcoRI and NcoI:

1 μg	Miniprep DNA
2 μL	10× Digesting buffer for EcoRI
0.8 μL	EcoRI
0.8 μL	NcoI
to 20 μL	ddH_2O

>Incubate at 37°C for 1–2 h.

(4) Run the digestion products on a 1% agarose gel. You should see two fragments, a 2 kb fragment and a 5 kb fragment.

(5) Sequence positive clones with pLKO.1 sequencing primer (5′ CAA GGC TGT TAG AGA GAT AAT TGG A 3′).

3.3. Producing Lentiviral Particles (see Note 17)

This protocol is for transfection with calcium phosphate solution in a 10 cm plate. The protocol can be scaled to produce different amounts of virus as needed.

- Day 1

  (1) Seed 293T cells in a 10 cm plate at 1.0×10^6/plate in 10 mL complete DMEM, incubate overnight in standard conditions (see Note 18).

- Day 2

  (2) Change medium (complete growth DMEM, 10 mL) 2.5–3 h prior to transfection (see Note 19).

  (3) Prepare transfection solution (start 1.5 h following medium change):

  Mix plasmids in a 2.5 mL tubes.

  25 μg pLKO.1-shRNA + 6.25 μg pCMV-VSV-G + 10 μg pCMV-dR8.2 dvpr (see Note 20).

  Add HEPES water into the tube, up to 250 μL.

  Add 250 μL (0.5 M) $CaCl_2$ and mix gently.

  Vortex for ≥2 min as 500 μL 2× HeBS is dropwise added (see Note 21).

  Incubate 20–25 min in room temperature (see Note 22).

  (4) Distribute phosphate-DNA mixture evenly over the 293T cells, incubate no longer than 18 h (overnight) in standard conditions (see Note 23).

- Day 3
  (5) Discard supernatant and add 6 mL new complete DMEM, incubate overnight.
- Day 4
  (6) Collect viral supernatant and store in 4°C. Add 6 mL new complete DMEM and incubate overnight.
  (7) Collect viral supernatant and pool with the supernatant from the day before.
  (8) Remove cell debris using a 0.45 μm filter, aliquot into 1.5 mL/tube, store at −80°C (see Note 24).

3.4. Titrating Lentiviral Suspensions (see Note 25)

For titration, serial dilutions of the unconcentrated viral suspensions are used to transduce 293T cells in 6-well plate. Cells are removed 72 h posttransduction and analyzed by FACS for the percentage of cells expressing GFP. Then, the titer can be calculated by a formula.

- Day 1
  (1) Seed 293T cells at 1.5×10^5 per well in six wells of 12-well plate in complete DMEM growth medium.
- Day 2
  (2) Remove medium, add 1 mL tenfold serial dilutions over a range of $1 \times (10)^{-2}$ to $1 \times (10)^{-6}$ into wells with your viral suspensions (e.g., 1, 10, 100, and 1,000 μL), each in duplicate. Add 8 μL (1 μg/μL) polybrene to each well (final concentration: 8 μg/mL) (see Note 26).
- Day 3
  (3) Replace the supernatants with fresh DMEM (without polybrene).
- Day 4–5
  (4) Change medium each day.
- Day 6
  (5) Detach and resuspend cells, pool the two wells of each concentration, measure the percentage of GFP-expressing cells by FACS.

Calculate the titer of viral suspension by the following formula: Titer (transducing units/mL) = (average cell number at the time of transduction) × (% of GFP-positive cells)/total volume of each well (in mL) × dilution factor (see Note 27).

3.5. Infecting Target Colon Cancer Cells with Lentivirus

(a) Determining the optimal puromycin concentration.

Since each cell line responds differently to puromycin selection, it is necessary to determine the optimal puromycin concentration for your cell line before initiating your experiment.

- Day 1
  (1) Plate target cells in ten 6 cm plates and grow at 37°C, 5% CO_2 overnight (see Note 28).
- Day 2
  (2) The target cells should be approximately 80–90% confluent.
  (3) Dilute puromycin in the preferred culture media for your target cells. The final concentration of puromycin should be from 1 to 10 μg/mL in 1 μg/mL increments.
  (4) Label plates from 1 to 10 and add appropriate puromycin-containing media to cells.
- Days 3+
  (5) Examine cells each day and change to fresh puromycin-containing media every other day.
  (6) The minimum concentration of puromycin that results in complete cell death after 3–5 days is the concentration that should be used for selection in your experiments (see Note 29).

(b) Lentiviral infection and selection.

- Day 1
  (1) Seed colon cancer cells in 6-well plate with complete growth medium. Duplicate wells for each lentiviral construct and controls should be used (see Note 30). Incubate 18–20 h at 37°C in a humidified incubator in an atmosphere of 5–7% CO_2.
- Day 2
  (2) Remove medium, and add 2 mL viral supernatant and 16 mL polybrene (1 mg/mL, final concentration: 8 mg/mL) to each well. Gently swirl the plate to mix (see Note 31).
  (3) Spin plates for 1.5 h (20°C, 1,200 g) and then incubate for 5 h (see Note 32).
  (4) 5 h later, replace the viral supernatant with 2 mL growth medium, and incubate overnight.
- Day 3
  (5) 19 h after infection, replace with 2 mL fresh medium to each well, and incubate (see Note 33).
- Day 4
  (6) 48 h after infection. Change to fresh puromycin-containing media as determined in Step (a).

- Day 5~
  (7) Change fresh puromycin-containing media every 3–4 days until resistant colonies can be identified.
  (8) Pick a minimum of five puromycin-resistant colonies and expand each clone (see Note 34).
  (9) Assay infected cells. Both the mRNA and protein expressions of PPARδ should be assayed to confirm the knockdown efficiency. The following recommendations are guidelines for the number of days you should wait until harvesting your colon cancer cells. However, you should optimize the time based on your cell line and assay:

Assay	Days postinfection
mRNA knockdown (quantitative PCR)	≥3 days
Protein knockdown (Western blot)	≥4 days
Phenotypic assay	≥4 days

4. Notes

1. The production of a high titer of infectious, replication-incompetent lentivirus is based on transient co-transfection of transfer vectors, packaging vectors, and envelope vectors into producer cells (293T cells). When expressed in 293T cells, the transfer vectors produce large numbers of transcripts that contain all of the functional elements (i.e., shRNA, Psi, RRE, and cPPT). Then these transcripts are efficiently packaged into vesicular stomatitis virus (VSV-G) pseudotyped particles produced by envelope vectors, with helper proteins produced by the packaging vectors. The pseudoviral particles generated in 293T cells within 48–72 h can be concentrated, frozen, and used in later experiments (Fig. 2).
2. The tricks for preparing pLKO.1 vector inserted with shRNA (pLKO.1-shRNA):

   When preparing pLKO.1 vector without insertion of shRNA, it is not hard to yield high copies of plasmids in bacteria (Escherichia coli, e.g., DH5α) using LB broth and ampicillin selection. However, low yields are a common problem when working with pLKO.1-shRNA plasmids. This seems to be because the bacteria have a very difficult time working through the hairpin structure (shRNA sequence) of the plasmid. We have found that there are two general problems that lead to low yields. The first is that the bacteria grow slower when the

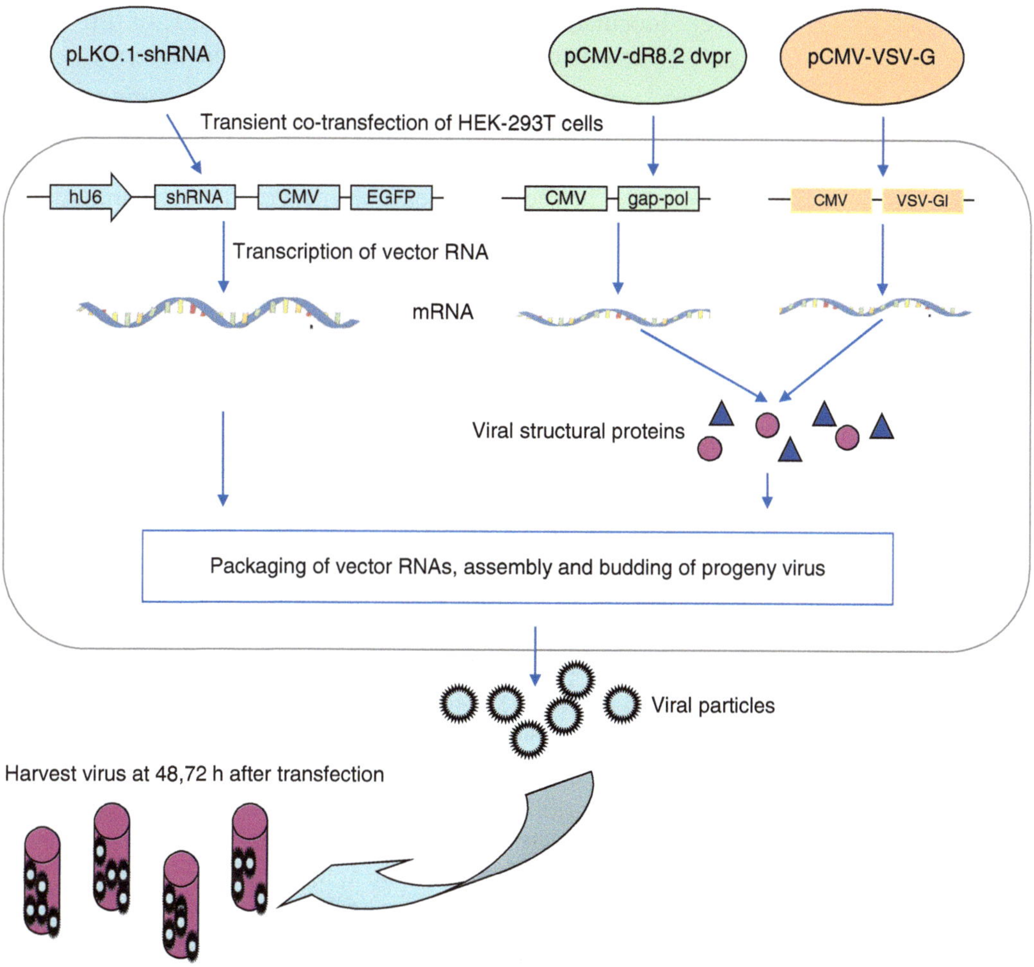

Fig. 2. The mechanism of lentivirus production. The transfer vectors, packaging vectors, and envelope vectors are cotransfected into 293T cells. When expressed in 293T cells, the transfer vectors produce large numbers of transcripts that contain all of the functional elements (i.e., shRNA, Psi, RRE, and cPPT). Then these transcripts are efficiently packaged into vesicular stomatitis virus (VSV-G) pseudotyped particles produced by envelope vectors, with helper proteins produced by the packaging vectors. The pseudoviral particles generated in 293T cells within 48–72 h can be concentrated, frozen, and used in later experiments.

plasmid contains a shRNA sequence. The second is that the number of plasmids per bacteria seems to be lower when the shRNA sequence is present (even though the pLKO.1 vector is a high-copy number vector). Followed are the suggestions we have found to greatly help handling this problem:

(a) Grow the cultures in Terrific Broth (TB) instead of LB, which can significantly increase the maxiprep yields.

(b) Use the ampicillin analog, carbenicillin, when growing bacteria with hairpins. Carbenicillin is more stable than ampicillin and prevents the outgrowth of "satellites."

(c) Inoculates at least 3 colonies into 1.3 mL of TB broth. This is because of the slow growth of the bacteria.

3. Three generations of HIV-based lentivirus packaging systems have been successively developed. The first generation system encompasses all HIV-1 genes except the envelope and thus is not safe. The second generation meets biosafety standard as it comprises self-inactivating (SIN) HIV-1-derived vectors, which looses the transcriptional capacity of the viral long terminal repeat (LTR) once transferred to target cells. The third generation offers maximal biosafety but is more cumbersome, involving the transfection of four different plasmids in producer cells. We prefer the second generation system as it is safe and satisfies most applications.

   Two pairs of second generation lentivectors have been recommended for use in conjunction with the pLKO.1 TRC-cloning vector: pCMV-dR8.2 dvpr and pCMV-VSV-G, and the psPAX2 and pMD2.G. In this chapter, we present the method using pCMV-dR8.2 dvpr and pCMV-VSV-G. Both of the lentivectors are high-copy plasmids usually provided in E. coli (DH5α). To prepare the two vectors, the DH5α can be cultured in LB broth and selected with ampicillin. Endotoxin-free plasmid isolation kit is recommended to extract the plasmids from DH5α.

4. Some other online programs are accessible for siRNA design. Whitehead Institute for Biomedical Research hosts an siRNA Selection Program that can be accessed after a free registration (http://jura.wi.mit.edu/bioc/siRNAext/). If you have MacOS X, another excellent program is iRNAi, which is provided free by the company Mekentosj (http://www.mekentosj.com/irnai/). TRC provides a free online access to the shRNA oligos designed for pLKO.1 cloning vector as well as the possible targets of your interest genes (http://www.broadinstitute.org/rnai/public/gene/search). You can get the information of the possible targets and the oligo designs for your interest genes by simply inputting the IDs of your genes or transcripts.

5. This strategy for choosing siRNA target sites is based on the observation that siRNAs with 3′ overhanging UU dinucleotides are the most effective. This is also compatible with using RNA Pol III to transcribe hairpin siRNAs because RNA Pol III terminates transcription at 4–6 nucleotide poly (T) tracts creating RNA molecules with a short poly (U) tail.

   In available publications, siRNAs with other 3′ terminal dinucleotide overhangs have been shown to effectively induce RNAi. If desired, you may modify this target site selection strategy to design siRNAs with other dinucleotide overhangs, but it is recommended that you avoid G residues in the overhang because of the potential for the siRNA to be cleaved by RNase at single-stranded G residues.

6. Human PPARδ gene produces five transcript variants that translate four isoforms of this protein (http://www.ncbi.nlm.nih.gov/gene/5467):
   (a) Transcript variant (1) (NM_006238.4) differs in the 5′ UTR compared to variant (3) (NM_001171818.1). Both variants (1) and (3) encode the same isoform (1) (NP_006229.1).
   (b) Transcript variant (2) (NM_177435.2) uses an alternate exon in the 3′ coding region and 3′UTR. It encodes isoform (2) (NP_803184.1), which is shorter and has a distinct C terminus, compared to isoform (1).
   (c) Transcript variant (4) (NM_001171819.1) differs in the 5′ UTR, lacks a portion of the 5′ coding region, and initiates translation at an alternate start codon, compared to variant (1). The encoded isoform (3) (NP_001165290.1) has a distinct N terminus and is shorter than isoform (1).
   (d) Transcript variant (5) (NM_001171820.1) lacks two in-frame exons in the coding region, compared to variant (1). The encoded isoform (4) is shorter than isoform (1).
   So, keep alert to find the common targets of the five transcript variants when you intend not to silence certain transcript variant or isoform.
   We have successfully selected and testified one target (5′-AAAGAAGGCCCGCAGCATCCT-3′) in PPARδ mRNA, which is highly effective and specific for RNAi in the colon cancer cell lines KM12C, KM12SM, and KM12L4a when infected at a multiplicity of infection (MOI) of lentivirus around 2 ~ 3 (22).
7. Do not change the ends; these bases are important for cloning the oligos into the pLKO.1 TRC-cloning vector.
   For your choices, TRC has published five sets of shRNA oligos verified for RNAi against human PPARδ mRNA (Table 2). These oligos are designed for cloning into the pLKO.1 TRC-cloning vector.
8. This will take a few hours, but it is important for the cooling to occur slowly for the oligos to anneal.
9. It is important to testify the efficacy of your restriction enzymes by single-enzyme cleave method before the double-enzyme cleave experiment, digesting the pLKO.1 vector with the enzyme separately and run electrophoresis for validation.
10. When visualizing DNA fragments to be used for ligation, use only long-wavelength UV light. Short wavelength UV light will increase the chance of damaging the DNA.

Table 2
The verified oligos for RNAi targeting human PPARδ

TRC No.	shRNA oligos (sense sequence: 5′→3′)
TRCN0000001661	CCGGGTATTATTTCACCAGCAGCATCTCGAGATGCTGCTGGTGAA ATAATACTTTTT
TRCN0000001662	CCGGCCTATTCATTGCGGCCATCATCTCGAGATGATGGCCGCAATG AATAGGTTTTT
TRCN0000001663	CCGGGATCAAGAAGACCGAAACCGACTCGAGTCGGTTTCGGTCTT CTTGATCTTTTT
TRCN0000001664	CCGGCCGCAAACCCTTCAGTGATATCTCGAGATATCACTGAAGGGT TTGCGGTTTTT
TRCN0000010647	CCGGGTGTGGAAGCAGTTGGTGAATCTCGAGATTCACCAACTGCTT CCACACTTTTT

11. If you are unable to measure the DNA concentration, use 1 μL.
12. Don't shake the tube in this operation.
13. Avoid too much LB (>200 μg/mL) that will decrease the contacts of plasmids with bacteria.
14. Remember to invert the plate when the bacterial solution seems absorbed by agarose.

    Plate the original bacteria (untreated) in a plate without antibiotics (control-1) and in another one with antibiotics (control-2) as controls. Plate the reaction bacteria solution in a plate without antibiotics (control-3). After incubation, there should have lots of colonies in control-1 and control-3, without colonies in control-2.

    Carbenicillin is preferable to ampicillin as it is more stable. Compared with ampicillin, carbenicillin will greatly reduce the false positive colonies although the colonies will grow slower and smaller.
15. Select at least three colonies for each tube because of the slow growth of the bacteria containing pLKO.1-shRNA. We recommend TB broth and carbenicillin instead of LB and ampicillin, which can significantly increase the plasmids yields.

    Set three controls: select 3 colonies from the plate incubating treated bacteria into TB broth without antibiotics (control-4); select 3 colonies from the plate of control-1 into the TB broth without antibiotics (control-5) or with antibiotics (control-6). After incubation, there should have lots of colonies in control-4 and control-5, without colonies in control-6.

Use a tube with a volume of at least four times the volume of the culture to avoid overflow during shaking.

16. To screen the insert, you can use miniprep kit to prepare the plasmids. However, as plasmids for transfection into eukaryotic cells must be very clean and free from endotoxin, we recommend preparing lentiviral plasmids using the endotoxin-free maxiprep Kit.
17. Before this step, you must contact your institution's Bio-Safety office to receive permission and institution-specific instructions. You must follow safety procedures and work in an environment (e.g., BL2+) suitable for handling HIV-derivative viruses.
18. Do not use 293T cells that are in culture for more than 30 passages.

    Although some literatures support antibiotics-free DMEM for this step, we have observed that antibiotics actually do not affect the transfection efficiency and the use of antibiotics-free medium may increase the risk of pollution.

    Optimize your cell density for plating so that cells should be ~80% confluent on the day of transfection.
19. Cautious medium replacement is necessary because 293T cells have a high tendency to detach.
20. You may vary the ratio of shRNA plasmid, packaging plasmid, and envelope plasmid to obtain the ratio that gives you the optimal viral production.
21. Vigorous vortexing will ensure the formation of a fine precipitate optimal for transfection.
22. Do not exceed 30 min.
23. The transfection can be started late in the afternoon and the medium is changed early the following morning. Medium needs to be changed within ~18 h of addition of the DNA.
24. Collect supernatant for the first time in the evening, around 8 h after medium replacement. The second harvest can take place early in the following morning around 12 h later.

    Do not use a 0.22 μm filter, as this is likely to shear the envelope of your virus.

    Supernatant can be harvested two or three times, every 8–12 h. Keep it at 4°C over the collecting period. The cleared supernatants can be kept at 4°C for 5–7 days.

    Supernatants can be used directly, stored at −80°C as aliquots, or concentrated if needed. Freeze/thaw cycles strongly reduce the titter.
25. The titer doesn't mean the real concentration of viral particles in suspension but the average transduction number of cells per 1 mL suspension.

We estimate the titer of the viral suspensions before proceeding with transduction experiments for the following reasons: (1) To ensure that pseudoviral stock is viable; (2) To determine the percentage of target cells that can be transduced with pseudoviral stock; and (3) To control the number of copies of integrated viral constructs per target cell.

26. Polybrene acts by neutralizing charge differences between the target cell and virion, facilitating localization and binding.
27. Only dilutions yielding to 1–20% of GFP-positive are considered for titer calculations. Below 1%, the FACS may not be accurate enough to reliably determine the number of GFP-positive cells. Above 20%, the chance for each GFP-positive target cell to be transduced twice significantly increases, resulting in underestimation of the number of transducing particles.
28. Growth rate of cells vary greatly. Adjust the number of cells plated to accommodate a confluency of 70% upon transduction. Also account for the length of time the cells will be growing before downstream analysis when determining the plating density does a pretest to seed your cells in different densities to see which one is the optimal for 3 days culture. In our experience, 3×10^5/well is good for HCT-116 cells, 1×10^5 is good for KM12C cells.
29. You may wish to repeat this titration with finer increments of puromycin for determining a more precise optimal puromycin concentration.
30. For the quality control of transduction and RNAi, the following controls should be used:

Recommended control	**Objective**
Negative control: untreated cells	Untreated cells will provide a reference point for comparing all other samples
Negative control: infected with lentivirus carrying nontargeting shRNA	The nontargeting lentivirus is a useful control that will activate RNAi pathway, but doesn't target any human genes. Cells infected with this virus will provide a useful reference for interpretation of knockdown
Positive control: infected with virus with shRNA targeting endogenous gene	The lentivirus with guaranteed effect of targeting endogenous gene is used for positive control. This control will provide clear and measurable knockdown of the human gene target

31. The indicated viral volume is for an MOI of ~2, which assumes typical viral yields from the 10 cm plate viral preparation method described in Subheading 3.3.

    To obtain optimal knockdown of PPARδ, you will need to transduce the lentiviral construct into cells using a suitable MOI. When transducing a lentivirus into a cell line for the first time, a range of volume or MOI should be tested. 2, 5, 10, and 15 mL of lentiviral suspensions or MOIs of 0.5, 1, and 5 should be used to determine the optimal transduction efficiency and knockdown for each cell line.

32. As some cell lines can't stand the high speed centrifugation, test your cell lines by centrifugation for 1.5 h (20°C, 1,200 × *g*) before the transduction experiment. By our observation, the colon cancer cell lines HCT-116, KM12C, KM12L4a, and KM12SM can grow well after this centrifugation.

    The efficiency of lentivirus infection can be substantially improved by centrifugal inoculation (spinoculation); the underlying mechanism of enhancement has not been defined. It is proposed that spinoculation works by depositing virions on the surfaces of target cells and that diffusion is the major rate-limiting step for viral adsorption under routine in vitro pulsing conditions (23).

33. Don't add puromycin in this step as the cells just infected by virus are too weak to stand the antibiotic.

34. Due to the random integration of the lentivirus into the host genome, varying levels of target gene knockdown may be seen with different puromycin-resistant colonies. Testing a number of colonies will allow the optimal degree of knockdown to be determined.

References

1. Kliewer SA, Forman BM, Blumberg B et al (1994) Differential expression and activation of a family of murine peroxisome proliferator-activated receptors. Proc Natl Acad Sci USA 91:7355–7359
2. Akiyama TE, Meinke PT, Berger JP (2005) PPAR ligands: potential therapies for metabolic syndrome. Curr Diab Rep 5:45–52
3. Desvergne B, Wahli W (1999) Peroxisome proliferator-activated receptors: nuclear control of metabolism. Endocr Rev 20: 649–688
4. Lee CH, Olson P, Evans RM (2003) Minireview: lipid metabolism, metabolic diseases, and peroxisome proliferator-activated receptors. Endocrinology 144:2201
5. Michalik L, Desvergne B, Wahli W (2004) Peroxisome-proliferator-activated receptors and cancers: complex stories. Nat Rev Cancer 4:61–70
6. Oliver WR Jr, Shenk JL, Snaith MR et al (2001) A selective peroxisome proliferator-activated receptor delta agonist promotes reverse cholesterol transport. Proc Natl Acad Sci USA 98:5306–5311
7. Barak Y, Liao D, He W et al (2002) Effects of peroxisome proliferator-activated receptor delta on placentation, adiposity, and colorectal cancer. Proc Natl Acad Sci USA 99: 303–308
8. Tan NS, Michalik L, Desvergne B et al (2004) Peroxisome proliferator-activated receptor-beta as a target for wound healing drugs. Expert Opin Ther Targets 8:39–48
9. Ding GL, Cheng LH, Qin QH et al (2006) PPARδ modulates lipopolysaccharide-induced

TNFα inflammation signaling in cultured cardiomyocytes. J Mol Cell Cardiol 40:821–828
10. Takahashi S, Tanaka T, Kodama T (2006) Peroxisome proliferator-activated receptor δ (PPARδ), a novel target site for drug discovery in metabolic syndrome. Pharmacol Res 53: 501–507
11. He TC, Timotly AC, Bert V et al (1999) PPARd is an APC regulated target of nonsteroidal anti-inflammatory drugs. Cell 99:335–345
12. Peters JM, Gonzalez FJ (2009) Sorting out the functional role(s) of peroxisome proliferator-activated receptor-β/δ (PPARβ/δ) in cell proliferation and cancer. Biochim Biophys Acta 1796:230–241
13. Mocellin S, Costa R, Nitti DM (2006) RNA interference: ready to silence cancer? J Mol Med 84:4–15
14. Scherer L, Rossi JJ (2003) Recent applications of RNA interference (RNAi) in mammalian systems. Lett Pept Sci 10:255–267
15. Sui G, Soohoo C, Affar E et al (2002) A DNA vector-based RNAi technology to suppress gene expression in mammalian cells. Proc Natl Acad Sci USA 99:5515–5520
16. Pan QW, Cai R, Liu XY et al (2006) A novel strategy for cancer gene therapy: RNAi. Chin Sci Bull 51:1145–1151
17. Tomar RS, Matta H, Chaudhary PM (2003) Use of adeno-associated viral vector for delivery of small interfering RNA. Oncogene 22: 5712–5715
18. Abbas-Terki T, Blanco-Bose W, Deglon N (2002) Lentiviral-mediated RNA interference. Hum Gene Ther 13:2197–2201
19. Rubinson DA, Dillon CP, Kwiatkowski AV et al (2003) A lentivirus-based system to functionally silence genes in primary mammalian cells, stem cells and transgenic mice by RNA interference. Nat Genet 33:401–406
20. An DS, Xie Y, Mao SH, Morizono K et al (2003) Efficient lentiviral vectors for short hairpin RNA delivery into human cells. Hum Gene Ther 14:1207–1212
21. Nishitsuji H, Ikeda T, Miyoshi H et al (2004) Expression of small hairpin RNA by lentivirus-based vector confers efficient and stable gene-suppression of HIV-1 on human cells including primary non-dividing cells. Microbes Infect 6:76–85
22. Yang L, Olsson B, Pfeifer D et al (2010) Knockdown of peroxisome proliferator-activated receptor-beta induces less differentiation and enhances cell-fibronectin adhesion of colon cancer cells. Oncogene 29: 516–526
23. O'Doherty U, Swiggard WJ, Malim MH (2000) Human immunodeficiency virus type 1 spinoculation enhances infection through virus binding. J Virol 74:10074–10080

Chapter 6

Dominant-Negative and Knockdown Approaches to Studying PPAR Activity

Sudhir Singh and Robert G. Bennett

Abstract

Manipulation of PPAR activity is often a valuable approach toward elucidation of the cellular effects of PPARs. The activity of specific PPARs can be decreased using chemical inhibitors, but these approaches can be affected by nonspecific interactions or cell toxicity. Alternative approaches include targeting PPAR gene expression or activity through molecular biology strategies. Here, we describe the targeting of PPARγ through dominant-negative and siRNA-mediated knockdown constructs.

Key words: Peroxisome proliferator-activated receptor, Dominant-negative, Adenovirus, siRNA

1. Introduction

Peroxisome proliferator-activated receptors (PPARs) are ligand-activated transcription factors that regulate expression of target genes (1). Paramount to the elucidation of biological effects of PPARs is the demonstration that loss-of-function of the particular PPAR results in perturbation of the effect. A common approach to reduce PPAR activity is the use of chemical inhibitors that block ligand binding and activation of PPAR transcriptional activity. While useful in many systems, chemical inhibitors can have nonspecific and cytotoxic effects that can perturb the results. Therefore, alternative approaches are often required, such as those involving the use of dominant-negative mutants or reduced PPAR expression.

The transcriptional activity of PPARs relies on ligand binding, association with coactivator or corepressor proteins, and DNA binding to target gene promoter regions. Numerous mutant forms of PPARγ have been produced that result in dominant-negative

Mostafa Z. Badr and Jihan Youssef (eds.), *Peroxisome Proliferator-Activated Receptors (PPARs): Methods and Protocols*, Methods in Molecular Biology, vol. 952, DOI 10.1007/978-1-62703-155-4_6, © Springer Science+Business Media New York 2013

forms of PPARs, as a result of impaired binding to coactivators or constitutive binding to corepressors, or inability to form a retinoid X- receptor-DNA complex (2–4). While dominant-negative contructs are of tremendous value in pinpointing the cellular effects of PPARs, many cell lines, especially primary cells, are difficult to transfect. Therefore, retroviral delivery of constructs can be a useful approach toward the use of dominant negative PPARs. Viral particles (such as adenovirus and lentivirus) carrying transcripts of interest are capable of infecting a wide number of cells without the need for lipofection or electroporation. Such approaches have been successfully employed frequently for PPARγ (2, 4–9), as well as PPARα (10) and PPARβ/δ (11, 12). Here, we describe adendovirus-mediated delivery of dominant-negative PPARγ (8) to reduce PPARγ transcriptional activity (13, 14).

The use of small inhibitory RNA (siRNA) to reduce the expression of target proteins is now a commonplace approach to the study of protein function. The method involves the design of small (19–23 bp) fragments of double-stranded RNA that bind to a protein complex known as RNA-induced silencing complex (RISC) (15). The RISC complex then binds to the target mRNA to form a triplex structure, followed by degradation of the mRNA complex by the nuclease Argonaute (16). Selected siRNA constructs are introduced into target cells using transfection reagents or electroporation approaches. Alternatively, the siRNA constructs can be delivered as short hairpin RNA (shRNA) inserted into plasmid or viral vectors for expression in cells. Because of possible global effects on expression caused by the response, it is critical to include controls using siRNA contructs targeting an unrelated protein (often housekeeping genes such as GAPDH) or nontargeting siRNA that will trigger the RISC machinery, target no mRNA species. Many suppliers now provide predesigned pools of 3–4 different siRNA constructs against a particular target to maximize the knockdown while minimizing off-target effects (17). We describe below the use of an siRNA pool (On-Target Plus, Thermo Scientific, Lafayette, CO) introduced into cells by an electroporation approach (Nucleofector, Lonza, Walkersville, MD), which we have used successfully to knockdown PPARγ (13, 14).

A number of methods can be employed to determine the effect of the above approaches on PPAR activity and expression. Because the activity of PPARs is employed at the protein level, the transcriptional activity or protein levels of PPARs should be determined whenever possible. However, the transcript level can provide a rapid means of assessing siRNA-mediated knockdown of PPAR expression for higher throughput screening methods. We describe below means to carry all of these determinations.

2. Materials

2.1. Adenovirus Amplification, Purification, and Infection

1. Healthy, low-passage HEK-293 cells (American Type Culture Collection, Manassas, VA).
2. Modified Eagle's Medium (MEM), supplemented with 10% fetal bovine serum.
3. Dominant-negative, wild-type, and control β-galactosidase (β-Gal) adenovirus stock.
4. Dry ice/ethanol bath.
5. Adenovirus Mini Purification Virakit (Virapure, San Diego, CA). Includes columns and loading, wash, and elution buffers.
6. Phosphate buffered saline 10× solution (PBS): 1.4 M NaCl, 27 mM KCl, 81 mM Na_2HPO_4, 15 mM KH_2PO_4. Dissolve 80 g NaCl, 2 g KCl, 11.5 g Na_2HPO_4 (anhydrous), 2 g KH_2PO_4 (anhydrous) in ~800 mL of deionized, endotoxin-free water. Adjust pH to 7.4 if necessary, bring to 1 L final volume. Sterile filter. Dilute with endotoxin-free water to 1× before use.
7. 100% ice-cold methanol fixative.
8. PBS +1% BSA: To 100 mL1× PBS, add 1 g bovine serum albumin (BSA). Stir until dissolved, sterile filter, and store at 4°C.
9. Adeno-X Rapid Titer kit (Clontech, Mountain View, CA). Includes mouse anti-hexon primary antibody, horseradish peroxidase (HRP) conjugated rat anti-mouse secondary antibody, DAB substrate, and peroxidase solution.
10. Experimental cells for adenoviral infection.
11. Fixative for β-galactosidase assay: 0.05% glutaraldehyde in 1× PBS. Prepare fixative fresh on the day of use by diluting 100 μL 25% glutaraldehyde solution (Sigma Chemical, St. Louis, MO) with 1× PBS.
12. X-Gal stock solution: 40 mg/mL X-gal (5-bromo-4-chloro-3-indolyl-b-galactopyranoside) in dimethylformamide. Protect from light. Store at –20°C.
13. X-Gal buffer: 2 mM $MgCl_2$, 35 mM potassium ferrocyanide, 35 mM potassium ferricyanide, 0.02% Nonidet P-40, 1× PBS.
14. X-Gal staining solution: Make fresh the day of use. Dilute X-Gal stock solution 1:40 in X-Gal buffer.

2.2. Knockdown of PPAR Expression with siRNA

1. Cultured cells of interest in log phase growth.
2. Nucleofector kit appropriate for cells of interest. Includes nucleofection solution and cuvettes.
3. Nucleofector II electroporation instrument.
4. Stock PPAR and control nontargeting siRNA solutions (20 nM).
5. Fluorescent siRNA for nucleofection efficiency (optional).

2.3. PPRE Luciferase Assay

1. PPRE reporter plasmid (e.g., ACO-PPRE-luc (18)).
2. *Renilla* luciferase reporter plasmid (e.g., pRL-TK, Promega, Madison WI).
3. Transfection reagent appropriate for cell line of interest.
4. Dual-Glo luciferase assay (Promega, Madison, WI).

3. Methods

3.1. Adenoviral Delivery of Dominant-Negative PPAR Constructs

3.1.1. Amplification and Purification of Adenoviral Vectors

1. Seed low-passage HEK-293 cells (see Note 1) onto a 100 mm^2 culture dish and grow to ~90% confluency. Infect the cells with experimental and control adenoviruses separately (see Note 2). The amount of adenovirus to use will depend on the nature and purity of the individual preparations. A general rule of thumb is 150–300 μL of crude virus, or 30–100 μL of purified virus.
2. Culture for 2–4 days, checking daily for signs of the cytopathic effect (CPE). Initially, the CPE is detectable as a loss of attached cells in a grape-shaped area, usually in the center of the dish. When 80–90% of the cells have detached, collect the suspended and attached cells by pipetting up and down over the growth surface until all cells are suspended.
3. Pellet the cells at 1,000 × *g* for 5–10 min. Remove and discard all but 1 mL of the supernatant. Resuspend the cell pellet in the remaining 1 mL of medium.
4. Lyse the cells by freezing the cell suspension in a dry ice/ethanol bath, followed by thawing in a 37°C water bath. Vortex to mix the cells. Repeat for a total of three freeze-thaw cycles. Centrifuge the suspension at 10,000 × *g* for 10 min, and collect and save the supernatant. The clarified crude lysate can be stored at −80°C, or processed to purification as below
5. The crude lysate is further purified using the Adenovirus Mini Purification Virakit (Virapure, San Diego, CA). Each purification column can purify up to one half the total crude lysate collected from a 100 mm^2 culture. All centrifugation steps are carried out at 2,200 × g for 5 min.
6. Condition the purification column by applying 0.4 mL loading buffer and centrifuge. Discard the flowthrough. Load up to 0.4 mL of crude lysate (see step 4 above) and centrifuge. Discard the flowthrough. If necessary, repeat the loading and centrifugation steps with additional lysate, up to half of the original crude lysate per column.
7. Wash the column by applying 0.4 mL of wash buffer to the column and centrifuging. Discard the flowthrough, and repeat the wash procedure two more times.

8. Place the column into a fresh sterile microcentrifuge tube. Add 0.4 mL of elution buffer to the column and centrifuge. The flowthrough contains the purified virus. Store the virus at −80°C, or proceed to determination of viral titer (see Note 3).

3.1.2. Determination of Viral Titer

1. Prepare serial dilutions of the purified viral stock in sterile PBS. The dilutions should range from 10^2–10^7-fold dilutions, for a total of six viral concentrations.
2. Seed low-passage log phase (recently passaged) HEK-293 cells into a 12-well plate at 5×10^5 cells per well in 1 mL of medium. Add 100 μL PBS (vehicle control) to one well, and 100 μL of the 10^2-fold dilution to a second well. Add 100 μL/well of each remaining diluted virus, two wells per dilution. Allow the cells to incubate in a culture hood for 48 h.
3. Carefully aspirate the media from the cells. Exercise caution, because the infected cells will detach easily. Leave the plate open in a laminar flow hood for 5 min to dry.
4. Fix the cells by gently adding 100% ice-cold methanol to each well along the well walls using a pipet. Incubate the plates at −20°C for 10 min. Aspirate methanol and wash the wells three times with 0.5 mL PBS + 1%BSA.
5. Add 250 μL of diluted mouse anti-hexon antibody (1:1,000 in PBS + 1% BSA) to each well. Incubate for 1 h at 37°C on an orbital shaker at low speed. Aspirate the anti-hexon antibody, and rinse the wells three times with 0.5 mL of PBS + 1% BSA. Add 250 μL of the secondary antibody (HRP-conjugated rat anti-mouse, 1:500 in PBS + 1% BSA), and incubate for 1 h at 37°C on an orbital shaker at low speed.
6. Prepare the 1× DAB working solution by diluting 600 μL DAB substrate solution with 6 mL peroxidase buffer. Bring working DAB solution to room temperature.
7. Remove the secondary antibody by aspiration, and rinse the wells three times with 0.5 mL of PBS + 1% BSA. Add 250 μL of 1× DAB working solution to each well, and incubate the plate at room temperature for 10 min. Protect the plate from light during the incubation. Remove the DAB solution by aspiration, and then add 1 mL PBS to each well.
8. Using an inverted microscope, observe a random field of cells under a 20× objective. Hexon-positive (infected) cells will appear as dark brown or black cells. The vehicle control well should have no hexon-positive cells. The wells with the higher concentration of virus may show morphological signs of cell death, and all cells may be hexon-positive. Looking at all of the wells, find the viral dilutions that resulted in 10% or fewer hexon-positive cells. Count the number of hexon-positive cells per field, a minimum of five fields per well. Calculate the mean positive cells per field for each well.

9. Calculate viral titer in infectious units (ifu)/mL for each well using the following formula:

$$\frac{\text{Infected cells per field} \times 573}{0.1\text{mL virus} \times \text{dilution factor}},$$

where 573 is the number of fields per well in a 12-well plate. The mean of the results from all of the counted wells is the ifu/mL for the viral preparation.

3.1.3. Determination of the Optimal Infectious Rate for the Target Cells Using a β-Gal Virus

1. Seed healthy, log phase cells in a 24-well plate at an appropriate density to reach 70–80% confluency at the time of infection, in a volume of 1 mL per well (see Note 4).
2. Dilute β-Gal adenovirus to appropriate concentrations to deliver 1–100 MOI (see Note 5). The diluted values should be tenfold higher than the final amount. Add 100 μL diluted virus to the cells. Incubate the cells for 4 h in a cell culture incubator. Remove the medium and replace with growth medium.
3. After 24–48 h, carefully remove the culture medium (see Note 6). Fix the cells in 0.05% glutaraldehyde for 5 min at room temperature. Add 1 mL per well PBS, aspirate. Add 1 mL PBS, incubate 10 min at room temperature. Aspirate, and then wash once more with 1 mL PBS.
4. Aspirate the final wash, and add 0.5 mL X-Gal staining solution. Incubate at 37°C for 1 h to overnight. Monitor frequently for development of blue color.
5. Choose the adenoviral MOI that produced good expression of β-Gal but with the least cytotoxicity (e.g., cell detachment). This adenoviral dilution will be the desired MOI for infection with the PPAR adenoviruses.

3.1.4. Infection of Cells with Wild-Type and Mutant PPARγ Adenoviruses

1. Calculate the infection strategy for the experiment. For experiments comparing different adenoviral constructs, the total amount of virus used for infection must be the same. A control virus (e.g., β-Gal) can be used to adjust the total (see Note 7).
2. Infect the cells as above, and incubate for 24 h (or an appropriate period depending on the cells) in a tissue culture incubator.
3. Assess the effect on PPARγ activity as described below.

3.2. Knockdown of PPAR Expression Using siRNA

The following protocol is for delivery of siRNA into HEK-293 cells by electroporation using the Nucleofector II instrument. The protocol can be easily adapted for use in many different cell lines and primary cells.

1. Passage healthy, low-passage cells 2 days before nucleofection, to maintain cells in log phase at the time of nucleofection.

Make sure the cells are not confluent at the time of harvest and are sufficient for the experiment. For HEK-293 cells, nucleofection requires 1.2×10^6 cells per reaction (see Note 8).

2. Prepare the following ahead of nucleofection: Power on the nucleofector instrument and select the program Q-001. Prepare nucleofector solution V by adding 4.5 parts to 1 part supplement, and allow it to reach room temperature. For each nucleofection, 100 μL of supplemented solution V will be needed (see Note 8). Prewarm 1 mL of culture medium per nucleofection reaction into wells of a 6-well plate to 37°C in a cell culture incubator. Similarly, prewarm 500 μL culture medium per reaction to 37°C in an appropriately sized tube (see Note 9).
3. Harvest the cells by standard trypsinization and perform a cell count. Collect the required number of cells by gentle centrifugation. Aspirate the supernatant and resuspend the pelleted cells in 100 μL supplemented nucleofector solution. Do not store the cell suspension in nucleofector solution for more than 20 min (see Note 10).
4. For each reaction, add 100 μL of the cell suspension to 3–30 pmol of control or PPARγ siRNA (see Note 11). Transfer the cell suspension to a nucleofection cuvette, being sure that the solution covers the bottom surface of the cuvette. Insert the cuvette into the instrument and press the start button. Remove the cuvette immediately after the program has finished and add 500 μL of the prewarmed medium. Transfer the sample to a well of the pre-warmed plate, and incubate the plate in a tissue culture incubator (see Note 12).
5. Assay for PPARγ knockdown as described below after 24–72 h. The efficiency of the siRNA transfer can be monitored using fluorescent siRNA (see Fig. 1).

3.3. Assessing PPARγ Activity and Expression

3.3.1. PPAR Response Element Luciferase Reporter Assay

Activation of PPARs results in increased expression of genes regulated by PPAR response elements (PPREs). A number of reporter plasmids have been described using multiple copies of PPREs upstream of luciferase genes (18–21). The effect on transcription can be easily measured by measurement of luciferase activity in the treated cells (see Fig. 2). The procedure below is a description of the assay we have used using a reporter plasmid firefly luciferase driven by the acyl-CoA oxidase PPRE (ACO-PPRE-luc, (21)) reporter plasmid with a *Renilla* luciferase reporter (pRL-TK, Promega, Madison, WI) as a transfection control.

1. Using appropriate conditions for the cell line of interest, transfect cells with ACO-PPRE-luc and pRL-TK at a ratio of 5:1–20:1, respectively, depending on the cell line. The transfection can take place before or after introduction of the

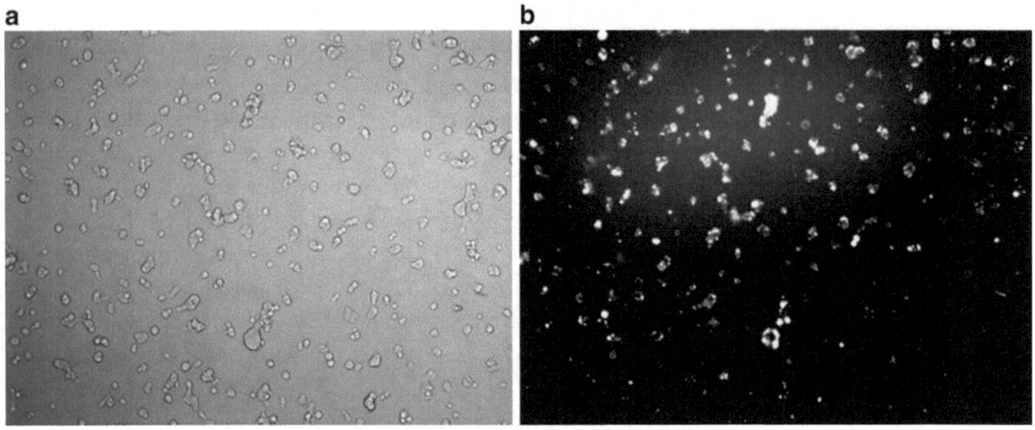

Fig. 1. Visualization of siRNA localization after a typical nucleofection reaction. HEK-293 cells were subject to nucleofection with fluorescent siRNA (siRNA-GLO Red, Thermo Scientific, Lafayette CO). After 24 h, the cells were visualized and brightfield (**a**) and fluorescent (**b**) images were captured. The fluorescence reveals efficient delivery of siRNA, predominantly localized to the cytosol and nucleus.

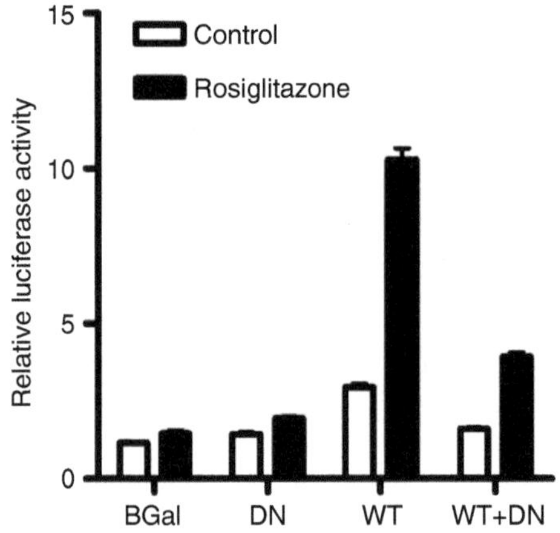

Fig. 2. Determination of PPARγ transcriptional activity in cells infected with wild-type (WT) and dominant-negative (DN) adenoviruses. Cells were infected with control (β-Gal), DN, WT, or both DN and WT adenoviruses at an MOI of 1.0 each. Control (β-Gal) adenovirus was used to adjust the total viral load to 2.0 MOI when necessary. After 4 h, the cells were transfected with ACO-PPRE-luc and pRL-TK reporter plasmids, and incubated for 24 h. Cells were treated with rosiglitazone (1 μM) or vehicle, and 24 h later firefly and *Renilla* luciferase activities were measured. Rosiglitazone caused a marked increase in PPARγ activity in the WT cells, which was reduced in presence of the DN mutant.

adenovirus or siRNA, or concomitantly with nucleofection of the siRNA (see Note 13). Place the cells in a cell culture incubator for 24 h.

2. Treat the cells with an appropriate PPAR activator (such as a thiazolidinedione for PPARγ, or clofibrate for PPARα) for 24 h.

3. Measure the firefly and *Renilla* luciferase activities using the Dual-Glo Luciferase Assay kit (Promega, Madison, WI) (see Note 14). Divide the firefly luciferase activity by the *Renilla* luciferase activity for each well to determine the relative luciferase (PPRE) activity. Successful dominant-negative inhibition or knockdown in the PPAR of interest should result in decreased relative luciferase activity compared to the appropriate controls.

3.3.2 Western Blotting

Western blotting is a standard protocol in virtually any biomedical research laboratory, and therefore the specific protocol will not be presented here. In general, from 24 to 72 h after application of the siRNA constructs, cell lysates are subject to western blotting to assess the protein levels of the PPAR of interest. In addition, this method can be used to assess the efficiency of viral infection of a PPAR transcript, as many of these retain antigenic properties for recognition by PPAR antibodies. The advantage of this approach is that it provides direct assessment of the protein level of the PPAR of interest. The disadvantage of this procedure is that, unlike the PPRE reporter assay, it provides no information about PPAR transcriptional activity.

3.3.3. Quantitative (Real-Time) PCR

This procedure determines the level of an mRNA of interest relative to that of a control (housekeeping) gene. This procedure has become fairly standard in research laboratories, and so specific procedures will not be provided here. However, because the changes in PPAR activity occur at the protein level, and because changes in protein levels are not always reflected by transcript levels, we do not commonly use this approach in out laboratory. The advantage of this approach is that it is a more rapid, high throughput procedure than PPRE activity assays or western blotting, and therefore is useful for screening potential treatments.

4. Notes

1. Most adenoviruses lack critical proteins to render them replication-incompetent. The HEK-293 cells are used because they express proteins that allow the viruses to replicate for amplification. Other cell lines can be used for this purpose, such as 911 or GH329.
2. All procedures with adenoviral particles should be performed by fully trained personnel in accordance with the local Institutional Biosafety Committee. In general, work should be performed in a BSL2 biological safety cabinet. All items that contact the virus (plasticware, surfaces, etc.) should be disinfected with 10% bleach solution for at least 10 min. Although most adenoviral vectors have been rendered replication-incompetent, with

extended use recombination events can occur to produce replication-competent virus, greatly increasing the hazardous potential. At each round of purification, the virus should be screened for replication-competent virus by standard approaches (22).

3. The effectiveness of some viral preparations will decrease with repeated freeze-thaw cycles. Therefore, it is recommended to store viral stocks in working aliquots.
4. Depending on the cell line, the presence of serum may greatly affect the efficiency of adenoviral infection. It is recommended to test serum at normal growth conditions (e.g., 10% serum), low-serum (e.g., 2%), and serum-free conditions to determine infection efficiency.
5. The optimal multiplicity of infection (MOI) will vary for each cell line, and is defined as the ifu per cell. Initial values should range from 1 to 100 MOI. For example, an MOI of 1.0 for a well containing 10^5 cells would require 10^5 ifu of virus. It is critical that the selected MOI does not cause cytotoxicity, and therefore the user must have a means of assessing cell death (e.g., cell morphology, lactate dehydrogenase leakage, etc.)
6. The time required for expression of the desired protein will vary depending on the cell line used. Efficient expression is detectable after 24 and 48 h in most cell lines, but shorter or longer times may be necessary.
7. When comparing the effects of different viral constructs, the total amount of virus must remain the same. Control or empty viruses can be used to adjust the total viral load. For example, if PPARγ activity is to be compared in cells overexpressing wt-PPARγ with various amounts of dn-PPARγ using an optimal MOI of 10, then cells could be infected with 5 ifu of wt-PPARγ+5 ifu β-Gal, 5 ifu of wt-PPARγ+5 ifu dn-PPARγ, or 10 ifu β-Gal.
8. The cell density and growth phase, nucleofection solution, oligonucleotide concentration, and nucleofector program are critical for the efficiency of the nucleofection procedure, and will vary depending on the cells used. An extensive list of optimized protocols for specific cells (both cell lines and primary cells), as well as a list of user-submitted protocols is available at the manufacturer's website (www.lonzabio.com).
9. Cells will be affected differently by the presence of serum and antibiotics in the growth medium. In our hands, the cell viability is often greater if the cells are placed into antibiotic-free medium immediately after the nucleofection reaction.
10. Exposure to nucleofector solution for more than 20 min can have greatly detrimental effects on cell viability. If multiple nucleofection reactions are planned, the cell collection and resuspension should be staggered to minimize exposure to the solution.

11. The total amount of oligonucleotide is critical to the success of the experiment. For HEK-293 cells, 3–30 pmol is recommended (30 pmol is 1.5 μL of a 20 μM siRNA stock) to cell suspension. Optimal siRNA concentration should be determined empirically. In addition, co-nucleofection of siRNA with other oligonucleotides (e.g., plasmids) can be accomplished, but the total oligonucleotide amount is limited. See instructions for the specific cell lines. It is critical that appropriate controls are included with each experiment. These may include nontargeting siRNA, or knockdown of unrelated transcripts. Transfection efficiency can be monitored using fluorescently labeled siRNA.
12. Immediately after the nucleofection reaction, the display on the instrument should read "OK". If an error message is displayed, refer to the instrument manual for an explanation. The most common causes of an error is failure to completely remove the cell growth medium after centrifugation, or failure to cover the bottom of the cuvette.
13. When using nucleofection, the reporter plasmids can often be introduced into the cells at the same time as the siRNA constructs, as long as the total oligonucleotide concentration does not exceed the usable level. In addition to *Renilla* luciferase, other kinds of reporter plasmids can be used for transfection normalization, such as fluorescent protein of β-Gal plasmids. For many cell lines, adherence to cell culture plasticware is decreased after transfection, and therefore cell cultureware coated with poly-d-lysine, collagen I, etc., can greatly improve performance.

Acknowledgments

Supported by funding through The Department of Veterans Affairs Biomedical Laboratory Research and Development Program, NIAAA (AA015509), and the Bly Memorial Research Fund (RGB), and by a University of Nebraska Medical Center Graduate College Fellowship (SS).

References

1. Tontonoz P, Spiegelman BM (2008) Fat and beyond: the diverse biology of PPARγ. Annu Rev Biochem 77:289–312
2. Agostini M, Schoenmakers E, Mitchell C et al (2006) Non-DNA binding, dominant-negative, human PPARγ mutations cause lipodystrophic insulin resistance. Cell Metab 4:303–311
3. Barroso I, Gurnell M, Crowley V et al (1999) Dominant negative mutations in human PPARγ associated with severe insulin resistance, diabetes mellitus and hypertension. Nature 402:880–883
4. Masugi J, Tamori Y, Kasuga M (1999) Inhibition of adipogenesis by a COOH-terminally truncated mutant of PPARγ2 in

3T3-L1 cells. Biochem Biophys Res Commun 264:93–99
5. Ferguson HE, Kulkarni A, Lehmann G et al (2009) Electrophilic peroxisome proliferator-activated receptor-γ ligands have potent antifibrotic effects in human lung fibroblasts. Am J Respir Cell Mol Biol 41:722–730
6. Gurnell M, Wentworth J, Agostini M et al (2000) A dominant-negative peroxisome proliferator-activated receptor γ (PPARγ) mutant is a constitutive repressor and inhibits PPARγ-mediated adipogenesis. J Biol Chem 275:5754–5759
7. Hata K, Nishimura R, Ikeda F et al (2003) Differential roles of Smad1 and p38 kinase in regulation of peroxisome proliferator-activating receptor gamma during bone morphogenetic protein 2-induced adipogenesis. Mol Biol Cell 14:545–555
8. Park Y, Freedman B, Lee E et al (2003) A dominant negative PPARγ mutant shows altered cofactor recruitment and inhibits adipogenesis in 3T3-L1 cells. Diabetologia 46:365–377
9. Wada K, Nakajima A, Katayama K et al (2006) Peroxisome proliferator-activated receptor γ-mediated regulation of neural stem cell proliferation and differentiation. J Biol Chem 281:12673–12681
10. Semple RK, Meirhaeghe A, Vidal-Puig A et al (2005) A dominant negative human peroxisome proliferator-activated receptor (PPAR) α is a constitutive transcriptional corepressor and inhibits signaling through all PPAR isoforms. Endocrinology 146:1871–1882
11. Bastie C, Luquet S, Holst D et al (2000) Alterations of peroxisome proliferator-activated receptor δ activity affect fatty acid-controlled adipose differentiation. J Biol Chem 275:38768–38773
12. Nahle Z, Hsieh M, Pieka T et al (2008) CD36-dependent regulation of muscle FoxO1 and PDK4 in the PPARδ/β-mediated Adaptation to Metabolic Stress. J Biol Chem 283: 14317–14326
13. Singh S, Bennett RG (2010) Relaxin signaling activates peroxisome proliferator-activated receptor gamma. Mol Cell Endocrinol 315:239–245
14. Singh S, Bennett RG (2009) Relaxin family peptide receptor 1 activation stimulates peroxisome proliferator-activated receptor gamma. Ann N Y Acad Sci 1160:112–116
15. Elbashir SM, Harborth J, Weber K et al (2002) Analysis of gene function in somatic mammalian cells using small interfering RNAs. Methods 26:199–213
16. Liu J, Carmell M, Rivas F et al (2004) Argonaute2 is the catalytic engine of mammalian RNAi. Science 305:1437–1441
17. Reynolds A, Leake D, Boese Q et al (2004) Rational siRNA design for RNA interference. Nat Biotechnol 22:326–330
18. Forman BM, Tontonoz P, Chen J et al (1995) 15-Deoxy-Δ12,14-prostaglandin J2 is a ligand for the adipocyte determination factor PPARγ. Cell 83:803–812
19. Palmer CNA, Hsu M, Griffin H et al (1995) Novel sequence determinants in peroxisome proliferator signaling. J Biol Chem 270: 16114–16121
20. Varanasi U, Chu R, Huang Q et al (1996) Identification of a peroxisome proliferator-responsive element upstream of the human peroxisomal fatty acyl coenzyme A oxidase gene. J Biol Chem 271:2147–2155
21. Jiang C, Ting AT, Seed B (1998) PPARγ agonists inhibit production of monocyte inflammatory cytokines. Nature 391:82–86
22. Dion LD, Fang J, Garver RI (1996) Supernatant rescue assay vs. polymerase chain reaction for detection of wild type adenovirus-contaminating recombinant adenovirus stocks. J Virol Methods 56:99–107

Chapter 7

Producing PPARgamma2 Knockdown in Mouse Liver

Tomomi Yamazaki and Osamu Ezaki

Abstract

A high saturated fat diet induces fatty liver in mice. The fatty liver is triggered initially by an increase in PPARγ2 protein in the liver, which leads to increased expression of lipogenic genes. Inactivation of PPARγ2 may improve fatty liver induced by a high saturated fat diet. RNA interference of PPARγ2 should be a powerful tool to elucidate the role of PPARγ2 in the development of fatty liver. Here, we describe our method for constructing the adenovirus vector containing short hairpin RNA targeted for PPARγ2. Adenovirus vector-mediated RNA interference for PPARγ2 should be useful for clarifying the biological role of the PPARγ2 pathway in various tissues and for therapeutic application in a variety of diseases.

Key words: RNAi, shRNA, Adenovirus, Cosmid, Plasmid, PPARγ2, Liver, Knockdown

1. Introduction

Among the viral vectors, adenovirus vectors have been used extensively to deliver foreign genes to a variety of cell types and tissues both in vitro and in vivo. They can be easily grown to high titer and efficiently transferred into both dividing and nondividing cells.

The adenovirus vectors are produced from human adenovirus serotype, adenovirus type 5 (Ad5) with the E1 and E3 genes deleted. The E1 deletion makes it impossible for the recombinant adenoviruses to replicate in normal cells; they can only replicate in HEK293 cells (an established cell line derived from human embryonic kidney cells), which express the E1 gene (1). The E3 gene is not essential for viral replication in vitro.

For studies of loss of functions via knockdown of target gene expression, RNA interference (RNAi) has been shown to have great promise for research and therapeutic use. RNAi is activated by introducing a double-strand RNA whose sequence is homologous

Mostafa Z. Badr and Jihan Youssef (eds.), *Peroxisome Proliferator-Activated Receptors (PPARs): Methods and Protocols*, Methods in Molecular Biology, vol. 952, DOI 10.1007/978-1-62703-155-4_7, © Springer Science+Business Media New York 2013

to the target gene transcript. The exogenous RNA is digested into 21–23 nucleotide small interfering RNAs (siRNAs), which bind a nuclease complex to form an RNA-induced silencing complex (RISC). The RISC then targets the endogenous gene transcripts by base-pairing and cleaves the mRNA (2, 3).

Several plasmid expression vectors have been designed to generate sustained production of siRNAs. Some of these vectors have been engineered to express short hairpin RNAs (shRNAs), which are processed in vivo into siRNA-like molecules capable of carrying out gene-specific silencing (4–6). These vectors contained a DNA sequence that encodes the shRNA cloned between a Pol III promoter and a transcription termination site. The transcript folds into a stem-loop structure. The ends of the shRNAs are processed in vivo, converting the shRNA into approximately 21-nucleotide siRNA-like molecules, which in turn initiate RNAi (4).

The full-length DNA method was developed for recombinant adenovirus preparation (7). This method allows for preparation of recombinant adenoviruses by digesting with *Bsp*T104 I the recombinant cosmid vectors in which the target gene is inserted and transfecting them into HEK293 cells. This method eliminates the chance of parent virus contamination and produces a high yield of the desired virus because it does not require homologous recombination with parent virus DNA within the cells (8).

Cellular entry of the adenovirus is initiated by binding of the fiber knob to the primary receptor, the coxsackie-adenovirus receptor (CAR). CAR mRNA expression is extremely variable among different tissues; the highest levels are found in the liver and adenovirus vector-based shRNA expresses highly in the mouse liver (9).

Thus, we generated adenovirus vector-based shRNA against PPARγ2 (Ad-shPPARγ2) and negative control (Ad-shNC). A single injection of Ad-shPPARγ2 in mice fed a high saturated fat diet for 4 weeks reduced expression of PPARγ2 mRNA in the liver (10).

2. Materials

1. pBAsi-hU6 Neo DNA (Catalog No. 3227) (Takara Bio Inc., Shiga, Japan).
2. *Bam*HI.
3. *Hind*III.
4. *Eco*RI.
5. *Eco*RV.
6. DNA Ligation Kit Ver. 2 (Catalog No. 6022) (Takara Bio Inc.).

7. *E. coli* DH5α competent cells.
8. LB agar plate containing 100 μg/mL ampicillin.
9. LB broth containing 100 μg/mL ampicillin.
10. TE buffer.
11. Adenovirus Dual Expression Kit (Catalog No. 6170) (Takara Bio Inc.).
12. 12. λ-packaging kit (Gigapack III XL Packaging Extract; Catalog No. 200207) (Agilent Technologies, La Jolla, CA, USA).
13. Transfection reagent (Trans IT 293; Catalog No. 2704) (Mirus Bio LLC, Madison, WI, USA).
14. Phenol/chloroform.
15. Ethanol.
16. Sterile distilled water.
17. EDTA solution.
18. Agarose for electrophoresis.
19. HEK293 cells (ATCC CRL-1573).
20. HeLa cells (ATCC CCL-2).
21. Dulbecco's Modified Eagle's Medium (DMEM) (glucose 1 g/L).
22. Fetal calf serum (FCS).
23. 0.02% EDTA/PBS (−); (−) means no calcium or magnesium.
24. RNase A.
25. *Cla*I.
26. DNA purification kit.
27. SOC medium (Catalog No. 15544–034) (Invitrogen, Carlsbad, CA, USA).
28. Cesium chloride.
29. HEPES.
30. SM buffer: 5.8 g NaCl, 2.0 g $MgSO_4 \cdot 7H_2O$, 50 mL 1 M Tris–HCl (pH 7. 5), 5. 0 mL 2% (w/v) gelatin in sterile distilled water, up to 1 L in sterile distilled water. Autoclave.

3. Methods

The mechanism of RNAi described in the Introduction is shown in Fig. 1. An overview of the protocol for generating Ad-shPPARγ2 and Ad-shNC is given in Fig. 2. Here, we describe the generation of Ad-shPPARγ2. Ad-shNC is prepared in the same way as Ad-shPPARγ2.

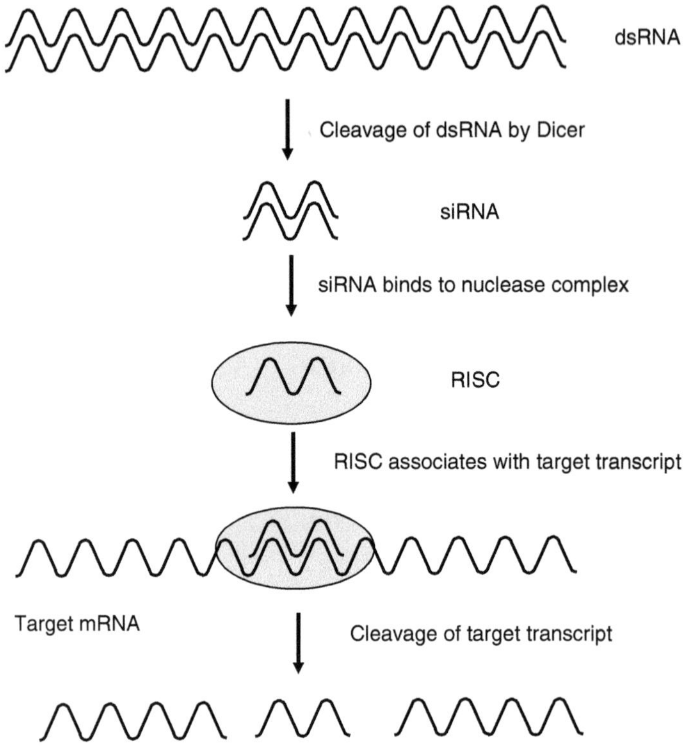

Fig. 1. Mechanism of RNAi. RISC, RNA-induced silencing complex.

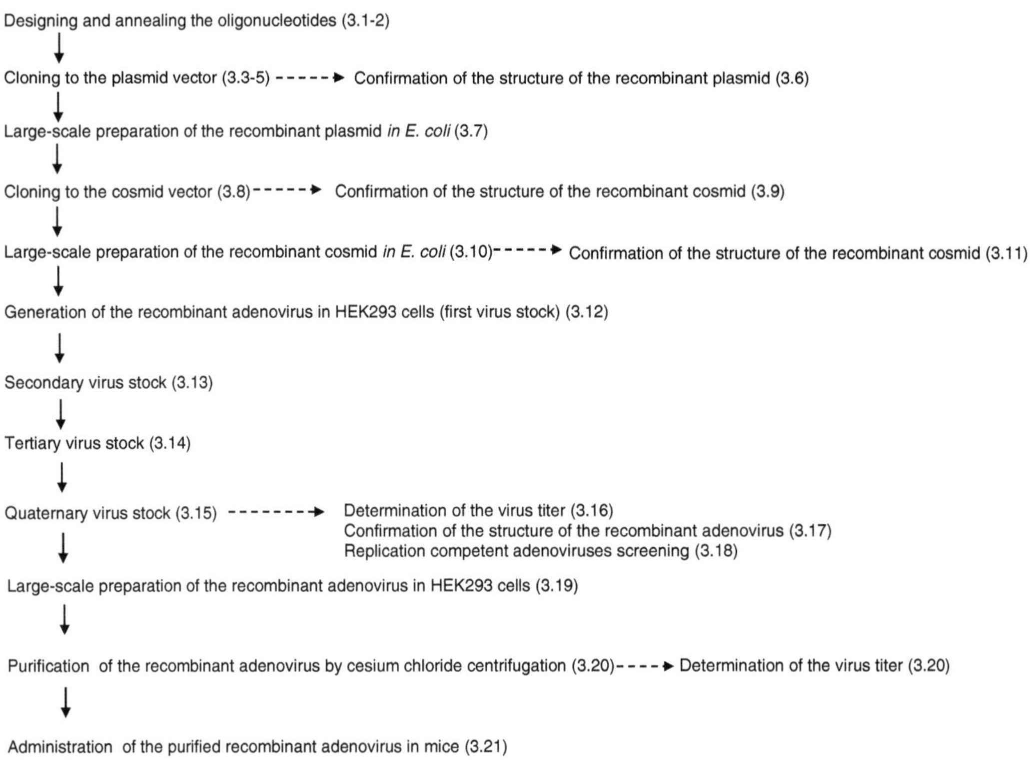

Fig. 2. Overview of the production of PPARγ2 knockdown in the mouse liver. Numbers in *parentheses* indicate the relevant step.

3.1. Designing the Oligonucleotides

Several plasmid shRNA expression vectors are available. We use pBAsi-hU6 Neo DNA (Takara Bio Inc., http://www.takara-bio.com). The structure of shRNA and the generation of siRNA from shRNA are shown in Fig. 3. The following sequences of the oligonucleotides should be included (Fig. 4).

1. A 5′-*Bam*HI restriction site overhang on the top strand and a 5′-*Hind*III restriction site overhang on the bottom strand.
2. A purine (G or A) residue located just downstream of the *Bam*HI site on the top strand (see Note 1).
3. The 19-base oligonucleotide sense sequence (target sense sequence) of the shRNA target site.
4. A nucleotide hairpin loop sequence.
5. The 19-base oligonucleotide antisense sequence (target antisense sequence) of the shRNA target site.
6. An RNA Pol III terminator sequence consisting of a six nucleotide poly (T) tract.

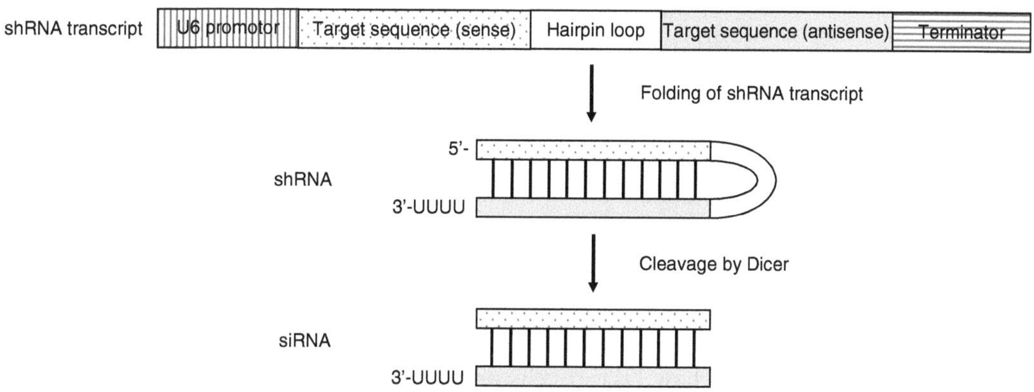

Fig. 3. Generation of siRNA.

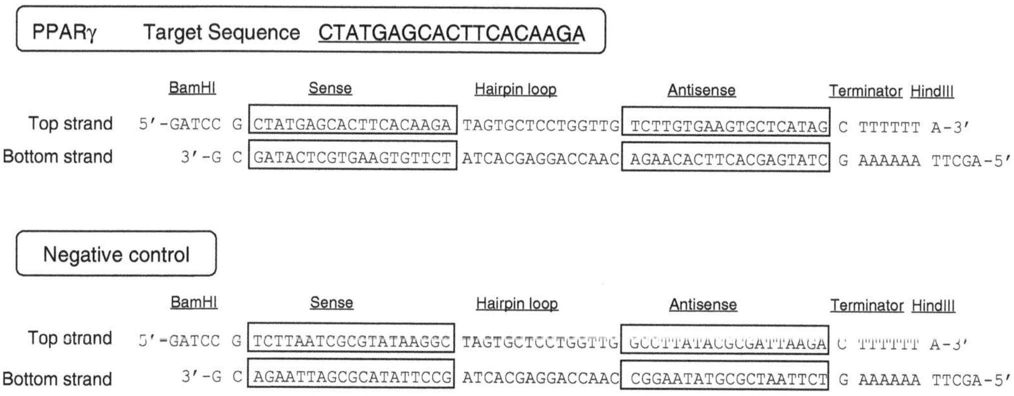

Fig. 4. Design of the shRNA oligonucleotide sequence.

3.2. Annealing the Oligonucleotides

1. Suspend each oligonucleotide in 10 mM Tris–HCl (pH 8.0) and 50 mM NaCl to a concentration of 40 μM.
2. Mix the oligonucleotides for the top strand and the bottom strand at a 1:1 ratio.
3. Heat the mixture to 95°C for 5 min.
4. Cool the mixture gradually to 25°C over 30 min.

3.3. Digesting the Vector

1. Combine the following reagents in a 1.5-mL tube: 4 μL (2 μg) pBAsi-hU6 Neo DNA vector, 10 units *Bam*HI, 10 units *Hind*III, 2 μL 10× buffer, up to 20 μL in sterile distilled water.
2. Incubate the reaction mixture for 1 h at 37°C.
3. Following ethanol precipitation, resuspend the mixture in 10 μL TE buffer.

3.4. Ligation of cDNA into a Plasmid Vector

A number of highly evolved DNA ligation kits are now available. We use DNA Ligation Kit Ver. 2 (Takara Bio Inc.)

1. Combine digested plasmid vector DNA and the annealed oligonucleotide to be inserted in a total volume of 5 μL. Recommended ratio of DNA vector:oligonucleotide = 0.03: 0.03–0.3 pmol.
2. Add 5 μL of Solution I (included in the DNA Ligation Kit Ver. 2) to the DNA solution and mix thoroughly.
3. Incubate at 16°C for 30 min.

3.5. Transformation

The ligation reaction mixture from Subheading 3.4, step 3 can be used directly for transformation immediately after ligation. We use DH5α competent *E. coli*.

1. Thaw DH5α cells on ice.
2. Place a 14-mL tube on ice.
3. Gently mix cells with the pipette tip and aliquot 50 μL of cells into the 14-mL tube on ice.
4. Add 3 μL of ligation reaction mixture to the cells and mix by swirling around once. Do not mix by pipetting up and down.
5. Incubate the tube on ice for 30 min.
6. Transfer the tube to a 37°C water bath. Store the tube in the water bath for exactly 20 s. Do not mix or shake.
7. Rapidly transfer the tube from the water bath to an ice bath. Allow cells to chill for 2 min.
8. Add 950 μL of pre-warmed (37°C) SOC medium to the tube.
9. Shake the tube at 37°C for 1 h at 225 rpm.
10. Spread 100 μL of the cell suspension on the pre-warmed LB plate containing ampicillin.

11. Invert the plate and incubate at 37°C overnight.
12. Pick up the colony and incubate it in 5 mL LB broth containing ampicillin at 37°C overnight.
13. Prepare plasmid DNA from 4 mL of the LB medium containing *E. coli* using a commercially available kit for mini-scale preparation of plasmid DNA.

3.6. Confirmation of the Recombinant Plasmid

Cut the plasmid DNA prepared in Subheading 3.5, step 13 (50 ng) with *Bam*HI and *Hind*III and confirm the approximately 60-bp band by 2.0% agarose gel electrophoresis.

3.7. Large-Scale Preparation of the Recombinant Plasmid

1. Add 1 mL of LB medium containing *E. coli* including recombinant plasmid prepared in Subheading 3.5, step 12 to 200 mL of LB broth containing ampicillin and incubate it at 37°C overnight.
2. Prepare the recombinant plasmid DNA using a commercially available large-scale plasmid DNA preparation kit (see Note 2).

3.8. Insertion of the Target DNA into the Cosmid Vector

To obtain recombinant adenovirus, we use the Adenovirus Dual Expression Kit (Takara Bio Inc.). The map of the cosmid vector pAxcwit2 is shown in Fig. 5.

1. Digest the plasmid DNA prepared in Subheading 3.7, step 2 with *Eco*RV.

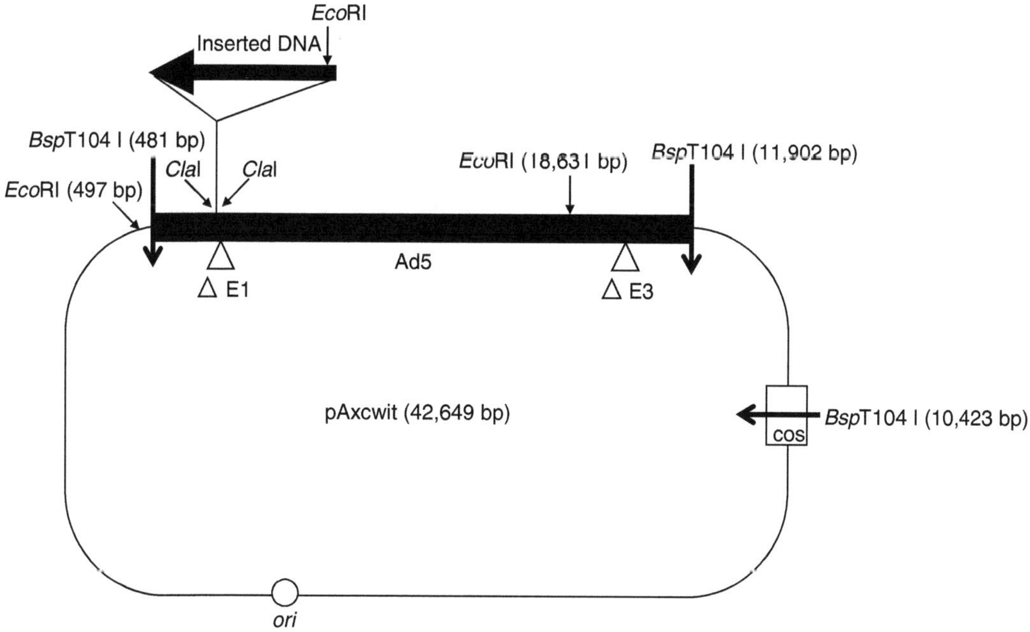

Fig. 5. Map of the cosmid vector containing adenovirus.

2. Purify the insert DNA fragment through phenol/chloroform extraction and ethanol precipitation (0.2 μg of DNA is used later in step 5 in this Subheading, 3.8).
3. Cut the cosmid vector completely with the restriction endonuclease *Swa*I as follows. Prepare the following reaction mixture in a 1.5-mL tube and incubate at 25°C for 2 h: 5 μL cosmid vector (pAxcwit2), 5 μL 10× *Swa*I digestion buffer, 2 μL *Swa*I (10 U/μL), up to 50 μL in sterile distilled water.
4. Following the addition of EDTA to a final concentration of 10 mM to the tube, purify the digested cosmid vector by phenol/chloroform extraction.
5. Add 0.2 μg of the insert DNA prepared in Subheading 3.8, step 2 to the digested cosmid vector in the tube prepared in Subheading 3.8, step 4.
6. Perform ethanol precipitation (see Note 3).
7. Dissolve the insert DNA fragment and digested cosmid vector by adding 5 μL of DNA Dissolution Buffer (included in Adenovirus Dual Expression Kit) and add 5 μL of the Ligation Solution (included in Adenovirus Dual Expression Kit). Incubate the mixture at 25°C for 10 min.
8. Perform ethanol precipitation. Do not allow DNA to dry.
9. Cut DNA with *Swa*I as follows. Add the following reagents to the ligated DNA prepared in Subheading 3.8, step 8, and incubate at 25°C for 2 h (see Note 4): 5 μL 10× *Swa*I digestion buffer, 2 μL *Swa*I (10 U/μL), up to 50 μL in sterile distilled water.
10. Package the cosmid DNA obtained from step 9 in this Subheading, 3.8 (both *Swa* I-digested and undigested cosmid DNA are included) using the λ-packaging kit. It will be easy to select the cosmid by the use of a kit that selectively packages relatively large DNA. We use Gigapack III XL Packaging Extract (Agilent Technologies; see Note 5).
11. Seed 1/100 (2 μL), 1/10 (200 μL), and the remaining infected DH5α obtained from Subheading 3.8, step 10 on three LB agar plates containing ampicillin and incubate them overnight at 37°C.
12. Pick up the colony into a 15-mL tube containing 5 mL LB broth containing ampicillin, shake it at 37°C overnight, and prepare cosmid DNA using a commercially available kit (see Note 6).

3.9. Confirmation of the Structure of the Recombinant Cosmid

Because recombinant virus preparation is a time-consuming process, it is important to make sure that a true recombinant cosmid has been constructed. Select the cosmid clone that contains the insert and confirm the orientation.

1. Digest the cosmid DNA prepared in Subheading 3.8, step 12 with *Cla*I. The insert can be cut out by *Cla*I digestion (Fig. 5).
2. Perform agarose gel electrophoresis. Expected sizes of the DNA fragments are 400 (including insert DNA) and 42,600 bp (see Note 7).
3. Digest the cosmid DNA prepared in step 12 in Subheading 3.8 with *Eco*RI. If the orientation is correct, the expected sizes of the fragments are 880, 18,000, and 24,000 bp. If the orientation is reversed, the DNA fragments of 530, 18,000, and 24,000 bp are separated (see Note 8).
4. Sequence the cosmid DNA prepared in step 12 in Subheading 3.8 and select the clone that does not have additional or deleted bases.

3.10. Large-Scale Preparation of the Recombinant Cosmid

Recombinant cosmid should be amplified to ready it for further recombinant adenovirus preparation.

1. Mix λ packaging extract and 0.5 μg of the recombinant cosmid DNA prepared in Subheading 3.8, step 12 and whose orientation and sequence were confirmed in Subheading 3.9 and perform packaging as described in step 10 of Subheading 3.8.
2. Plate a 1/100 (20 μL) volume of the mixture onto an LB agar plate containing ampicillin and incubate overnight at 37°C. Add the remaining mixture to 50 mL of LB broth containing ampicillin and incubate overnight at 37°C with shaking.
3. On the following day, if ten or more colonies per plate are detected, prepare cosmid DNA from 50 mL of the liquid culture using a commercially available preparation kit. If less than ten colonies per plate are detected, discard the liquid culture and repeat steps 1 and 2 in this Subheading, 3.10 because this indicates low transformation efficiency. Since cosmids will be deleted in *E. coli* during extended passaging, they should be stored as DNA after amplification.

3.11. Confirmation of the Insert of the Recombinant Cosmid

Using a portion of recombinant cosmid DNA from Subheading 3.10, step 3, confirm the insert as described in steps 1 and 2 in Subheading 3.9.

3.12. Generation of Recombinant Adenovirus (Full-Length DNA Transfer Method)

1. Digest the recombinant cosmid DNA prepared in Subheading 3.10, step 3 with the restriction enzyme *Bsp*T104 I (Fig. 5) as follows. Prepare the following reaction solution and incubate at 37°C for 2 h: 10 μg recombinant cosmid DNA, 5 μL *Bsp*T104 I (10 U/μL), 10 μL 10× L Buffer (included in Adenovirus Dual Expression Kit), up to 100 μL in sterile distilled water.

2. Extract once with phenol/chloroform and twice with chloroform, precipitate with ethanol, and dissolve the DNA pellet in 30 μL of sterile distilled water.
3. Run 1 μL of the DNA on an agarose gel electrophoresis to confirm complete digestion with *Bsp*T104 I. Three bands with sizes of 1,500, 10,000, and 32,000 bp will be visible if digestion is complete.
4. Prepare two 60-mm cell culture plates of HEK293 cells. One plate will be used in step 5 in this Subheading, 3.12 and the other in step 7 also in this Subheading, 3.12. Prepare the cells so that they will be 100% confluent at the time of each use.
5. Transfect 10 μg of the *Bsp*T104 I-digested cosmid prepared in step 2 in this Subheading, 3.12 into HEK293 cells on the 60-mm plate using TransIT-293 (Mirus Bio; see Note 9).
6. On the following morning, harvest the cells using EDTA-PBS (−) (see Note 10).
7. Plate each of the cell suspension from step 6 in this Subheading, 3.12, and a tenfold diluted cell suspension prepared as follows onto collagen-coated 96-well plates. To seed a consistent number of cells onto the tenfold dilution plate, mix the non-transfected HEK293 cells grown on the 60-mm plate prepared in step 4 in this Subheading, 3.12, and the transfected HEK293 cells from step 6 in this Subheading, 3.12, at the following ratio. Suspend the transfected HEK293 cells in the 60-mm plate in 11 mL of medium (A). Suspend the non-transfected HEK293 cells in the 60-mm plate in 11 mL of medium (B). Prepare tenfold diluted cell suspension from 1 mL of A and 9 mL of B and add 100 μL of the cell mixture per well. Prepare undiluted cell suspension from 10 mL of A and add 100 μL of the cell mixture per well.
8. Add 50 μL of 10% FCS-DMEM to each well after 5–6 days and after 10–11 days, changing tips for each well.
9. Cytopathic effects (CPE) resulting from virus propagation are observed in the wells in 7–15 days (see Note 11). Transfer cells with medium from each well in which all cells have degenerated into 1.5-mL sterile tubes under sterile conditions, freeze quickly in dry ice or liquid nitrogen, and store at −80°C.
10. Preferentially select six tubes containing cultures collected from wells that showed complete degeneration of cells in a relatively late period (at 10 days or later) (see Note 12) and lyse cells by six freeze-thaw cycles. Freeze quickly in dry ice or liquid nitrogen and thaw in a 37°C water bath.
11. Collect the supernatant by centrifugation at 1,500 ×*g* for 10 min at 4°C and store at −80°C as the first virus stock (6 clones from 6 tubes).

3.13. Preparation of the Secondary Virus Stock

1. Prepare cultures of HEK293 cells that are 70–100% confluent in 5 mL of medium per flask in six collagen-coated 25-cm^2 flasks.
2. Gently add 150 μL of the first virus stock per flask into one of the six flasks. Do the same for the other five first virus stocks.
3. Slowly swirl the flasks to disperse the virus evenly to all cells.
4. Incubate the cells in a CO_2 incubator (37°C, 5% CO_2).
5. After 3–4 days, the cells will show CPE. Collect the cells with medium and lyse the cells with a sealed-type sonicator to release the viruses (see Note 13). In our work, cells in two of the six flasks showed CPE, so we could obtain two clones.
6. Collect the supernatant by centrifugation at 1,500×*g* for 10 min at 4°C. Freeze and store at −80°C (secondary virus stock).

3.14. Preparation of the Tertiary Virus Stock

1. Prepare cultures of HEK293 cells that are 70–100% confluent in 15 mL of medium per flask in two collagen-coated 75-cm^2 flasks.
2. Gently add 600 μL of the secondary virus stock per flask into one of the flasks. Do the same for the other secondary virus stock.
3. After 3–4 days, collect the cells that show CPE with medium and lyse the cells as described in step 5 in Subheading 3.13.
4. Collect the supernatant as described in Subheading 3.13, step 6. Freeze and store at −80°C (tertiary virus stock).

3.15. Preparation of the Quaternary Virus Stock

1. Prepare cultures of HEK293 cells that are 70–100% confluent in 40 mL of medium per flask in two collagen-coated 225-cm^2 flasks.
2. Gently add 3 mL of the tertiary virus stock per flask into one of the flasks. Do the same for the other tertiary virus stock.
3. After 3–4 days, collect the cells that show CPE with medium and lyse the cells as described in step 5 in Subheading 3.13.
4. Collect the supernatant as described in step 6 in Subheading 3.13.
5. Dispense 1 mL each into microcentrifuge tubes, freeze quickly in dry ice, and store at −80°C (quaternary virus stock). We could obtain two clones as quaternary virus stock. Quaternary virus stocks are actually used in in vitro experiments (see Note 14).

3.16. Determination of Viral Titer (End-Point Dilution Assay)

1. Prepare HEK293 cells in one 96-well plate per virus stock. Carefully seed all wells at the same density (10^4 cells per well) in 100 μL of medium per well.
2. On the following day, prepare serial dilutions of virus stock as follows. Make a 1:100 dilution by adding 12 μL of virus stock to 1,188 μL of growth medium. Starting with the 1:100

dilution, prepare serial 1:10 dilutions by transferring 120 μL diluted virus to 1,080 μL medium. The range of dilutions for testing is 10^{-3} to 10^{-10}.

3. Take the 96-well culture plates from the incubator and inspect the wells to ensure that the cells have attached to form an even monolayer.
4. Add 100 μL of diluted virus to each well in columns 1–10 and rows 1–8 (A–H) as follows. The diluted virus whose titer is 10^{-3} is added in the uppermost ten wells, i.e., row 1 (A) and columns 1–10. The diluted virus whose titer is 10^{-10} is added in the bottommost 10 wells, i.e., row 8 (H) and columns 1–10. Add 100 μL of virus-free medium to wells in columns 11–12, i.e., the remaining wells. These wells serve as controls for the viability of noninfected cells.
5. Cover the plates and incubate in a CO_2 incubator for 10 days at 37°C.
6. Using a microscope, check each well for CPE. For each row, count the number of wells having CPE. A well is scored as CPE-positive even if only a few cells show CPE. To ascertain the CPE, compare the infected well with the noninfected control wells.
7. Calculate the fraction of CPE-positive wells in each row.
8. Calculate viral titer (see Note 15, Table 1).

Table 1
Calculation of the viral titer

Dilution	Fraction of CPE-positive wells
10^{-3}	10/10 = 1
10^{-4}	10/10 = 1
10^{-5}	10/10 = 1
10^{-6}	10/10 = 1
10^{-7}	10/10 = 1
10^{-8}	8/10 = 0.8
10^{-9}	4/10 = 0.4
10^{-10}	0/10 = 0

$x = 1 + 1 + 1 + 1 + 1 + 1 + 1 + 0.8 + 0.4 + 0 = 8.2$
Titer $= 10^{(x+0.8)} = 10^{(8.2+0.8)} = 10^{9}$ (pfu/mL)
CPE, cytopathic effects

Viral titer (pfu/mL) = $10^{(x+0.8)}$, where x = the sum of the fractions of CPE-positive wells.

3.17. Confirmation of the Recombinant Adenovirus

1. Prepare cultures of HEK293 cells that are 70–100% confluent in 0.5 mL of medium per well on a collagen-coated 24-well plate. Use one well per virus stock.
2. Add 10 μL of quaternary virus stock to the well. Infect cells by swirling the plate gently to disperse the virus stock solution evenly. The cells should be kept in a CO_2 incubator at 37°C.
3. After 3 days, collect cells and medium from a well containing HEK293 cells to a 2-mL tube, remove the supernatant after centrifugation at 1,500 × *g* for 5 min at 4°C, and store the cell pellet at −80°C (cell pack). Prepare one cell pack per virus stock.
4. Extract all DNA from each cell pack and confirm the structure of the recombinant adenovirus DNA as follows. Add the following reagent to the cell pack: 40 μL 10× TNE (included in Adenovirus Dual Expression Kit), 4 μL proteinase K (20 mg/mL) (included in Adenovirus Dual Expression Kit), up to 400 μL in sterile distilled water.
5. Suspend the cell pack by vortexing.
6. Add 4 μL of 10% SDS (included in Adenovirus Dual Expression Kit) per cell pack suspension and mix well by vortexing.
7. Incubate at 50°C for 1 h.
8. Extract DNA with phenol/chloroform twice.
9. Extract DNA with chloroform twice.
10. Precipitate with ethanol and dissolve the DNA pellet in 50 μL of TE buffer containing 20 μg/mL RNase A per tube.
11. Cut the insert DNA by digesting 15 μL of DNA suspension per cell pack with the restriction enzyme *Cla*I and confirm the desired restriction pattern (400, 460, and 30,800 bp) by agarose gel electrophoresis.

3.18. Replication Competent Adenoviruses Screening (PCR Method)

Even virus samples that did not contain replication competent adenoviruses (RCAs) initially may still acquire the E1 gene during the course of multiple passages in HEK293 cells. It is therefore recommended to check for RCAs (11).

1. Prepare cultures of HeLa cells that are 70–100% confluent in 0.5 mL of medium per well on a collagen-coated 24-well plate. Use one well per quaternary virus stock.
2. Add 10 μL of quaternary virus stock to the well. Infect cells by swirling the plate gently to disperse the virus stock solution evenly. The cells should be kept in a CO_2 incubator at 37°C.
3. Culture for 3 days and verify microscopically that no CPE are visible.

4. Aspirate the medium, collect the cells from the plate using trypsin, and centrifuge at 150×*g* for 5 min at room temperature. Wash the cell precipitate with 1 mL of PBS three times.
5. Extract DNA from cell precipitate as described in Subheading 3.17, steps 4–10.
6. Prepare the PCR mixture as follows: 2 μL DNA sample (see Note 16), 0.5 μL TaKaRa Taq (5 U/μL), 4 μL dNTP mix (2.5 mM/dNTP), 5 μL 10× PCR buffer (Mg^{2+} plus), 0.5 μL sense primer* (20 pmol/μL), 0.5 μL antisense primer* (20 pmol/μL), up to 50 μL in sterile distilled water.

   *Primer set designed to amplify 240 bp beginning at the start codon of E1A.

   Sense: 5′-ATGAGACATATTATCTGCCAC-3′ (Ad5; 560–580 bp).

   Antisense: 5′-GTAAGTCAATCCCTTCCTGCAC-3′ (Ad5; 800–779 bp).
7. Carry out the reaction using the following conditions: 30 cycles for 94°C for 30 s, 55°C for 30 s, and 72°C for 30 s.
8. Run 7 μL of PCR product on agarose gel. The absence of 240-bp bands suggests that there is no RCA contamination. The presence of 240-bp bands indicates a high probability of RCA contamination.

3.19. Large-Scale Preparation of Recombinant Adenovirus

1. Prepare cultures of HEK293 cells that are 70–100% confluent in ten collagen-coated 225-cm^2 flasks.
2. After selection of one quaternary virus stock that shows the predicted pattern after *Cla*I digestion and has higher titer and no RCA contamination, dilute the selected quaternary virus stock with 5% FCS-DMEM to prepare 50 mL of diluted virus stock containing 1×10^8 pfu/mL.
3. Remove the medium from the flasks and gently add 5 mL of the diluted virus stock prepared in Subheading 3.19, step 2 per flask.
4. Slowly swirl the flasks to disperse the virus stock solution evenly over all cells. Perform this step 3–4 times every 15–20 min. The cells should be kept in a CO_2 incubator.
5. After 1 h, add 30 mL of 5% FCS-DMEM per flask (35 mL in total per flask).
6. After 3 days, all the cells will show CPE. Transfer the cells with medium into 10 sterile 50-mL tubes and centrifuge at 1,500×*g* for 10 min at 4°C.
7. Discard the supernatant leaving 10 mL and disrupt the cells using a sealed-type sonicator to liberate the viruses.

8. Collect the supernatant as concentrated virus stock by centrifugation at 3,000 rpm for 10 min at 4°C (approximately100 mL in total).

3.20. Purification of Recombinant Adenovirus

The use of purified recombinant adenovirus is needed to perform in vivo studies (12).

1. Perform the first cesium chloride density gradient-based virus purification by layering the following reagents in the order shown in a tube compatible with a swing-out rotor with a maximum speed of 141,000 × *g*: 10 mL 4.0 M cesium chloride/10 mM HEPES, 5 mL 2.2 M cesium chloride/10 mM HEPES, 20 mL concentrated virus stock.
2. Centrifuge at 100,000 × *g* for 2 h at 4°C using the swing-out rotor.
3. Collect the virus band using a capillary tube (see Note 17).
4. Add an equal volume of saturated cesium chloride to the recovered virus stock obtained in step 3 in this Subheading, 3.20.
5. Perform the second cesium chloride density gradient-based virus purification by layering the following reagents in the order shown in a tube compatible with a swing-out rotor with a maximum speed of 285,000 × *g*: 4–5 mL recovered virus + saturated cesium chloride, 2 mL 4.0 M cesium chloride/10 mM HEPES, 3–4 mL 2.2 M cesium chloride/10 mM HEPES.
6. Centrifuge at 200,000 × *g* for 3 h at 4°C using the swing-out rotor.
7. Collect the virus band using a capillary tube (see Note 18).
8. Transfer to a sterilized dialysis tube under sterile conditions and dialyze against 1 L of 10% glycerol in PBS (−). Replace the outer solution after 2 h and dialyze overnight.
9. Collect the dialyzed virus stock, aliquot for each use, and store at −80°C.
10. Thaw one of the tubes and determine the titer as described in Subheading 3.16 prior to use in the experiments.

3.21. PPARγ2 Knockdown in the Mouse Liver

1. A single intravenous injection of Ad-shPPARγ2 obtained from step 10 in Subheading 3.20 in a single dose (2×10^9 pfu in 200 μL) was administered into each mouse fed a high saturated fat diet for 4 weeks.
2. Five days later, the Ad-shPPARγ2-injected mice fed a high saturated fat diet for 4 weeks showed reductions of PPARγ2 mRNA and protein in the liver of 47 and 43% (Fig. 6), respectively, with a reduction in liver triglyceride concentration of 38% in 5 days (10).

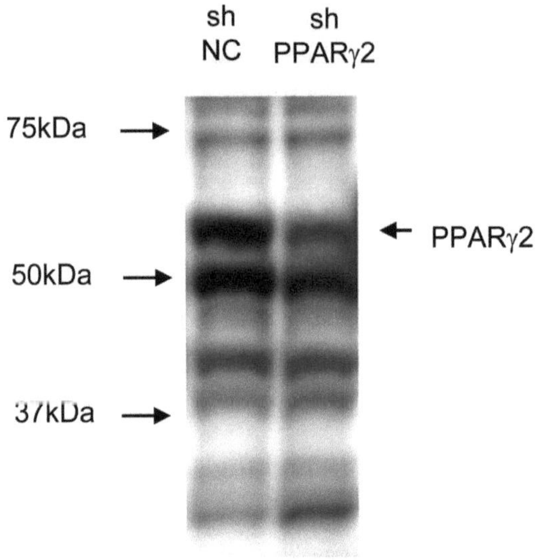

Fig. 6. Western blot analysis of PPARγ2. Mice fed a high saturated fat diet for 4 weeks were injected with Ad-shPPARγ2 or an Ad-negative control and were killed 5 days later. Nuclear protein from liver was extracted and separated by SDS-PAGE (7.5% gel). Typical result of western blot analyses of PPARγ is shown. Nonspecific bands are shown as markers of protein loading.

4. Notes

1. RNA Pol III prefers to initiate transcription with a purine.
2. If possible, check the ability of the recombinant plasmids to reduce the mRNA expression of PPARγ2 in the cells. We use HEK293 cells infected with PPARγ2 adenovirus that overexpress mRNA of PPARγ2. It is also possible to use other cells that abundantly express mRNA of PPARγ2.
3. Because the cosmid DNA becomes insoluble once it is dried, immediately perform the next process simply by removing residual ethanol.
4. This reaction is important to prevent the appearance of cosmids without inserts. Because the recombinant cosmids no longer contain the *Swa*I site, they are not digested. The cosmids not containing the insert are digested with *Swa*I and become linear DNA. The DNA digested with *Swa*I (linear DNA) is less efficiently packaged than undigested DNA.
5. Streak the DH5α glycerol stock onto the LB agar plate. Incubate the plate overnight at 37°C. The next day, pick up a single colony and culture it in 5 mL of LB broth overnight at 37°C with shaking. On the following day, inoculate the 5 mL of LB broth supplemented with 0.2% maltose with 50–100 μL

of the overnight culture and incubate at 37°C with shaking until the OD_{650} reaches approximately 1.0 (4–6 h). Take 1 mL of the culture in a sterile tube and discard the supernatant after centrifugation. Suspend the pellet in 500 μL of 10 mM $MgSO_4$ to prepare 2× bacterial cultures. Mix 1 μL of the cosmid DNA obtained from step 9 in Subheading 3.8 and 12.5 μL of packaging extract and incubate at 22°C for 1.5 h (mixture 1). Add 100 μL of SM buffer to mixture 1 (mixture 2). Mix 100 μL of mixture 2 and 100 μL of the 2× bacterial cultures and incubate at room temperature for 10 min.

6. We picked up 18 colonies. Do not prepare stock in the *E. coli* because the cosmid is lost when it is subcloned over generations in *E. coli*.
7. Four of the 18 clones had the inserted DNA.
8. The *Eco*RI restriction site is on the upstream side of the U6 promoter in the inserted DNA (Fig. 5). The smallest DNA fragment from the cosmid vector that has inserted DNA in the right direction is longer than that in the reverse direction. Four of the 18 clones were properly inserted.
9. Remove the medium from the 60-mm plate and add 2.5 mL of Opti-MEM (Catalog No. 31985070) (Invitrogen). Mix 15 μL of TransIT-293 Reagent with 250 μL of Opti-MEM by vortexing and incubate at room temperature for 5 min. Add the *Bsp*T104 I-digested cosmid solution to the mixture prepared above, mix gently, and incubate at room temperature for 5 min. Add this *Bsp*T104 I-digested cosmid suspension to the plate in a dropwise fashion and swirl the plate gently to evenly disperse the *Bsp*T104 I-digested cosmid suspension. Incubate the plate at 37°C in a CO_2 incubator.
10. If the cells do not detach readily, use trypsin at a lower concentration than usual (approximately 0.025%).
11. Infected cells typically become round and detach from the bottom. These changes are collectively referred as the CPE.
12. Do not select cultures collected from wells in which virus propagation occurred in a relatively early period because they may contain multiple virus clones.
13. Otherwise disrupt the cells with three consecutive freeze-thaw cycles. Freeze cells in a dry ice bath and thaw cells by placing the tubes in a 37°C water bath. Do not allow the suspension to reach 37°C. Vortex the cells after each thaw.
14. This virus is adequate for use in the in vitro study after steps in Subheadings 3.16, 3.17, 3.18. Further scaling up and purification are not needed.
15. The formula given is based on the Spearman–Kaber method. The assay is a reliable indicator of viral titer only if the following

three conditions are met: The negative control wells show no visible signs of CPE or growth inhibition, the wells infected with the least dilute virus (10^{-3}) are all CPE-positive, and the wells infected with the most dilute virus (10^{-10}) are all CPE-negative. When the titer is too high and wells infected with the most dilute virus (10^{-10}) are not all CPE-negative, use virus that is more highly diluted (for example, from 10^{-6} to 10^{-13}) and properly calculate the titer again.

16. Negative control (reaction mixture containing 2 μL of sterile distilled water in place of DNA sample) and positive control (reaction mixture containing 2 μL of DNA sample extracted from uninfected HEK293 cells in the same manner) should also be used.
17. The virus band will be readily visible against a black background. The upper white band contains cellular proteins. The viruses are in the lower white band.
18. If two bands are visible, the bottom band will be the virus band, and the upper band will contain proteins.

References

1. Graham FL, Smiley J, Russell WC, Nairn R (1977) Characteristics of a human cell line transformed by DNA from human adenovirus type 5. J Gen Virol 36:59–74
2. Hammond SM, Caudy AA, Hannon GJ (2001) Post-transcriptional gene silencing by double-stranded RNA. Nat Rev Genet 2:110–119
3. Nykanen A, Haley B, Zamore PD (2001) ATP requirements and small interfering RNA structure in the RNA interference pathway. Cell 107:309–321
4. Brummelkamp TR, Bernards R, Agami R (2002) A system for stable expression of short interfering RNAs in mammalian cells. Science 296:550–553
5. Paddison PJ, Caudy AA, Bernstein E, Hannon GJ, Conklin DS (2002) Short hairpin RNAs (shRNAs) induce sequence-specific silencing in mammalian cells. Genes Dev 16:948–958
6. Paul CP, Good PD, Winer I, Engelke DR (2002) Effective expression of small interfering RNA in human cells. Nat Biotechnol 20:505–508
7. Berkner KL, Sharp PA (1983) Generation of adenovirus by transfection of plasmids. Nucleic Acids Res 11:6003–6020
8. Miyake S, Makimura M, Kanegae Y, Harada S, Sato Y, Takamori K, Tokuda C, Saito I (1996) Efficient generation of recombinant adenoviruses using adenovirus DNA-terminal protein complex and a cosmid bearing the full-length virus genome. Proc Natl Acad Sci USA 93: 1320–1324
9. Fechner H, Haack A, Wang H, Wang X, Eizema K, Pauschinger M, Schoemaker R, Veghel R, Houtsmuller A, Schultheiss HP, Lamers J, Poller W (1999) Expression of coxsackie adenovirus receptor and alphav-integrin does not correlate with adenovector targeting in vivo indicating anatomical vector barriers. Gene Ther 6:1520–1535
10. Yamazaki T, Shiraishi S, Kishimoto K, Miura S, Ezaki O (2011) An increase in liver PPARgamma2 is an initial event to induce fatty liver in response to a diet high in butter: PPARgamma2 knockdown improves fatty liver induced by high-saturated fat. J Nutr Biochem 22(6):543–53
11. Lochmuller H, Jani A, Huard J, Prescott S, Simoneau M, Massie B, Karpati G, Acsadi G (1994) Emergence of early region 1-containing replication-competent adenovirus in stocks of replication-defective adenovirus recombinants (delta E1 + delta E3) during multiple passages in 293 cells. Hum Gene Ther 5: 1485–1491
12. Kanegae Y, Makimura M, Saito I (1994) A simple and efficient method for purification of infectious recombinant adenovirus. Jpn J Med Sci Biol 47:157–166

Chapter 8

Adipose Tissue-Specific PPARγ Gene Targeting

Weimin He

Abstract

The nuclear receptor peroxisome proliferator activated receptor gamma (PPARγ) is most abundantly expressed in adipose tissue and has been shown to play imperative roles in controlling adipogenesis and lipogenesis in cultured cell systems in vitro as well as in mice and humans. However, it is unclear how important the role this receptor plays in regulating physiological functions of mature adipocytes in vivo. The Cre-loxP gene targeting strategy is employed to specifically disrupt PPARγ in mature adipocytes in mice. In this chapter, I will describe generation of "floxed" PPARγ mice, which bear loxP sequences in the introns of PPARγ gene locus flanking the coding exons 1 and 2 of PPARγ and creation of the aP2-Cre transgenic mice, which express Cre recombinase under the control of the promoter of adipocyte fatty acid binding protein (aP2). Crossing of the two mouse lines results in deletion of PPARγ gene only in differentiated adipocytes in Cre positive mice.

Key words: Adipose tissue, PPARγ, Gene targeting, Embryonic stem cells, lox*P*, Cre transgenic mice

1. Introduction

Adipose tissue was once regarded as a tissue that only stores lipids as triglycerides in feeding states and releases nonesterified fatty acids as energy sources for peripheral tissues in fasting states. Studies in the last several decades, however, have established adipose tissue as an important endocrine organ/tissue (1). This tissue secretes many endocrine factors called adipocytokines (or adipokines) and some proinflammatory factors (in obese condition) to critically regulate many important pathophysiological activities, including food intake (2), immunity and inflammation (3, 4), systemic glucose and lipid metabolism (5), and bone homeostasis (6), to name a few. Lack of sufficient amount of functional adipose tissue (known as lipodystrophy) is associated with adverse health

Mostafa Z. Badr and Jihan Youssef (eds.), *Peroxisome Proliferator-Activated Receptors (PPARs): Methods and Protocols*, Methods in Molecular Biology, vol. 952, DOI 10.1007/978-1-62703-155-4_8, © Springer Science+Business Media New York 2013

problems including dyslipidemia, insulin resistance, and diabetes (7). In contrast to relatively uncommon lipodystrophy, however, the last three decades have witnessed an alarming increase in the rate of obesity in humans (8), a state of massive adipose tissue expansion (excess lipid accumulation in adipocytes with or without a significant increase in adipocyte numbers) in most developed countries as well as many developing countries (9). Obesity has now become a major risk factor for dyslipidemia, hypertension, insulin resistance, diabetes, cardiovascular complications, cancers, and even aging, collectively called metabolic syndrome or Syndrome X (10).

Adipose tissue formation requires the activities of several key regulatory factors. One of these factors is the nuclear hormone receptor peroxisome proliferator activated receptor gamma (PPARγ) that is highly expressed in adipose tissue (11). Dominant negative mutations of PPARγ have been identified in human patients with partial lipodystrophy and metabolic disorders (12, 13). In line with this, loss of function of PPARγ results in lack of discernible adipose tissue in experimental animals (14).

In addition to its essential role in promoting adipocyte differentiation and lipid storage in adipose tissue, PPARγ also controls adipose tissue functions. Thus, pharmacological activation of PPARγ by its agonists thiazolidinediones (TZDs), a class of synthetic compounds that are used to treat diabetes, sensitizes insulin action in a variety of insulin resistant animals and diabetic patients, even at the expense of further increasing adipose tissue weights in overweight or obese patients (15). This effect of PPARγ activation is associated with: (1) lipid repartitioning (increasing lipid accumulation in adipocytes and reducing lipid contents in circulation and in insulin targeting tissues such as the liver and skeletal muscle, thereby decreasing lipid-induced insulin resistance), (2) changes in the expression of adipokines and proinflammatory cytokines in adipose tissue (increasing adipokines that sensitize insulin action and decreasing adipokines and proinflammatory cytokines that cause insulin resistance), and (3) significant reduction in oxidative stress in adipose tissue (16, 17). Interestingly, a relatively common polymorphism in an adipose tissue-specific isoform of PPARγ, namely PPARγ2^{Pro12Ala} that exhibits decreased receptor-mediated transcriptional activity, is associated with lower body mass index (BMI) and improved insulin sensitivity in human subjects (18, 19), decreased risk of type 2 diabetes in healthy offspring of type 2 diabetic patients (20), and reduced incidence of myocardial infarction in diabetic patients (21). This finding is counterintuitive to our understanding of the PPARγ activation in enhancing insulin sensitivity in humans and animals. Yet this beneficial effect of reducing PPARγ activity in adipose tissue on insulin sensitivity can be recapitulated in PPARγ heterozygous (PPARγ$^{+/-}$) mice (22, 23). Thus, manipulating PPARγ activity in adipose tissue in different states appears to be beneficial to the systemic insulin action in peripheral insulin target organs and tissues.

It was unclear how supraphysiological activation of PPARγ by its agonists and a decrease in PPARγ transcriptional activity in adipose tissue both result in increased insulin sensitivity. Two research groups have used either a dominant negative PPARγ deletion mutant (24) or knockdown of PPARγ (mediated by small inhibitory RNA (25)) to study PPARγ function in 3T3-L1 adipocytes and found reduced insulin-induced glucose intake. However, these studies can only partially address the PPARγ function in adipose tissue and may not be totally relevant to the function of PPARγ in adipose tissue in vivo. To investigate the PPARγ function in adipose tissue in vivo, we employed cre-loxP-mediated gene targeting (26) to disrupt PPARγ specifically in adipose tissue. This chapter contains details of materials and methods used to knock out PPARγ in adipose tissue in experimental animals.

2. Materials

2.1. Buffer Solutions

1. 10× TBE buffer: 890 mM Tris–borate, pH 8.3, 20 mM EDTA.
2. Phosphate buffered saline: 137 mM NaCl, 2.7 mM KCl, 8.1 mM Na_2HPO_4, 1.76 mM KH_2PO_4, pH 7.4.
3. Hepes-buffered saline: 121 mM NaCl, 5.4 mM KCl, 0.44 mM KH_2PO_4, 5.56 mM glucose, 20 mM Hepes, pH 7.3.
4. ES cell electroporation buffer: 20 mM Hepes, pH 7.0, 137 mM NaCl, 5 mM KCl, 0.7 mM Na_2HPO_4, 6 mM dextrose, 0.1 mM β-Mercaptoethanol.
5. ES cell lysis buffer: 10 mM Tris–HCl, pH 7.5, 10 mM EDTA, 10 mM NaCl, 0.5% Sarcosyl, 1 mg/mL proteinase K.
6. Depurination buffer: 0.25 M HCl.
7. Denaturation solution: 1.5 M NaCl, 0.5 M NaOH.
8. Neutralizing solution: 1.5 M NaCl, 1 M Tris–HCl, pH 7.5.
9. 20× SSC solution: 3 M NaCl, 300 mM sodium citrate, pH 7.4.
10. Washing buffer 1: 2× SSC, 0.1% SDS.
11. Washing buffer 2: 0.5× SSC, 0.1% SDS.
12. 50× Denhardts solution: 1% ficoll, 1% polyvinylpyrrolidon (PVP), 1% BSA.
13. Pre-hybridization buffer (Southern blot): 3× SSC, 50% formamide, 5× Denhardts, 0.5% SDS.

14. Hybridization buffer (Southern blot): 3× SSC, 50% formamide, 5× Denhardts, 0.5% SDS, 5% Dextran Sulfate, 20 μg/mL salmon sperm DNA.
15. Mouse-tail lysis buffer: 134 mM Tris, pH 8.8, 33 mM $(NH_4)_2SO_4$, 20 mM β-Mercaptoethanol, 20% DMSO, 6 mM $MgCl_2$, 1 mM dNTP (each), and 40 μg/mL proteinase K.

2.2. Enzymes and Reagents

1. Restriction enzymes: BamHI, NotI, HindIII, XhoI, SalI, SacI, Asp718, KpnI, SmaI, EcoRI.
2. T4 DNA ligase (1 U/μL).
3. LB medium and agar.
4. Agarose.
5. Jetsorb Gel Extraction Kit.
6. DNA purification kits.
7. Plasmid miniprep kits (Qiagen).
8. Plasmid maxiprep kits (Qiagen).
9. Phenol, phenol/chloroform, and chloroform.
10. DMEM (Invitrogen).
11. Fetal bovine serum.
12. Nonessential amino acids (NEAA) (100× solution).
13. Penicillin–Streptomycin (100× solution).
14. Trypsin–EDTA (0.25%, 1 mM).
15. 14.4 M β-Mercaptoethanol
16. Leukemia inhibitory factor (LIF).
17. 50 mg/mL G418.
18. 2 mM FIAU.
19. Taq DNA polymerase.
20. dNTPs (10 mM each).
21. Proteinase K.

2.3. Vectors

1. pBluescript.
2. pSP72.
3. pGEM72-KJ1.

2.4. Oligonucleatides

1. DNA oligonucleotides containing lox*P* sequences.
2. DNA oligonucleotides for genotyping.

2.5. Cells

1. *Escherichia coli* competent cells, DH5α.
2. *E. coli* Ultra-competent cells, DH5α.
3. J1 Mouse embryonic stem (ES) cells.

3. Methods

3.1. Generation of Floxed PPARγ Mice

3.1.1. Creation of Floxed PPARγ Targeting Construct

1. The procedure of constructing floxed PPARγ targeting construct is summarized in Fig. 1a. A single genomic fragment encompassing the coding exons 1 and 2 of PPARγ and surrounding intron regions is obtained and restriction digested into the three fragments (a-c). Three pairs of oligos containing 34-bp lox*P* sequences are designed (see Table 1). Each pair of oligos is annealed into a double strand DNA. The first pair of oligos is annealed to result in a DNA with a 5′-HindIII site and a 3′-Asp718 compatible site at either end; annealing of the

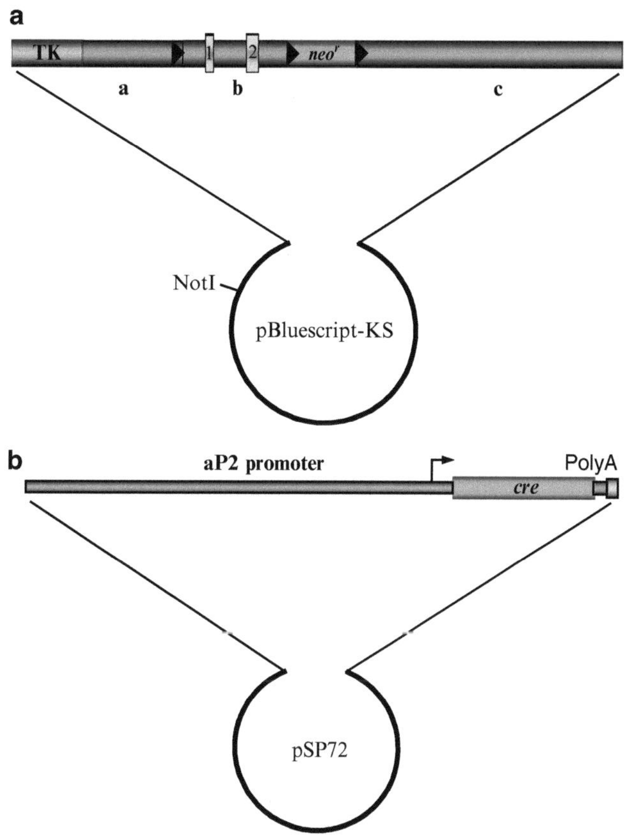

Fig. 1. Schematic representation of floxed PPARγ targeting construct for generation of floxed PPARγ mice and aP2-*cre* transgene construct for creation of aP2-Cre transgenic mice. (a) Floxed PPARγ targeting construct. Three *black triangles* represent lox*P* sites. *Yellow boxes* represent the coding exons 1 and 2 of PPARγ. TK represents herpes simplex virus *thymidine kinase* gene. *neo^r* indicates a *neo^r* gene that is controlled by the PGK promoter. (a–c) Genomic fragments used for cloning of lox*P* sequences and the *neo^r* cassette. (b) aP2-*cre* transgene construct. A 5.4 kb fatty acid binding protein 4 (FABP4)/aP2 gene promoter is cloned into pSP72 cloning vector containing the P1 bacteriophage *cre* recombinase gene. The *cre* gene contains a nuclear localization signal and a *myc*-tag sequence at the 5′ end and is followed by an intron sequence and a poly(A) signal at the 3′ end.

Table 1
Oligonuceleotide sequences for cloning of three loxP sites and for mouse genotyping

lox*P* oligos Pair 1 (lox*P* site upstream of exons 1 and 2 of PPARγ)
5′-AGCTAGCATAACTTCGTATAGCATACATTATACGAAGTTATAAGCTT-3′
5′-GTACAAGCTTATAACTTCGTATAATFTATGCATTACGAAGTTATGCT-3′
lox*P* oligos Pair 2 (lox*P* site downstream of exons 1 and 2 of PPARγ)
5′-TCGAGCTCATAACTTCGTATAGCATACATTATACGAAGTTAT-3′
5′-AATTATAACTTCGTATAATGTATGCATTACGAAGTTATGAGC-3′
lox*P* oligos Pair 3 (lox*P* site upstream of *neo*^{r} cassette)
5′-TCGAGATAACTTCGTATAGCATACATTATACGAAGTTATAAGCTT-3′
5′-ATAACTTCGTATAATGTATGCATTACGAAGTTATC-3′
Genotyping oligos for *cre*^{r}
5′-GCATTACCGGTCGATGCAACGAGTG-3′
5′-GAACGCTAGAGCCTGTTTTGCACGTTC-3′
Genotyping oligos for 5′ lox*P* site
5′-CTGGTGAAGTATACTATACACTGTGCAGCC-3′
5′-GTGTCATAATAAACATGGGAGCATAGAAGC-3′

second pair results in a DNA fragment having a 5′-XhoI site and a 3′-EcoR1 compatible site at either end; the third pair is annealed to give rise to a DNA fragment with a 5′-XhoI site and a 3′-SacI compatible site at either end.

2. Annealed short DNA fragments are cloned into the vector (pBluescript) containing either fragment **a** or **d** or a *neo* gene cassette that is driven by a PGK promoter and tailed with a PGK polyA signal. The first lox*P* sequence is cloned downstream of genomic fragment **a**; the second lox*P* sequence is cloned upstream of the *neo* cassette; and the third lox*P* sequence is cloned upstream of genomic fragment **c** (see Fig. 1a). A herpes simplex virus *thymidine kinase* gene (TK) is then cloned into the vector containing the **a**-lox*P*. The fragment **b** is cloned into the vector containing the lox*P*-*neo* cassette. Finally, TK-**a**-lox*P*, **b**-lox*P*-*neo* cassette, and lox*P*-**c** are assembled into a full-length targeting construct. This cloning strategy results in a targeting construct containing a TK gene, a short 5′ recombination arm, three lox*P* sequences flanking either the coding exons 1 and 2 of PPARγ or the neo cassette, and a long 3′ recombination arm.

3. During cloning, the restricted DNA fragments are separated by 10% agarose gels in 1× TBE buffer, purified using DNA fragment purification kits, and then ligated into the cloning vector cleaved with appropriate restriction enzymes. Once the ligation reaction is completed, half of ligation reactions is incubated with *E. coli* DH5α on ice for 30 min. Heat-shock bacterial cells at 37°C for 1 min and then incubate in 1 mL LB medium at 37°C with agitation.
4. Plate 100 μL of cells on prewarmed LB agar plates containing 50 μg/mL ampicillin.
5. Incubate plates overnight at 37°C.
6. Pick colonies from the LB agar plates, grow cells in 2 mL of LB broth overnight at 37°C and purify plasmids using Qiagen miniprep kits. Positive clones are identified by analytical restriction digestion and each lox*P* site is sequenced to confirm correct sequences. For purification of full-length targeting construct plasmid, Qiagen maxiprep kids are used to provide high quality plasmid DNA.
7. Preparation of targeting fragment is as follows. Full-length targeting construct DNA (50 μg) is digested with NotI (200 U) in 200 μL 1× reaction solution for 6 h. Separate the DNA fragment from the vector DNA in a 0.8% agarose DNA gel. Cut out the 18 kb fragment and purify the DNA with the DNA fragment purification kits. After elution from columns, the DNA is further cleaned by phenol, phenol/chloroform, and chloroform extraction. DNA is then precipitated, washed, resuspended at 1 μg/μL, and stored at −20°C.

3.1.2. Preparation of Mitotically Inactivated Mouse Embryonic Fibroblasts

1. Mate male and female mice containing *neo* gene in their genome and check vaginal plugs.
2. At E12.5-13.5, female mice are dislocated and their abdomens are rinsed well with 70% ethanol.
3. Pull out uteruses free of adjacent mesenchyme into a 10 cm plate with Hepes solution.
4. Remove embryos from extraembryonic tissue, cut umbilical cord and transfer embryos to a new dish with Hepes solution.
5. Tear off head with forceps, scoop out innards with forceps and scissors and transfer carcasses to a new dish with Hepes solution.
6. Transfer carcasses with fine tip forceps to a new sterile dish with Hepes solution. Repeat this step three times to make sure carcasses are completely sterile.
7. Remove carcasses to a new empty Petri dish, touch carcasses to side of the dish to wick off Hepes solution. Pile up carcasses in a stack along side of dish.

8. Add 0.5 mL trypsin solution to cover pile, mince the pile of tissue with autoclaved fine sharp scissors and add another 0.5 mL of trypsin solution. Put plate (tilted) into a 37°C incubator for 30 min.
9. Add 5 mL DMEM containing 10% FBS and transfer cells to a 50 mL tube. Rinse the plate with 5 mL media and pipette media up and down repeatedly to disrupt cells till media is turbid. Transfer turbid solution to a new tube.
10. Add 15 mL new media to the pellet, pipette up and down, let big tissue chunks to sink, and transfer supernatant into the previous tube with turbid supernatant. Repeat this step one more time.
11. Aliquot cells to T175 culture flasks (2 embryos equivalents of cells per 175 flask) and add 25 mL culture media (DMEM-Glutamax plus 10% FBS, 1× NEAA, 1× penicillin–streptomycin) to each flask. Culture cells at 37°C with 5% CO_2 in a humidified incubator specifically designated for ES cell culture (see Note 1).
12. When confluent, cells are split at 1:4 and cultured again. Collect confluent cells by trypsinization and wash cells with PBS. Count total cells and γ-irradiate cells in a 50 mL tube with cobalt-60 irradiator (2,500 rad). Mitotically inactivated cells can also be obtained by treating proliferating cells with mitomycin C (see Note 2). Centrifuge at 200 × *g* for 5 min, aspirate, and resuspend cells in fresh medium. Add 2× freezing media (40 mL DMEM, 40 mL FBS, and 20 mL DMSO) to make it 5×10^6 cells/m. Aliquot cells into cryovials (0.5 or 1 mL/vial) and place them in a Styrofoam box. Place Styrofoam box in a –80°C freezer at least for 24 h; then store cells in liquid nitrogen.

3.1.3. Introducing Targeting DNA into Mouse ES Cells Through Electroporation

1. Gelatinize a T25 culture flask for 10 min and aspirate.
2. Thaw a vial of EF cells (2.5×10^6) in 37°C water bath, transfer cells to 15 mL tube, add 5 mL of DMEM, and centrifuge for 5 min at 200 × *g*.
3. Resuspend cells in 3 mL fresh medium. Plate cells in the flask and incubate cells overnight in the humidified 37°C incubator with 5% CO_2. It is important to have all culture space covered by feeder cells (see Note 3).
4. Thaw a vial of J1 ES cells in 37°C water bath, transfer cells into a 15 mL tube, add 5 mL EF cell medium, and pipette up and down to disperse cells. Centrifuge for 5 min at 200 × *g* and aspirate.
5. Resuspend cells in 3 mL ES cell culture medium (DMEM-Glutamax supplemented with 15% FBS, 1× NEAA, 1× penicillin–streptomycin, 500 units/mL LIF, and 100 μM β-Mercaptoethanol). Adding LIF to the medium will help to maintain cells in undifferentiated state (see Note 4). Plate cells in the flask and incubate in the 37°C incubator with 5% CO_2.

6. Change medium every 24 h (see Note 5). One day prior to cell passage (desired cell density ~75%), gelatinize five 10-cm culture dishes and plate feeder cells overnight.
7. Change medium early in the morning (2–3 h before cell handling) (see Note 5). Wash cells with 5 mL PBS twice and add 0.5 mL of trypsin–EDTA. Incubate cells at 37°C for 5 min. Pat the flask gently to dislodge cells and incubate for another 3 min at 37°C.
8. Add 5 mL medium, pipette up and down to disperse to single cells. Transfer cells to a 15 mL tube and centrifuge for 5 min at 200 × *g*.
9. Resuspend cells in 15 mL media. Plate 5 mL of cells into each dish, add another 5 mL fresh medium, mix well, and let cell grow in the 37°C incubator.
10. Change medium every 24 h. A day prior to electroporation of ES cells, gelatinize five 10-cm cell culture dishes and plate EF cells.
11. At the day of electroporation. Change medium 2–3 h before cell handling.
12. Wash cells with 10 mL PBS twice, and trypsinize cells with 1 mL of trypsin–EDTA at 37°C for 5 min. Rock the dishes gently to dislodge cells and trypsinize for 3 more min at 37°C.
13. Add 7 mL medium, pipette up and down to disperse to single cells. Pool all three dishes of cells to a 50 mL tube and count cells. Centrifuge cells for 5 min at 200 × *g*.
14. Resuspend cells in electroporation buffer (10^7 cells/mL). Add DNA (20 μg) into a cuvette first and then 0.8 mL of cells (8×10^6). Mix well with DNA and cap the cuvette. Electroporate cells with BioRad Gene Pulsar (400 V, 25 μF, time constant 0.4). Let cells sit in the hood for 10 min.
15. Collect all cells into a 50 mL tube, add 35 mL of ES cell medium and resuspend cells well. Plate 9 mL of cell suspension in a 10-cm dish (1.6×10^6/dish) prepared in Step 10. Incubate the cells in the 37°C incubator with 5% CO_2. Change medium the second morning.
16. Start selection by changing cells to selection medium (ES cell medium plus G418 at 175 μg/mL or ES cell medium plus G418 at 175 μg/mL and FIAU at 0.2 μM) 48 h after electroporation. One plate is selected with G418 only and the other four plates are put under G418/FIAU double selections.
17. Replace medium daily. Swirl plates gently to dislodge dead cells before medium change.
18. After 6 days of selection (8 days after electroporation), replace with regular ES cell medium (without G418 and FIAU) and culture for 1 last day. The colonies are ready to be picked.

3.1.4. Picking ES Colony and Freezing ES Cells in 96-Well Plates

1. Change medium 2–3 h before picking colonies. Place a converted microscope into the tissue culture hood and UV-irradiate for 1 h.
2. Gelatin-coat two 96-well plates. Prepare another two 96-well U-bottomed plates by adding 20 μL of trypsin–EDTA solution per well (48 wells at a time).
3. Aspirate medium from one 10-cm $G418^r$/$FIAU^r$ plate and wash cells with PBS twice. Add 10 mL PBS to cover the cells.
4. Place the dish on the inverted microscope and pick colonies using long multi-Flex tips with round ends. Gently push and tease up an entire colony and suck it up along with little volume of PBS (10 μL maximal). Each colony is transferred to the trypsin–EDTA solution in a well of the plate prepared in step 2.
5. After finishing one 10-cm dish (usually 48 colonies in 30 min). Incubate the plate at 37°C for 10 min. Pat the plate gently and incubate at 37°C for another 5 min.
6. During the incubation, add ES medium to the gelatinized plates (150 μL/well) prepared in step 2.
7. Retrieve the trypsinized colonies from the incubator and add 80 μL ES medium per well. Pipette up and down 8–9 times to disperse colonies into single cells.
8. Transfer 50 μL of cells per well to the gelatin-coated plate containing ES medium (step 6) and mix well. Put the plate in the incubator and grow cells until confluent. Change the medium as necessary. These cells will be used for genomic DNA extraction.
9. Transfer the remaining 60 μL of cells into each well of a fresh 96-well plate and place on ice.
10. Repeat the steps 4–9 until all 4 dishes of good colonies are picked (2× 96-well plates).
11. Add 60 μL ice-cold 2× freezing medium into each well. Add 30 μL of sterile light paraffin oil per well to prevent degassing and evaporation during storage at –80°C.
12. Seal the 96-well plates with Parafilm and put into a Styrofoam box and store the cells at –80°C.

3.1.5. Isolation and Restriction Digestion of Genomic DNA from ES Cells

1. Rinse confluent ES cells in the gelatinized plates twice with PBS.
2. Add 50 μL of ES lysis buffer per well. Seal the plates with Parafilm.
3. Incubate the plates at a conventional 60°C oven in a moisturized condition (by putting one sealed plate facing up on top of wet paper towels in a Ziploc bag) overnight.

4. Add 100 μL per well of a mix of NaCl and ethanol (150 μL of 5 M NaCl into 10 mL of cold absolute ethanol) to precipitate genomic DNA using a multichannel pipettor.
5. Allow the plates to sit on the bench for 30 min at room temperature without disturbing.
6. Invert the plates carefully to discard the solution; the DNA remains attached to the wall of the wells. Blot the excess liquid on paper towels.
7. Rinse the DNA three times by dripping 150 μL of 70% ethanol per well using the multichannel pipettor. Discard the alcohol by inverting the plate each time. After the final wash, invert the plates and allow it to dry on the bench (Do not let DNA dry completely or it is very difficult to dissolve).

3.1.6. Identification of Positive ES Clones in Southern Blot Analyses with a 5′ Probe

1. Digest ES cell DNA with restriction enzyme NheI (usually a 96-well plate is enough to get several positive clones, so wrap and store the other plate at 4°C). Prepare a 3 mL master digestion mix: 300 μL 10× restriction buffer, bovine serum albumin (100 μg/mL), RNase (100 μg/mL), and NheI (10 units/well). Bring to 3 mL with ddH_2O.
2. Aliquot 30 μL of restriction digest mix per well, mix with DNA and incubate the reactions at 37°C overnight in the humidified incubator.
3. Prepare six 1% agarose gels and add ethidium bromide (4 μL of 10 mg/100 mL gel). Add gel-loading buffer to the samples. Load samples into wells of the gels and run the gels in 1× TBE buffer at 80 V for 4–5 h.
4. Transfer the gels to plastic containers (one gel in one container). Rinse with ddH_2O twice. Add 250 mL of depurination buffer to the gels and shake gently for 15 min.
5. Decant depurination buffer. Rinse the gels with ddH_2O, slosh gently, rinse again with ddH_2O, and decant. Then add 250 mL of denaturation solution to the gels for 20 min with gentle shake. Repeat this step with new solution for 20 min.
6. Pour off denaturing solution; rinse gel with ddH_2O twice, and add ~250 mL neutralizing solution. Shake for 15 min.
7. Pour off neutralizing solution; replace with fresh neutralizing solution; shake again for 15 min.
8. Decant neutralizing solution; rinse gel with ddH_2O, slosh gently, rinse again with ddH_2O, pour off, and add ~250 mL 20× SSC; shake again for 30 min.
9. Cut six Hybond nitrocellulose membranes slightly larger than the size of gels. Pre-wet the blots in ddH_2O and in 20× SSC for 5 min. Transfer DNA from the gels to the blots using a conventional transfer method (with 2× SSC) (3 MM paper/gel/blot/3 MM paper/paper towels) overnight (with the

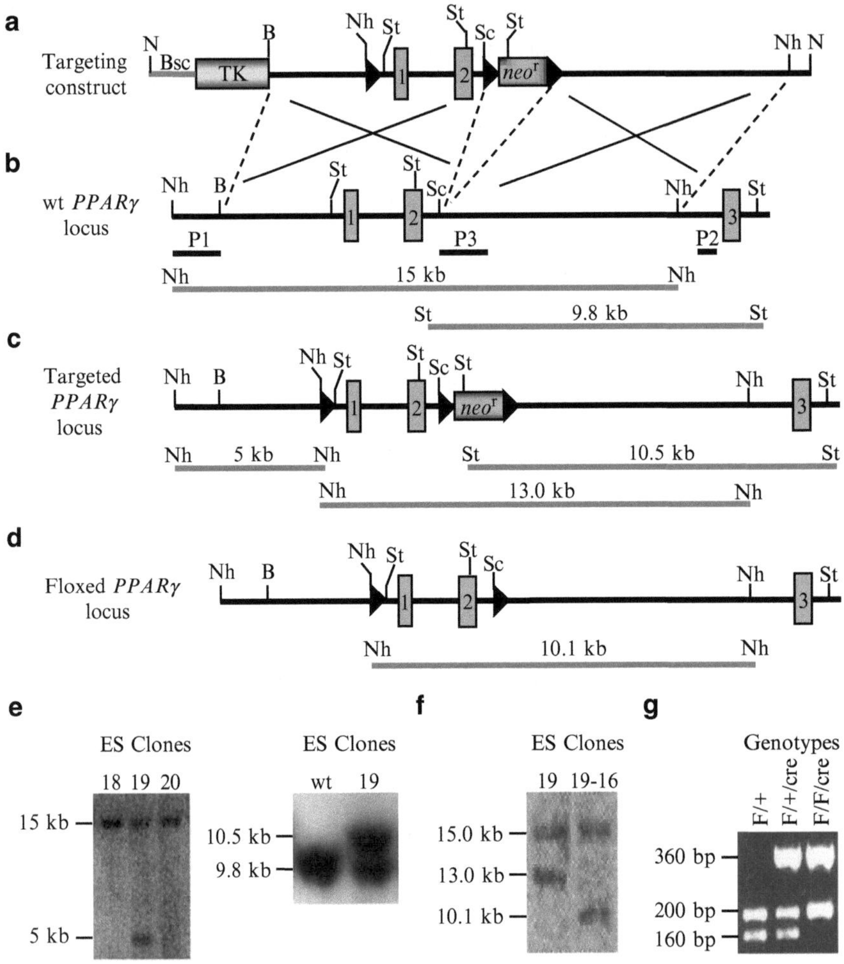

Fig. 2. PPARγ targeting strategy. (*a*) Floxed PPARγ targeting construct. (*b*) Wild-type (wt) PPARγ locus. (*c*) Targeted PPARγ locus. (*d*) Floxed PPARγ locus. (*e*) Southern blot analysis of targeted ES cells. (*Left*) Hybridization with 5′ probe (P1). (*Right*) Hybridization with 3′ probe (P2). (*f*) Southern blot analysis with an inside probe (P3) in ES cells transiently transfected with a *cre* plasmid. Restriction enzymes: N, NotI; Nh, NheI; B, BamHI; St, StuI; Sc, SacI. (*g*) PCR genotyping of animals with a pair of primers detecting cre gransgene (360 bp) and a pair of the other primers detecting either floxed allele (200 bp) or a wild type allele (160 bp).

capillary effect of 2× SSC). Squeeze away any bubbles between the gel and 3 MM papers.

10. Place the blots in the UV Cross-linker and cross-link on the automatic setting.
11. Wet the blots in 2× SSC, roll and place them into hybridization tubes.
12. Warm pre-hybridization solution to 42°C. Add pre-hybridization solution to the tubes. Place tubes in 42°C hybridization oven for 3 h.
13. During pre-hybridization, prepare ^{32}P-labeled probes with a 5′ genomic fragment (P1) upstream of the recombination arm (see Fig. 2*B*). Boil the DNA (100 ng/12 μL) for 10 min and

cool on ice. Add random priming reagent, Klenow enzyme, and [α-^{32}P]-dCTP to the boiled DNA. Incubate at 37°C for 60 min.

14. Separate the nonincorporated label with a G-25 Sephadex spin column. Determine the specific activity of the probe by reading 1 μL of the probe in a scintillation counter.
15. Boil the hot probe for 10 min and cool immediately on ice.
16. Warm the hybridization solution to 42°C. Add the hot probe (10^7 cpm/10 mL) to the hybridization solution and mix. Pour off pre-hybridization solution, add hybridization solution, and incubate the tubes with the blots at 42°C overnight.
17. Preheat the wash solution 1–56°C. Decant the hybridization solution (discard properly). Add the wash solution 1 to the tubes and wash the blots at 56°C for 30–60 min.
18. Preheat the wash solution 2–56°C. Pour off the wash solution 1 and add the wash solution 2 and wash at 56°C for 30 min.
19. Check the blot with a Geiger counter. The background on blank areas of the membrane should be 200 cpm or less. If the background is too high, repeat wash with the wash solution 2.
20. Wrap the blots with Saran Wrap and expose to Kodak films in autoradiography cassettes at −80°C overnight. Develop the films and identify positive clones with correct sizes of DNA bands (15 kb for wild-type locus and 5 kb for targeted locus) (see Fig. 2*C*, *E*, *left*). Total eight positive clones are obtained.

3.1.7. Expansion of Positive ES Clones and Confirmation of Correct Targeting with a 3′ Probe

1. Coat a 24-well plate with gelatin and plate feeder cells to nine wells. Let cells to attach overnight.
2. Thaw ES clones by placing the 96-well plate containing positive clones in the 37°C incubator.
3. Remove paraffin oil; then transfer one wild-type clone and eight positive clones to each well of the 24-well plate. Add 1.5 mL ES medium to each well and mix.
4. Change medium daily. Prepare two 6-well plates by coating with gelatin and inoculating with EF cells 24 h before cells are ready for expansion.
5. Change medium 2–3 h prior to expansion. Trypsinize cells and add 1 mL of ES cell medium to disperse cells.
6. Transfer cells to the 6-well plates, add another 2 mL ES cell medium per well, and mix. Grow cells at 37°C and change medium daily.
7. When cells are 75% confluent, gelatinize a 12-well plate. Change medium 2–3 h prior to expansion. Trypsinize cells, transfer cells to a 15 mL tube, add 5 mL of ES medium, and spin at 200 × *g* for 5 min.

8. Resuspend cells in 2.5 mL ES medium. Plate 0.5 mL ES cells per clone in the 12-well plate. Change medium as necessary until cells are confluent. Add 300 μL of lysis buffer to each well and lyse cells for 30 min at 37°C. Transfer lysates to microcentrifuge tubes, add 600 μL of 100% ethanol, centrifuge, and wash with 70% ethanol twice. Air-dry DNA and resuspend in 1× TE buffer.
9. Freeze the remaining 2 mL of cells into two vials till correct recombination at 3′ side is confirmed.
10. Digest 10 μg of genomic DNA with StuI for 3 h. Separate DNA fragments in a 0.8% agarose gel. Perform Southern analysis with ^{32}P-labeled probes prepared from a 3′ genomic fragment (P2) (see Fig. 2*B*) using random priming kit as described in Subheading 3.1.7. Seven out of 8 clones are correctly recombined at the long arm (3′ side) (see Fig. 2*C, E right*).

3.1.8. Generation of Floxed PPARγ ES Clones Without Neo Cassette

This is how I made *neo*-less floxed PPARγ ES clones. However, readers are strongly encouraged to use an alternate way to delete *neo* cassette from germ cells in vivo (see Note 6)

1. Thaw 1 vial of one of the positive ES clones (see #19 Fig. 2*E*) and plate cells in a T25 flask coated with gelatin and plated with EF cells. Incubate cells at 37°C with 5% CO_2. Change medium daily.
2. Coat six 6-cm culture dishes with gelatin and plate EF cells 24 h before cells are 75% confluent.
3. Change medium 2–3 h prior to cell handling. Wash cells with PBS twice, add trypsin–EDTA and incubate at 37 C for 5 min.
4. Add 5 mL of medium, pipette up and down to disperse cells, and centrifuge cells in a 15 mL tube.
5. Wash cells with PBS twice. Resuspend cells in PBS at 1×10^6/mL. Add 25 μg of pCMX-*cre* plasmid (*cre* gene fused with a nuclear localization signal) into a cuvette. Add 0.8 mL of ES cells into the cuvette and mix well with DNA. Electroporate cells (800 V, 3 μF, time constant 0.1). Let cells sit in the hood for 10 min.
6. Dilute cells with ES medium and plate cells in the 6-cm dishes prepared in step 2 (3×10^4/dish×2, 1.5×10^4/dish×2, and 7.5×10^3/dish×2). Incubate cells at 37°C. Change medium daily.
7. The day before colonies are ready to be picked, gelatinize a 96-well plate and plate EF cells.
8. Change medium in the morning. Pick colonies in the early afternoon and grow ES cells in the 96-well plate as described in Subheading 3.1.4.

9. When clones are 75% confluent, split cells with trypsin, and freeze half of each clone into a 96-well plate.
10. Grow another half in a 96-well plate coated with gelatin. Change medium as necessary until the cells are confluent.
11. Extract DNA, digest with NheI, perform DNA gel electrophoresis, transfer DNA to nitrocellulose membranes, and perform Southern blot analyses using ^{32}P-labeled probes prepared with a genomic fragment (P3) (see Fig. 2*B*) as described in Subheading 3.1.6. Two clones (see 19–16 Fig. 2*F* and 19–52 not shown) show deletion of *cre* cassette from the targeted PPARγ locus (see Fig. 2*F*).
12. Both clones are expanded and prepared for microinjection into blastocysts (C57Bl/6J). Injection of the clone 19–16 ES cells gives rise to good chimeras (90% agouti coat color) and 19–52 clone gives rise to chimeras with 50–60% agouti coat color.
13. The male chimeric mice are crossed to C57Bl/6J mice.
14. Majority of F1 mouse offsprings from 19 to 16 clone-injected chimeras show good agouti coat color. Wean mice at 21 days postnatal, clip tails (0.5 cm) and place tails in microcentrifuge tubes. Add 300 μL of mouse-tail lysis buffer per tube and incubate at 55°C overnight.
15. Add 600 μL of absolute alcohol to each tube and shake tubes vigorously. DNA can easily be seen in the solution. Precipitate DNA for 10 min. Wash the DNA pellets with 70% alcohol twice and air-dry DNA. Resuspend DNA in 100 μL of 1× TE buffer.
16. Determine genotypes of the mice using regular PCR with the primers designed to detect 5′ lox*P* site (see Table 1). PCR condition: 94°C, 1 min, (94°C, 20 s, 60°C, 30 s, 71.5°C, 70 s) × 35 cycles.
17. Run 2% DNA gels to show PCR products. The 200 bp PCR product is floxed PPARγ allele while the 160 bp DNA product represents the wild-type PPARγ allele (see Fig. 2*G*). Consistent with coat color, 19–16 clone-derived chimeras give rise to F1 mice bearing a floxed PPARγ allele (PPAR$\gamma^{f/+}$), suggesting successful germ line transmission from this clone of ES cells.

3.2. Generation of aP2-cre Transgenic Mice

1. A plasmid containing the promoter of mouse fatty acid binding protein 4 (FABP4, also named aP2) gene (pBluescript-aP2p) is obtained.
2. A plasmid containing the P1 bacteriophage *cre* recombinase cDNA fused to a nuclear localization signal and a myc-tag (pSP72-*cre*) is obtained.
3. The 5.4 kb aP2p is digested with KpnI/SmaI, purified, and cloned into the pSP72 cloning vector (restricted with the same enzymes) upstream of the *cre* gene as shown in Fig. 1*B*.

4. Digest the high quality plasmid DNA (20 μg) with HindIII.
5. Separate the whole digest on an agarose gel (0.8%). Add crystal violet to the gel buffer (1:1,000 dilution of 2% stock) instead of ethidium bromide, which is toxic even at low concentrations.
6. Cut the band containing the linearized aP2p-*cre* DNA fragment from the gel and purify the DNA using Jetsorb Gel Extraction Kit according to the manufacturer's protocol.
7. Ultrafiltrate the eluted DNA through a 0.45 μm Ultrafree-MC spin column. Dilute the DNA to 3 ng/μL in MI buffer and pass through a 0.2 μm Ultrafree-MC spin column.
8. The DNA is sent for microinjection of pronuclei. Four positive transgenic lines are obtained, two of which have 3–4 copies of *cre* transgene in their genome, respectively.

3.3. Creation of Adipose Tissue-Specific PPARγ Knockout Mice

1. Cross the floxed PPARγ mice to the aP2-cre transgenic mice. This results in $PPAR\gamma^{f/+}$/aP2-Cre, $PPAR\gamma^{f/+}$, $PPAR\gamma^{+/+}$/aP2-Cre, and $PPAR\gamma^{+/+}$ offsprings.
2. $PPAR\gamma^{f/+}$/aP2-Cre mice are backcrossed to C57Bl/6J mice for 4–5 generations.
3. Backcrossed $PPAR\gamma^{f/+}$/aP2-Cre mice are intercrossed to their $PPAR\gamma^{f/+}$ littermates. This results in creation of the following six genotypes of mice: $PPAR\gamma^{f/f}$/aP2-Cre (adipose tissue-specific PPARγ knockout mice, FKOγ mice), $PPAR\gamma^{f/f}$ (double-floxed PPARγ mice), $PPAR\gamma^{f/+}$/aP2-Cre (adipose tissue-specific PPARγ heterozygous mice), $PPAR\gamma^{f/+}$ (floxed PPARγ mice), $PPAR\gamma^{+/+}$/aP2-Cre (aP2-Cre transgenic mice), and $PPAR\gamma^{+/+}$ (wild-type mice). Figure 2*G* shows 3 genotypes of mice: $PPAR\gamma^{f/+}$ (f/+), $PPAR\gamma^{f/+}$/aP2-Cre (f/+/cre), and $PPAR\gamma^{f/f}$/aP2-Cre (f/f/cre).
4. $PPAR\gamma^{f/f}$/aP2-Cre mice and $PPAR\gamma^{f/f}$ mice are then intercrossed to generate cohorts of FKOγ and $PPAR\gamma^{f/f}$ mice for investigation of PPARγ function in mature adipocytes. $PPAR\gamma^{f/f}$ mice, which do not show difference compared to $PPAR\gamma^{+/+}$/aP2-Cre mice or $PPAR\gamma^{+/+}$ mice, are used as controls.
5. Characterize the FKOγ mice (27).

4. Notes

1. It is recommended that a specific tissue culture incubator and a hood are designated for ES cells culture only since any mycoplasma, bacterial or fugal contamination will compromise the effort of carrying out gene targeting project. The incubator

should be humidified (it is best to use autoclaved water for this purpose). Change water every week and monitor potential fungal growth in the water tray during ES culture. In case fungal contamination happens, bomb the incubator with paraformaldehyde overnight and clean the inside completely with 70% alcohol and then autoclaved water.

2. The author prepared mitotically inactivated mouse embryonic fibroblast feeder cells by γ-irradiation. In case a γ-irradiator (Such as Cobalt-60 irradiator) is not available, mitotically inactivated mouse embryonic fibroblasts can be prepared by treating proliferating cells with mitomycin C (10 μg/mL final concentration) for 3 h. The treated cells need to be washed with PBS several times to get rid of mitomycin C completely so that ES cell growth will not be compromised.
3. Whenever possible, plate feeder cells 1 day before ES cells are plated. This will let feeder cells to attach and spread completely to cover the culture space of a flask or a dish so that that ES cells will attach and grow only on the feeder layer rather than on the uncovered space.
4. ES cells depend on the cytokine LIF to maintain undifferentiated while proliferating. It is not crucial to add LIF into the ES medium before the step of electroporation (since feeder cells can supply some LIF). But it is worth to "waste" some LIF to guarantee that ES cells stay healthy and undifferentiated.
5. It is absolutely required that ES cells are refreshed with fresh medium every 24 h. It is also very important to refresh ES cells with fresh medium 2–3 h prior to cell split/passage.
6. The presence in an intron of the floxed *neo* cassette often results in interference with expression of a targeted gene. Consequently, many investigators selectively remove the floxed *neo* cassette by transient expression of Cre in ES cells. Although the author has no difficulty in obtaining good ES clones for germ line transmission after many rounds of freezing, thawing and culturing, the added manipulation of the ES cells may reduce the likelihood that a clone be able to transmit through the germ line. Therefore, the author recommends deleting *neo* resistant gene cassette in vivo. EIIa-Cre transgenic mice (Jackson Laboratory) can be used to achieve this purpose. Crossing of the EIIa-Cre mice to the mice bearing lox*P*-Gene X-lox*P*-*neo*-lox*P* allele will result in mosaic mice with partial and complete excision of loxP-flanked sequences (including *neo* cassette). The mosaic mice can then be mated, and the *neo*-less conditional knockout allele can be found in the offspring after screening.

Acknowledgments

The author thanks Dr. Yaacov Barak for the PPARγ genomic fragments and pBluescript-TK, pGEM72-KJ1-*neo*^{r}, and pSP72-cre plasmids, J1 ES cells, and his precious help during the project. The author also appreciates Dr. Peter Tontonoz for the pBluscript-aP2-promoter plasmid.

References

1. Kershaw EE, Flier JS (2004) Adipose tissue as an endocrine organ. J Clin Endocrinol Metab 89:2548–2556
2. Flier JS (2004) Obesity wars: molecular progress confronts an expanding epidemic. Cell 116:337–350
3. Fantuzzi G (2005) Adipose tissue, adipokines, and inflammation. J Allergy Clin Immunol 115:911–919
4. Ouchi N, Parker JL, Lugus JJ, Walsh K (2011) Adipokines in inflammation and metabolic disease. Nat Rev Immunol 11:85–97
5. Lazar MA (2005) PPAR gamma, 10 years later. Biochimie 87:9–13
6. Takeda S, Elefteriou F, Karsenty G (2003) Common endocrine control of body weight, reproduction, and bone mass. Annu Rev Nutr 23:403–411
7. Garg A, Misra A (2004) Lipodystrophies: rare disorders causing metabolic syndrome. Endocrinol Metab Clin North Am 33:305–331
8. Spiegelman BM, Flier JS (1996) Adipogenesis and obesity: rounding out the big picture. Cell 87:377–389
9. Kopelman PG (2000) Obesity as a medical problem. Nature 404:635–643
10. Reaven GM (2005) Why Syndrome X? From Harold Himsworth to the insulin resistance syndrome. Cell Metab 1:9–14
11. Rosen ED, Walkey CJ, Puigserver P, Spiegelman BM (2000) Transcriptional regulation of adipogenesis. Genes Dev 14:1293–1307
12. Barroso I, Gurnell M, Crowley VE, Agostini M, Schwabe JW, Soos MA, Maslen GL, Williams TD, Lewis H, Schafer AJ et al (1999) Dominant negative mutations in human PPARgamma associated with severe insulin resistance, diabetes mellitus and hypertension. Nature 402:880–883
13. Agostini M, Schoenmakers E, Mitchell C, Szatmari I, Savage D, Smith A, Rajanayagam O, Semple R, Luan J, Bath L et al (2006) Non-DNA binding, dominant-negative, human PPARgamma mutations cause lipodystrophic insulin resistance. Cell Metab 4:303–311
14. Barak Y, Nelson MC, Ong ES, Jones YZ, Ruiz-Lozano P, Chien KR, Koder A, Evans RM (1999) PPAR gamma is required for placental, cardiac, and adipose tissue development. Mol Cell 4:585–595
15. Olefsky JM (2000) Treatment of insulin resistance with peroxisome proliferator-activated receptor gamma agonists. J Clin Invest 106:467–472
16. Evans RM, Barish GD, Wang YX (2004) PPARs and the complex journey to obesity. Nat Med 10:355–361
17. Tontonoz P, Spiegelman BM (2008) Fat and beyond: the diverse biology of PPARgamma. Annu Rev Biochem 77:289–312
18. Deeb SS, Fajas L, Nemoto M, Pihlajamaki J, Mykkanen L, Kuusisto J, Laakso M, Fujimoto W, Auwerx J (1998) A Pro12Ala substitution in PPARgamma2 associated with decreased receptor activity, lower body mass index and improved insulin sensitivity. Nat Genet 20:284–287
19. Altshuler D, Hirschhorn JN, Klannemark M, Lindgren CM, Vohl MC, Nemesh J, Lane CR, Schaffner SF, Bolk S, Brewer C et al (2000) The common PPARgamma Pro12Ala polymorphism is associated with decreased risk of type 2 diabetes. Nat Genet 26:76–80
20. Ostergard T, Ek J, Hamid Y, Saltin B, Pedersen OB, Hansen T, Schmitz O (2005) Influence of the PPAR-gamma2 Pro12Ala and ACE I/D polymorphisms on insulin sensitivity and training effects in healthy offspring of type 2 diabetic subjects. Horm Metab Res 37:99–105
21. Doney AS, Fischer B, Lee SP, Morris AD, Leese G, Palmer CN (2005) Association of common variation in the PPARA gene with incident myocardial infarction in individuals

with type 2 diabetes: A Go-DARTS study. Nucl Recept 3:4

22. Kubota N, Terauchi Y, Miki H, Tamemoto H, Yamauchi T, Komeda K, Satoh S, Nakano R, Ishii C, Sugiyama T et al (1999) PPAR gamma mediates high-fat diet-induced adipocyte hypertrophy and insulin resistance. Mol Cell 4:597–609
23. Miles PD, Barak Y, He W, Evans RM, Olefsky JM (2000) Improved insulin-sensitivity in mice heterozygous for PPAR-gamma deficiency. J Clin Invest 105:287–292
24. Masugi J, Tamori Y, Kasuga M (1999) Inhibition of adipogenesis by a COOH-terminally truncated mutant of PPARgamma2 in 3T3-L1 cells. Biochem Biophys Res Commun 264:93–99
25. Liao W, Nguyen MT, Yoshizaki T, Favelyukis S, Patsouris D, Imamura T, Verma IM, Olefsky JM (2007) Suppression of PPAR-gamma attenuates insulin-stimulated glucose uptake by affecting both GLUT1 and GLUT4 in 3T3-L1 adipocytes. Am J Physiol Endocrinol Metab 293:E219–227
26. Marth JD (1996) Recent advances in gene mutagenesis by site-directed recombination. J Clin Invest 97:1999–2002
27. He W, Barak Y, Hevener A, Olson P, Liao D, Le J, Nelson M, Ong E, Olefsky JM, Evans RM (2003) Adipose-specific peroxisome proliferator-activated receptor gamma knockout causes insulin resistance in fat and liver but not in muscle. Proc Natl Acad Sci USA 100:15712–15717

Chapter 9

Site-Directed Mutagenesis to Study the Role of Specific Amino Acids in the Ligand Binding Domain of PPARs

Nico Mitro, Federica Gilardi, Marco Giudici, Cristina Godio, Elena Scotti, and Maurizio Crestani

Abstract

The role of certain amino acids in the interactions of ligands with their cognate nuclear receptors is usually achieved by the resolution of the crystal structure of the receptor complexed with the ligand. As a complementary functional approach, site-directed mutagenesis, a technique broadly used in molecular biology, allows the assessment of the role of a specific amino acid in determining the interaction with a specific ligand. This method makes it possible to evaluate several mutations of a key amino acid for ligand binding and to determine the relationship between protein structure and ligand interaction. Here, we describe an application of this technique to evaluate different point mutations on the transcriptional activity of peroxisome proliferator-activated receptor γ (PPARγ) in the absence or presence of chemically different ligands.

Key words: Nuclear receptors, Peroxisome proliferator-activated receptors, Transcriptional activity, Mutagenesis

1. Introduction

In vitro site-directed mutagenesis is a technique used to characterize the relationships between protein structure and function by introducing mutations in a given protein. Different methods of this technique have been published and several kits to introduce mutation in single- or double-stranded DNA are commercially available (1–4). The following protocol is useful to introduce point mutations, delete or insert single or multiple adjacent mutations and consequently change amino acids in a given protein sequence such as the ligand binding domain of PPARs (5, 6).

Mostafa Z. Badr and Jihan Youssef (eds.), *Peroxisome Proliferator-Activated Receptors (PPARs): Methods and Protocols*, Methods in Molecular Biology, vol. 952, DOI 10.1007/978-1-62703-155-4_9, © Springer Science+Business Media New York 2013

This method was first described in 1978 by Michael Smith (7). The procedure is based on the use of a supercoiled double-stranded DNA vector with the gene of interest that has to be mutated and employs two oligonucleotide primers carrying the desired mutation each complementary to opposite strands of the vector (Fig. 1). By using polymerase chain reaction (PCR) with a high fidelity DNA polymerase the protocol allows to generate a plasmid containing the desired mutation. After the PCR cycle, the product vector is treated with the endonuclease *Dpn*I. The *Dpn*I is a restriction enzyme whose target sequence is 5′-GmethylatedATC-3′ specific for methylated DNA. The use of *Dpn*I is important to digest and degrade the parental DNA template for the isolation of the mutagenized vector, since virtually all the DNA purified from *E. coli* (bacteria used for plasmid vector preparations) is dam methylated and sensible to *Dpn*I action (8). Ultimately, the mutagenized plasmid is transformed into ultra-competent *E. coli* cells to

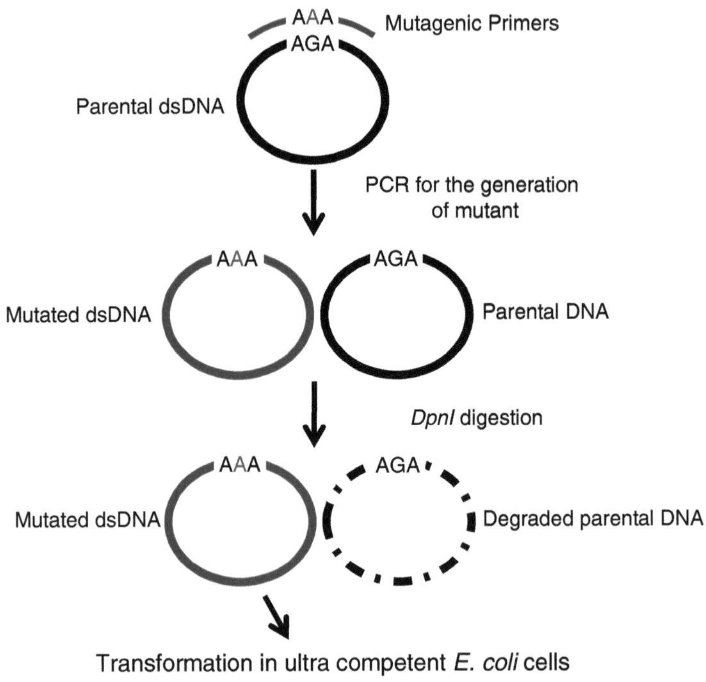

Fig. 1. Cartoon depicting the principle of site-directed mutagenesis. The method is based on a parental plasmid DNA prepared from a dam positive *E. coli* strain (see Note 1). Parental DNA is double stranded (indicated as parental dsDNA). Design appropriate mutagenic primers carrying the mutation, perform the PCR cycles and obtain several copies of mutated double strand parental DNA (indicated as mutated dsDNA). The PCR cycles generate for the majority mutated dsDNA and some hybrid of parental DNA forms with mutated DNA. *Dpn*I digestion (that selectively target methylated adenin generated by the dam positive *E. coli* cells) degrades the parental DNA and the mixture is now enriched in mutated dsDNA. Finally, the mutated dsDNA can be introduced by transformation in ultra-competent *E. coli* cells for amplification, colony screening, and DNA sequencing to verify that the plasmid carries the desired mutation. The figure depicts the mutation in the parental dsDNA of arginin (codon: AGA) into lysine (codon: AAA).

screen for positive colonies. After sequencing, the plasmid is ready to be amplified and then to be used for example in transfection assays or mutated protein purification.

2. Materials

1. Plasmid vector containing the gene of interest to be mutated.
2. High fidelity DNA polymerase.
3. 10× reaction buffer (usually provided with the DNA polymerase) and deoxynucleotides mix (dNTPs).
4. *Dpn*I restriction enzyme.
5. Two oligonucleotide primers carrying the desired mutation (see Subheading 3.1).
6. Ultra-competent *E. coli* cells (store at −80°C).
7. Luria–Bertani (LB) broth with antibiotic to select the vector: 10 g tryptone/L, 5 g yeast extract/l, 10 g NaCl/L, pH 7. Dispense 200 mL in 2 L flasks. Autoclave 20 min at 1 atmosphere to sterilize the medium, cool down, and add the antibiotic to a desired final concentration.
8. LB agar plate: add 20 g of agar to 1 L of LB broth, autoclave 20 min at 1 atmosphere to sterilize the medium, cool down to 55°C, add the antibiotic to a desired final concentration. Aliquot into Petri dishes (~15 to 20 mL/100-mm plate).

3. Methods

This protocol allows the introduction of point mutations, deletions, or insertions in double-stranded DNA to test the different behavior of a mutated protein as compared to a wild type protein. The following protocol has been successfully used to generate point mutants of PPARγ ligand binding domain. These mutants have been used to pinpoint key amino acids involved in ligand–receptor interaction (see Fig. 2). However, this protocol could be adapted to introduce mutations in other nuclear receptors or proteins.

3.1. Design of Mutagenic Primers

3.1.1. Experimental Protocol

The mutagenic primers to be used with the following protocols should be designed by inserting the desired mutation. In order to have successful mutagenic primers both oligonucleotides must contain the desired mutation and anneal to the same sequence on

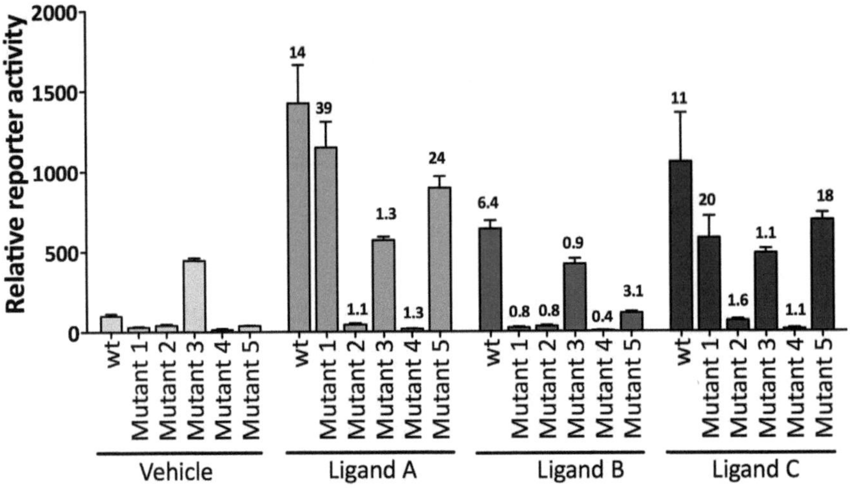

Fig. 2. Example of the effects of different point mutations on the transcriptional activity of PPARγ in the absence or presence of chemically distinct ligands. Five point mutants of the ligand binding domain of PPARγ were cotransfected into HEK293 cells along with the reporter vector bearing the luciferase cDNA driven by multiple Gal4 binding sites cloned in front of the *Herpes simplex* virus thymidine kinase minimal promoter. Transfected cells were treated with the indicated ligands added at a final concentration of 5 μM in dimethyl sulfoxide. Results are expressed as percentage of the value in cells transfected with wt receptor in the absence of ligands and each point is the mean ± SD of three independent experiments each performed in triplicate wells. The fold activation of each mutant by the added ligands as compared with the vehicle (DMSO) is indicated above the bars.

opposite strands of the target vector carrying the gene of interest that should be mutated.

The length of the primers should be between 20 and 40 bases and their melting temperature should be at least 78°C. Oligonucleotides longer than 40 bases may be used, but their use may increase the risk to form secondary structures or primer dimers, which may affect the mutagenesis reaction and ultimately the desired mutation.

Usually, the mutation (base replacement, deletion, or insertion) should be in the middle of the primer.

To calculate the melting temperature of the primers for introducing a point mutation this formula is usually used:

$$\text{Melting temperature } (mT):$$
$$81.5^{\circ}\text{C} + 0.41^{\circ}\text{C} \times (\%GC) - (675/N) - \%\text{ mismatch},$$

where

N is the base length of the primer.

%GC is the percent of guanine and cytosine present in the primers. Ideally, the primers should have a minimum GC content of 40% and should terminate in one or more C or G bases.

% mismatch is the percent of bases changed in the primers.

However, %GC and % mismatch are whole numbers.

To calculate the melting temperature of the primers for introducing insertions or deletions, use this modified version of the previous formula:

$$\text{Melting temperature (mT)}: 81.5 + 0.41(\%\text{GC})(675/N)$$

In this case N does not include the bases, which are inserted or deleted.

Nowadays, several free online softwares are available to design mutagenic primers and to calculate the corresponding melting temperature, however the melting temperature calculated with online software may differ from the one calculated with the previous formula.

To obtain optimal performance and mutation efficiency of the primers they need not be 5′ phosphorylated and must be purified either by polyacrylamide gel electrophoresis (PAGE) or by fast performance liquid chromatography (FPLC).

Finally the concentration of the primers should be kept in excess and it is advisable to vary the amount of template while keeping the concentration of the primer constantly in excess to ensure a successful mutagenesis.

3.2. Preparation of the Mutated Strand

3.2.1. Experimental Protocol

Before starting the reaction to prepare the mutated strand, make sure that the plasmid DNA template vector has been prepared from a dam + *E. coli* strain (see Note 1). To maximize the performance of temperature cycling, it is better to run the PCR using thin 0.2 mL tubes, which ensure optimal contact with the thermal cycler-heating block.

1. Synthesize two complementary primers carrying the mutation, flanked by unmodified nucleotide sequence (see Subheading 3.1).
2. Prepare the mutagenic reaction as indicated below (see Note 2):

   5 μL of 10× reaction buffer (usually provided with the DNA polymerase).
   X μL (10–50 ng) of plasmid DNA template.
   X μL (150 ng) of mutagenic primer 1 (see Note 3).
   X μL (150 ng) of mutagenic primer 2 (see Note 3)
   1 μL of dNTP mix.
   Then add:

   1 μL of high fidelity DNA polymerase (typically 2.5 U/μL).
   Add bi-distilled water to a final volume of 50 μL.
3. Once the reaction is ready, place it in a thermal cycler and run the following thermal protocol:

Step 1:	95°C for 1 min
Step 2:	95°C for 1 min 60°C for 1 min 68°C for 1 min for kb of vector length (i.e., vector length 5 kb = 5 min) Cycle this step 20 times
Step 3:	68°C for 7 min

4. At the end of the PCR, place the tube on ice for 2 min to cool the reaction.

3.3. Digestion with DpnI

3.3.1. Experimental Protocol

1. Add 1 μL of the *Dpn*I restriction enzyme (10 U/μL) to each amplification reaction.
2. Gently mix each reaction mix by pipetting the solution up and down, spin down the reaction mix in the tube with a benchtop centrifuge for 1 min, and incubate the reactions at 37°C for 1 h to digest the parental DNA plasmid.

3.4. Transformation of Ultra-competent E. coli Cells

3.4.1. Experimental Protocol

1. Thaw the ultra-competent *E. coli* cells on ice. Include a positive and a negative control for transformation to make sure that this procedure is working properly.
2. Aliquot 2 μL of the *Dpn*I-treated DNA from each sample (for negative control add water while for positive control add, e.g., the parental plasmid) to a prechilled 14-mL polypropylene round-bottom tube.
3. For each control and sample reaction to be transformed, aliquot 50 μL of the ultra-competent cells. Gently swirl the contents of the tube and incubate the cells on ice for 20–30 min, swirling every 5 min.
4. Preheat LB broth in a 42°C water bath.
5. Heat-pulse the tubes in a 42°C (do not exceed 42°C) water bath for 30–45 s. The duration of the heat-pulse is critical for obtaining the highest efficiencies.
6. Incubate the tubes on ice for 2 min.
7. Add 0.5 mL of preheated (42°C) LB broth to each tube, and then incubate the tubes at 37°C for 1 h with shaking at 225–250 rpm.
8. Plate 250–300 μL of each transformation reaction on agar plates containing the appropriate antibiotic for the plasmid vector (see Note 4).
9. Incubate the agar plates overnight at 37°C.

3.5. Expected Results after the Transformation

The expected colony number should be between 10 and 100 depending upon the base composition and length of the DNA vector used. The negative control of the transformation should give no colonies while the positive control (e.g., parental DNA vector) should range between 100 and 2,000 colonies.

Once the colonies from the mutated plasmid are obtained, a mini DNA preparation and a sequencing analysis of the insert of interest should be performed to verify that selected clones contain the desired mutation(s).

This protocol has been successfully applied to generate several mutants in PPAR ligand binding domain.

4. Notes

1. The majority of the commonly used *E. coli* strains are dam positive. Plasmid DNA isolated from dam negative strains cannot be used with this protocol.
2. Set up an initial mutagenic reaction using 10 ng of plasmid DNA template. If this initial reaction is unsuccessful, set up a series of sample reactions using various concentrations of the vector template ranging from 5 to 50 ng and keep the mutagenic primer concentration constant.
3. The mutagenesis protocol uses 150 ng of each oligonucleotide primer. To convert nanograms to picomoles of oligo, use the following equation:

$$\text{X pmoles of oligo} = [(\text{ng of oligo})/(330\times\text{number of bases of the primer})]\times 1{,}000$$

For example, for 150 ng of a 30 base long primer:

$$[(150\text{ng of oligo})/(330\times 30\text{ bases})]1{,}000 = 15\text{pmoles}.$$

4. Make sure that the ultra-competent cells carry an antibiotic resistance different from that of the mutated plasmid. For example, if the mutated plasmid carries the ampicillin resistance cassette, the ultra-competent cells should not be resistant to ampicillin. This is in order to allow the selection of transformed colonies containing the mutated plasmid from the parental, non-transformed ultra-competent cells.

Acknowledgments

This work was supported by grants from the European Commission (LSHM-CT2006-037498), the Giovanni Armenise-Harvard Career Development Award and the Cariplo Foundation (2008.2511 and 2009.2727). We are grateful to Miss Elda Desiderio Pinto for invaluable administrative support.

References

1. Kunkel TA (1985) Rapid and efficient site-specific mutagenesis without phenotypic selection. Proc Natl Acad Sci U S A 82:488–492
2. Sugimoto M, Esaki N, Tanaka H et al (1989) A simple and efficient method for the oligonucleotide-directed mutagenesis using plasmid DNA template and phosphorothioate-modified nucleotide. Anal Biochem 179:309–311
3. Taylor JW, Ott J, Eckstein F (1985) The rapid generation of oligonucleotide-directed mutations at high frequency using phosphorothioate modified DNA. Nucleic Acids Res 13:8765–8785
4. Vandeyar MA, Weiner M, Hutton C et al (1988) A simple and rapid method for the selection of oligodeoxynucleotide-directed mutants. Gene 65:129–133

5. Pochetti G, Godio C, Mitro N et al (2007) Insights into the mechanism of partial agonism: Crystal structures of the peroxisome proliferator-activated receptor γ ligand-binding domain in the complex with two enantiomeric ligands. J Biol Chem 282:17314–17324
6. Pochetti G, Mitro N, Lavecchia A et al (2010) Structural insight into peroxisome proliferator-activated receptor gamma binding of two ureidofibrate-like enantiomers by molecular dynamics, cofactor interaction analysis, and site-directed mutagenesis. J Med Chem 53: 4354–4366
7. Hutchison CA III, Phillips S, Edgell M et al (1978) Mutagenesis at a specific position in a DNA sequence. J Biol Chem 253:6551–6560
8. Nelson M, McClelland M (1992) Use of DNA methyltransferase/endonuclease enzyme combinations for megabase mapping of chromosomes. Methods Enzymol 216:279–303

Chapter 10

PPAR SUMOylation: Some Useful Experimental Tips

Benoit Pourcet, Bart Staels, and Corine Glineur

Abstract

Studies on the regulation of nuclear receptors, such as the peroxisome proliferator-activated receptors (PPARs), are important to enhance our understanding of their molecular, cellular, and physiological behavior. A decade ago, it was shown that the SUMOylation pathway plays a very important role in the regulation of transcription factor activity. The SUMOylation process involves the covalent binding of SUMO protein to the target protein. However, experimental procedures to demonstrate that low-expressed proteins, such as PPARs, are SUMOylated, remain tricky, and require specific optimization for each protein.

Here, we provide a simple and useful experimental method to investigate the SUMOylation of PPARs in a cellular context. The procedure for studying SUMOylation in living cells is based on the purification under denaturating conditions of total SUMOylated proteins followed by the specific detection of the PPAR proteins. For that purpose, cells are transfected with both 6xHistidine-tagged SUMO and PPAR expression vectors. Since the polyHistidine tag binds to nickel cationic ion-linked agarose matrix (Ni-NTA matrix), His-tagged SUMO proteins covalently linked to the protein substrate can be specifically precipitated and separated from the unSUMOylated proteins. The SUMO-modified PPAR proteins can subsequently be visualized by western blotting using anti-PPAR antibodies. Many questions relative to the regulation of PPAR SUMOylation can be appropriately addressed by adapting this protocol.

Key words: SUMOylation assays, SUMO, nuclear receptor, Peroxisome proliferator-activated receptor, Nickel affinity purification, Histidine tag

1. Introduction

The small ubiquitin-like modifier (SUMO) proteins, SUMO-1, SUMO-2, and SUMO-3 belong to the ubiquitin-like protein (UBL) family (1) and participate in a post-translational modification process called SUMOylation. This reversible pathway involves the formation of an isopeptide bond, which occurs between the C-terminal carboxyl group of the SUMO protein and the ε-amino group of a lysine side chain (K) in the target protein (2) (see Fig. 1).

Mostafa Z. Badr and Jihan Youssef (eds.), *Peroxisome Proliferator-Activated Receptors (PPARs): Methods and Protocols*, Methods in Molecular Biology, vol. 952, DOI 10.1007/978-1-62703-155-4_10, © Springer Science+Business Media New York 2013

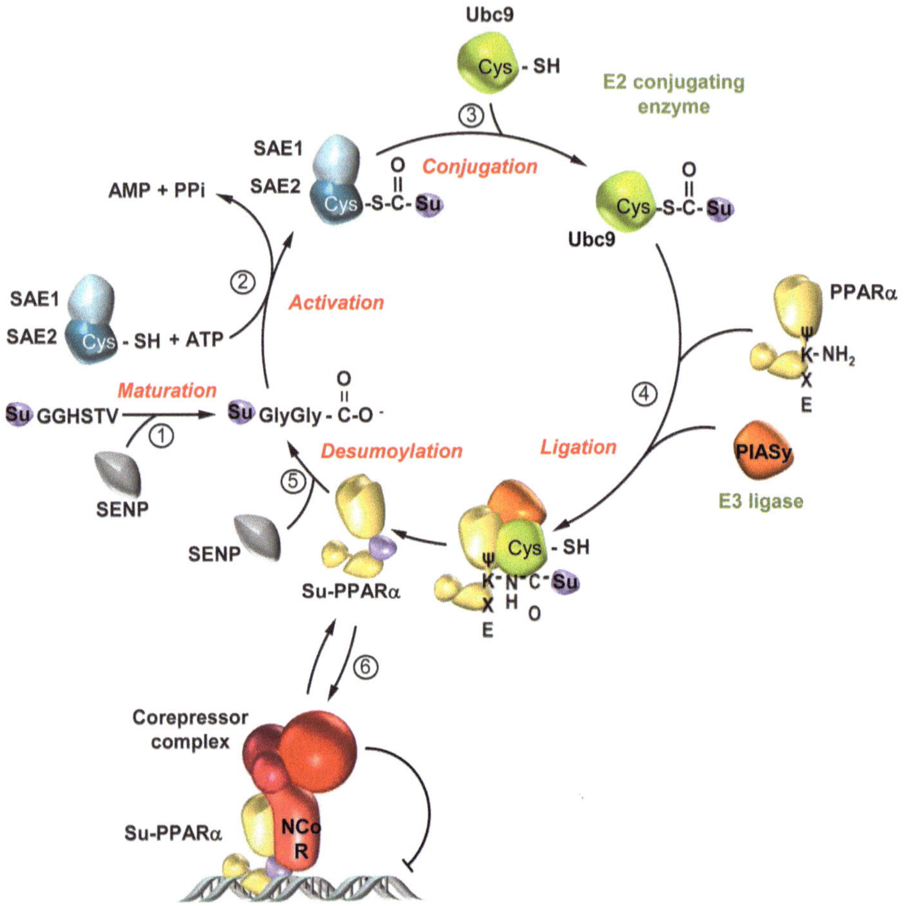

Fig. 1. The SUMOylation pathway. This post-translation modification process can be divided into different steps: the maturation (1), the activation (2), the conjugation (3), the ligation (4), and the deSUMOylation (5). The SUMOylation of a target such as the nuclear receptor PPARα can lead to the recruitment of factor such as the nuclear corepressor NCoR to PPARα target gene (6). This event triggers the selective inhibition of a PPARα target genes subset. *AMP* adenosine monophosphate; *ATP* adenosine triphosphate; *Cys* cysteine; *PIASy* protein inhibitor of activated STAT y; *PPARα* peroxisome proliferator-activated receptor alpha; *SENP* sentrin-specific protease; *Ubc9* Ubiquitin conjugating enzyme 9; *SAE* SUMO activating enzyme; *Su* SUMO; *Su-PPARα* SUMO-modified PPARα.

This lysine is a common part of a consensus motif ΨKXE/D (Ψ is an hydrophobic amino acid, K the modified lysine, X any amino acid, and E or D are acidic amino acids) (see Fig. 1 (1–5)) (2).

The SUMO pathway is an enzymatic cascade beginning with the maturation of the SUMO protein and ending with the deSUMOylation of the substrate. This process requires the sequential activity of several enzymes (see Fig. 1). The first step called maturation step, consists in the digestion of the C-terminal extremity of the inactive SUMO precursor by a C-terminal hydrolase (see Fig. 1 (1–1)) (3). This reaction thus releases the carboxyl group of the glycine residue involved in the isopeptide bond (1 (1–1)). The following activation step leads to the ATP-dependent transfer of the SUMO protein to the E1 activating complex formed by the SUMO

Activating Enzyme 1 and 2 (SAE1 and 2) (1 (1–2)) (3). SUMO is then conjugated to the E2 conjugating enzyme Ubc9 through a thioester bond between the C-terminal glycine of SUMO and the side chain of the Cys 93 residue in Ubc9 (1 (1–3)) (4). Although SUMO-linked Ubc9 can interact directly with the target and transfer the SUMO protein to the acceptor lysine residue, E3 protein ligases, such as protein inhibitor of activated STAT (PIAS), are often required to facilitate this process (1 (1–4)) (2). Finally, sentrin-specific (SENP) or ubiquitin-like protein-specific (Ulps) proteases perform deconjugation and the free SUMO can be recycled in another conjugation round (5) (1 (1–5)). Usually, the SUMO protein bound to the target protein serves as an anchor for the recruitment of additional modulators containing a SUMO interaction motif (SIM) (2).

The SUMOylation pathway plays a critical role in numerous biological processes, such as DNA repair, regulation of nuclear translocation, signal transduction, chromosomal segregation, and gene expression (6). Indeed, the SUMO pathway regulates the activity of several factors taking part in the transcriptional process (7). For instance, the nuclear receptors peroxisome proliferator-activated receptors (PPARs) are key transcriptional regulators of genes implicated in the control of energy expenditure, lipid and glucose homeostasis and in the anti-inflammatory response (8). Current studies aim at understanding the mechanisms involved in the regulation of their activity. It is well established that PPAR activity is modulated by molecular events such as ligand binding (9, 10) or post-translational modifications such as phosphorylation and ubiquitination (11). Additionally, it has been shown that the SUMOylation of PPARγ affects its transcriptional activities (12–14). Recently, we have demonstrated that the human PPARα is also SUMOylated, which selectively favors the recruitment of the nuclear corepressor NCoR to PPARα. In the same study, the data using NCoR siRNA and PPARα siRNA show that this selective recruitment results in the inhibition of the expression of a subset of PPARα target genes (15) (see Fig. 1 (1–6)).

If a protein is suspected to be SUMOylated, a first step is implemented to verify the covalent binding of SUMO to the putative target, another step is to identify the lysine acceptor residue(s) and finally to determine the functional relevance of such modification. A simple methodology can be followed to address the first point (see Fig. 2). Previous high throughput studies have already identified numerous proteins as targets for SUMOylation by using proteomic mass spectroscopy methodology (16–20). A literature search could provide information whether the protein of interest is among the hundreds of SUMO substrates described so far. In addition, bioinformatics analyses by using the SUMOsp (http://sumosp.biocuckoo.org/online.php) and SUMOplot (http://www.abgent.com.cn/doc/sumoplot/login.asp) softwares can be used to predict

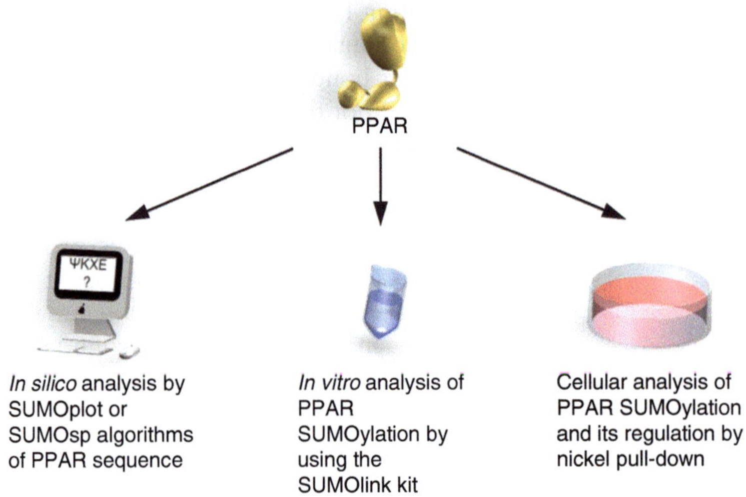

Fig. 2. A methodology to study PPARs SUMOylation.

the existence of a SUMO consensus site in the protein sequence. These preliminary analyses may suggest the existence of SUMO modification on a protein substrate. However, since a SUMO consensus motif is not necessarily a target for the SUMOylation pathway and since SUMOylation can also occur on lysine residues outside the SUMO consensus motif (16), use of such prediction softwares can only be suggestive. To determine SUMOylation of a protein target, numerous techniques have been optimized so far. Although none of them alone are sufficient to affirm that a protein is SUMOylated, together they can provide sufficient evidences to reasonably conclude that a protein is a substrate of the SUMO pathway. The most accurate, trustful (and also expensive) method is probably the use of mass spectroscopy to identify SUMO embranchment in purified endogenous proteins. Another valuable approach is immunoprecipitation of an endogenous protein followed by a western blotting for endogenous SUMO(s) proteins or immunoprecipitation of SUMOylated proteins followed by a western blotting for the endogenous protein of interest. Unfortunately, for most substrates, and especially for low-expressed proteins such as nuclear receptors, the level of SUMOylated proteins is very low, making their detection extremely difficult in the absence of overexpression of the substrate, SUMO(s) or both. Regarding this, many experimental procedures have been developed to examine protein SUMOylation. Since the binding of Ubc9 enzyme to the substrate is a prerequisite to its SUMOylation (2), the identification of a complex Ubc9-target protein by co-immunoprecipitation or by GST pull-down should be a first indication for the substrate SUMOylation (15, 21–23). Moreover, the availability of purified SUMO machinery factors allowed the identification of SUMO substrates by *in vitro* SUMOylation assays (24). Although this technique is very informative and quite easy to realize with the commercially available *in vitro* SUMOylation assay kits, its biological relevance is very restricted. For that reason, it must be complemented by an *in vivo* or a cellular approach. For that reason, we have developed a cellular PPARα SUMOylation assay.

Here, we provide a detailed protocol, which has been used to investigate the SUMOylation of human PPARα in cells (15). The overall method is based on the purification of SUMOylated protein from cell lysates and, subsequently, the specific detection of the SUMOylated PPARα (see Fig. 3). For that purpose, the human PPARα and the SUMO-1 protein tagged with six histidines in tandem are both over-expressed in the human hepatoma cell line HuH-7. In this cellular context, SUMO-target proteins, including PPARα, are covalently linked to His-tagged SUMO-1 by the endogenous SUMO machinery. Different cell such as NIH 3T3

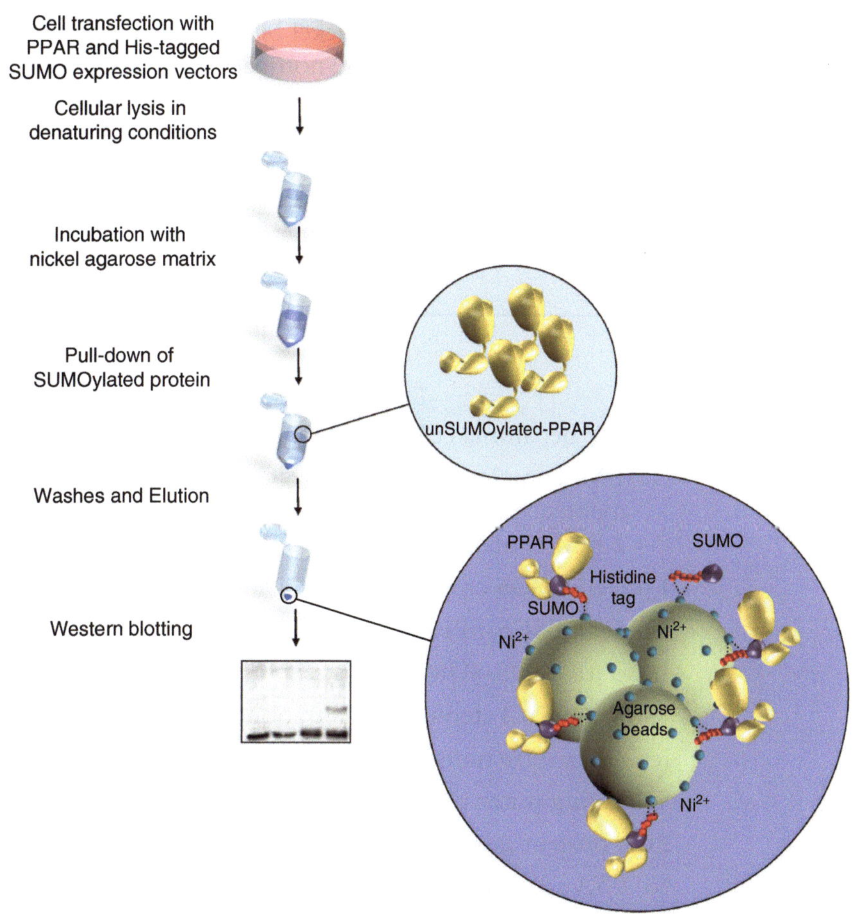

Fig. 3. An overview of the nickel affinity purification. Nickel ion (Ni^{2+}) (in *blue*) is linked to a nitrilotriacetic acid agarose beads (NTA) (in *green*) and forms a chelation complex with the imidazole ring of two histidines. SUMO protein is represented as a purple ball, histidine tag as a chain of six *red* spherical items. Hydrogen bond between Ni^{2+} and His are shown as *interrupted line*.

and HEK293T or Hela cells types have successfully been used to demonstrate PPARs SUMOylation for murine PPARα (25) for PPARγ (13, 14, 26). To demonstrate protein SUMOylation, the cell lysates are incubated with a nickel cationic (Ni^{2+}) ion-linked agarose matrix (nickel-nitrilotriacetic acid agarose Ni-NTA). Ni^{2+} ion establishes high affinity hydrogen bonds with two histidine imidazole rings (see Fig. 3). In this method, unlike the others, cells are lysed in denaturing conditions (6 M guanidine-HCl or 8 M urea) throughout a procedure, which does not disrupt the isopeptide bond but reduces deconjugation by SUMO proteases and decreases non-covalent interactions (27). Furthermore, unfolded proteins facilitate the access of the polyhistidine tag to the Ni^{2+}-agarose matrix enhancing their recovery. Since Ni^{2+} ions are linked to agarose beads, the complex protein-SUMO-His_6-Ni^{2+}-Agarose can be easily precipitated by centrifugation and separated from non-SUMOylated proteins including non-modified PPARα (see Fig. 3). At low concentrations (10–20 mM), imidazole competes with imidazole loops of histidines for nickel binding, and under these conditions it can be used to release nonspecifically bound proteins from the Ni^{2+}-agarose matrix. At higher concentrations of imidazole (200–500 mM) altogether with the protein loading buffer (28), polyHis-tagged proteins are eluted. SUMOylated proteins are then recovered from the pellet and protein content is specifically analyzed by western blotting (see Fig. 3).

2. Materials

HuH-7 hepatoma cell line (29).

DMEM complete medium: 37°C pre-heated DMEM supplemented with 10% fetal bovine serum, 20 μg/mL Gentamycin, 2 mM Glutamine.

1 μg/μL pSG5 empty vector.

1 μg/μL pSG5-hPPARα expression vector (30).

1 μg/μL pSG5-SUMO1-His_6 expression vector (see Note 1).

150 mM NaCl sterilized by filtration.

JetPEI *in vitro* transfection reagent.

Sonicating water bath *or* 23G needle and associated syringe.

BCA protein assay kit.

Ni-NTA agarose.

RIPA buffer: 10 mM Tris–HCl, pH 8.0; 150 mM NaCl; 0.5% Na Deoxycholate; 1 mM EDTA; 1% Triton X-100; 0.1% SDS; protease inhibitor cocktail complete EDTA-free (see Note 2).

Lysis buffer: 6 M Guanidine-HCl; 10 mM Tris–HCl, pH 8.0; 100 mM NaH_2PO_4; 10 mM imidazole; protease inhibitor cocktail complete EDTA-free (see Note 2). The pH must be adjusted to 8.0 with NaOH.

Wash buffer 1: 6 M Guanidine-HCl; 1 mM Tris–HCl, pH 8.0; 100 mM NaH_2PO_4; 10 mM imidazole; 300 mM NaCl; 0.5% Triton X-100; protease inhibitor cocktail Complete EDTA-Free (see Note 2). The pH needs to be adjusted to 8.0 with NaOH.

Wash buffer 2: 8 M Urea; 10 mM Tris–HCl, pH 8.0; 100 mM NaH_2PO_4; 20 mM imidazole; 300 mM NaCl; 0.5% Triton X-100; protease inhibitor cocktail complete EDTA-free (see Note 2). The pH must be adjusted to 6.4 with HCl.

Wash buffer 3 (PBS): 137 mM NaCl; 2.7 mM KCl; 10 mM Na_2HPO_4; 1.8 mM KH_2PO_4. This buffer must be cooled down to ice temperature.

500 mM Imidazole.

Laemmli buffer 3×: 187.5 mM Tris–HCl, pH 6.8; 6% SDS; 30% Glycerol; 6% β-mercaptoethanol; 0.3% w/v Bromophenol blue.

Polyacrylamide gels and electrophoresis equipment: as per user preference.

Here, we use the Protean II XL system (Biorad) with a 15-well comb and a 1.5 mm spacer.

40% (w/v) solution of acrylamide/Bis-acrylamide mixture (37.5:1 (w/w)).

Resolving buffer: 1.5 M Tris–HCl, pH 8.8.

Stacking Buffer: 1 M Tris–HCl, pH 6.8.

10% w/v SDS (sodium dodecyl sulfate) diluted in deionised water.

10% w/v APS (ammonium persulfate) diluted in deionised water.

TEMED (tetramethylethylenediamine).

Western-blotting (WB) materials and transfer equipment: as per user preference.

We have used the Trans-blot Cell (Biorad) as blotting system.

WB blocking solution: 5% w/v skimmed milk, PBS, 0.1% Tween 20.

Primary antibody anti-human PPARα (Santa Cruz, sc-9000×) diluted 1/10,000 in 3% w/v skimmed milk, PBS, 0.1% Tween 20.

Primary antibody anti-6xHis (Clontech, 631212) diluted 1/1,000 in 3% w/v skimmed milk, PBS, 0.1% Tween 20.

Secondary antibody coupled to horseradish peroxidase (HRP) diluted 1/10,000 in 10% w/v skimmed milk, PBS, 0.1% Tween 20.

WB washing solution: PBS, 0.1% Tween 20.

Immobilon western chemiluminescent HRP substrate.

Amersham hyperfilm ECL.

3. Methods

The protocol is divided into three parts: transfection, nickel purification, and western blotting analysis. These can be performed in one flow; however, it can also be stopped at several points, as highlighted in Fig. 4. From cultured cells, this method will allow to generate 90 μL of nickel affinity purified SUMO1-branched PPARα and 100 μL of crude lysate as input control before the

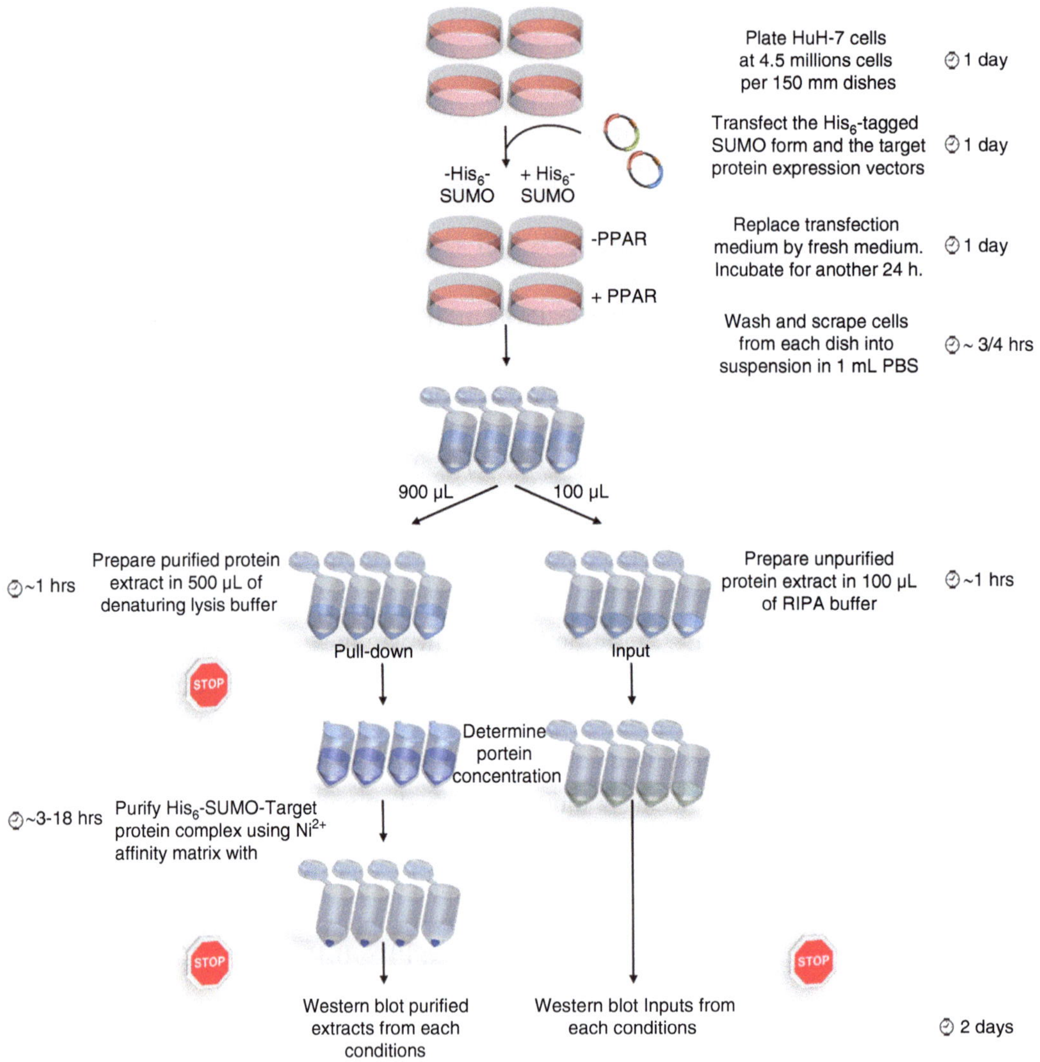

Fig. 4. A timeline of a cellular analysis of protein SUMOylation. The overall procedure of the nickel affinity purification from the transfection to the western-blotting analysis takes around 5 days. However, it can be stopped after different steps where the samples should be stored at −20°C (*Stop panel*). Purified His_6-tagged proteins and the input proteins are finally analyzed by western blotting.

Table 1
Composition of transfection mixes for each cell culture dish

	Ø	SUMO-1	Ø	SUMO-1
Conditions	**Ø (μL)**	**Ø (μL)**	**PPARα (μL)**	**PPARα (μL)**
DNA mix				
pSG5	32.6	16.3	16.3	
pSG5-hPPARα			16.3	16.3
pSG5-SUMO1-His_6		16.3		16.3
NaCl 150 mM	1597.4	1597.4	1597.4	1597.4
Transfectant mix				
JetPEI	65.2	65.2	65.2	65.2
NaCl 150 mM	1564.8	1564.8	1564.8	1564.8
Add the Transfectant mix to the DNA mix. Incubate 20 min at room temperature. Add this transfection mix drop by drop to the cells in 13 mL of DMEM complete medium as following				
DMEM complete medium	13 mL	13 mL	13 mL	13 mL

purification. Both extracts can be analyzed simultaneously or separately.

3.1. Transfection

1. HuH-7 human hepatoma cells are plated at 4.5 millions of cells per 150 mm Ø cell culture dish in 20 mL of DMEM complete medium and left to settle overnight at 37°C, 5% CO_2 atmosphere.
2. Transfection mix is prepared into a 15 mL conic tube according to the conditions shown in Table 1 and added to cells. However, some intermediary mixes are prepared to avoid dif ferences in plasmid content between each condition as shown in Table 2.
3. Cells are incubated with the transfection mix for 24 h at 37°C, 5% CO_2 atmosphere (see Note 3).
4. Medium is then replaced with fresh DMEM complete medium, and cells are incubated for another 24 h at 37°C, 5% CO_2 atmosphere (see Note 4).

3.2. Nickel Pull-Down

1. Wash cells twice with ice-cold PBS. Aspirate the remaining buffer.
2. Add 1 mL of ice-cold PBS to the dish, harvest cells by scraping them from each dish and transfer the suspension to a 1.5 mL microcentrifuge tube (see Note 5).
3. Save 100 μL from these cell suspensions as input.

Table 2
Composition of transfection premixes

Mix	DNA intermediary mix Ø (μL)	DNA intermediary mix SUMO-1 (μL)	Transfectant mix (μL)
pSG5	40.75		
pSG5-SUMO1-His_6		40.75	
NaCl 150 mM	3,993.5	3,993.5	7,041.6
JetPEI			293.4
Vortex and add DNA intermediary mix in 2 new tubes for each as following			
Volume per tube	2 tubes with 1,613.7 μL	2 tubes with 1,613.7 μL	
Add 16.3 μL per tube of pSG5 or pSG5-hPPARa following conditions shown in Table 1			
Add 1,630 μL of Transfectant mix per tube, vortex and spin down			
Incubate for 20 min at room temperature			
Add this DNA+JetPEI mix to the cells in 13 mL of DMEM complete medium			

Centrifuge at 800 × *g* for 5 min, discard the supernatant.

Add 100 μL of RIPA buffer and incubate 30 min at 4°C while agitating.

Clear the lysate by centrifuging at 20,000 × *g*, 4°C, for 15 min.

Store the supernatant at −80°C for further analysis.

4. Centrifuge the remaining cells from step 2 (900 μL) at 800 × *g*, 4°C, for 5 min.

   Discard the supernatant and add 500 μL of ice-cold lysis buffer to the cell pellet.

5. Shear DNA (see Note 6):

   Either by passing the lysate 10 times through a syringe-associated 23G needle.

   Or by sonicating with a Bioruptor using cycles of 2 min (30 s on, 30 s off) at high level.

6. Incubate for 10 min at 4°C while agitating. Centrifuge 15 min, at 4°C, 20,000 × *g*. Transfer the supernatant to a new 1.5 mL microcentrifuge tube. Samples can be stored at −80 or −20°C after this step.

7. Protein dosage with the Biorad protein assay kit in a 96-well plate (optional). For that purpose, use 10 μL of protein lysate per reaction and add 200 μL of Biorad protein assay solution 1×. Incubate at room temperature for 5 min. Determine the optical density of your sample at a 595 nm wavelength

($OD_{\lambda=595nm}$). Transfer an equal amount of protein per condition in a new 1.5 mL microcentrifuge tube (see Note 7). Complete protein lysate volume with lysis buffer to 500 μL.

8. Wash Ni-NTA matrix twice with lysis buffer (see Note 8). Centrifuge the total required volume of matrix for 2 min, at 4°C, 5,000 × *g*. Aspirate the storage buffer (see Note 8), add 1 mL of lysis buffer and centrifuge with the same conditions. Do it twice and then, resuspend the matrix in the same volume of lysis buffer in order to prepare the slurry 50%.
9. Add 70 μL of slurry 50% Ni-NTA matrix to the cell lysate. Incubate at 4°C, for 2–18 h while rotating.
10. Wash the beads three times with the Wash buffer 1 (see Note 9), three times with the Wash buffer 2 (see Note 9) and two times with the Wash buffer 3 as described below:

    Centrifuge for 2 min, at 4°C, 5,000 × *g*.

    Aspirate the supernatant (see Note 8).

    Add 1 mL of Wash buffer.

    Mix by vortexing.

    Centrifuge for 2 min, at 4°C, 5,000 × *g*.

    Remove the supernatant.
11. Elution is performed as described below:

    Add 60 μL of imidazole 500 mM (see Note 10).

    Vortex for 30 sec.

    Incubate 5 min, at room temperature.

    Centrifuge for 1 min, at 4°C, 20,000 × g.

    Add 30 μL of Laemmli 3× to the matrix.

    Vortex for 30 sec.

    Centrifuge for 2 min, at 4°C, 20,000 × g.

    (After this step, samples can be stored at −80°C)

    Boil samples for 5 min at 95°C.

    Spin down droplets for 30 sec with a bench centrifuge.

3.3. Western Blotting (see Note 11)

1. Use your favorite SDS-polyacrylamide gel electrophoresis system (see Note 12).
2. Load 45 μL of purified protein sample or 20–50 μg of input protein sample per well.
3. Use your favorite blotting system to transfer protein onto the membrane (see Note 13).
4. Block the membrane with the WB blocking solution for 30 min to 1 h at room temperature while agitating.

5. Then, incubate on a rocking platform with the primary antibody solution for 18 h, at 4°C.
6. Wash 6 times 5 min with the WB washing solution at room temperature on a rocking platform.
7. Incubate the membrane with the secondary antibody solution for 30 min to 1 h at room temperature on a rocking platform.
8. Wash 6 times 5 min with the WB washing solution at room temperature on a rocking platform.
9. Incubate the membrane with immobilon western substrate for 5 min at room temperature.
10. Develop the film according to the manufacturer manual.
11. Adjust the exposure time to the expected signal intensity (see Notes 14–17).

4. Notes

1. In some rare cases, such as the nuclear receptor Retinoid X Receptor alpha (RXRα) (21), the SUMOylation of the modified protein has been assessed by immunoprecipitating the endogenous SUMO-modified protein. However, even in this study (21), the use of a Histidine tagged-SUMO protein and a FLAG tagged-Retinoid X Receptor alpha (RXRα) has been used to confirm the SUMOylation of the nuclear receptor. Tags must be wisely chosen since they can lead to huge differences in results. For instance, the murine myc-PPARα cannot be SUMOylated by the CFP-SUMO1 in the study of Yamashita et al. (26), whereas mPPARα is modified by the myc-SUMO1, the HA-SUMO2, and the myc-SUMO3 in Leuenberger et al. study (25). In most successful studies, small tags such as 6xHistidine, myc, FLAG or HA (12–15, 21, 25, 31, 32) have been used and should be chosen instead of bigger tags such as CFP or GFP.
2. Some protease inhibitor cocktails on the market can have some EDTA in their composition. In our protocol, this must be avoided since EDTA washes the nickel out from the beads. Some protease inhibitor cocktails such as the Complete EDTA-free tablets from Roche can be used in replacement.
3. In some cases, a Ubc9 expression vector might be used to enhance the SUMOylation pathway (32, 33). Furthermore, the presence of the E3 ligase might be not required in *in vitro* SUMOylation test, but is essential in cellular assays (34). In most cases, cells already express E3 ligases regulating the SUMOylation of the protein of interest. However, it can be a

critical issue. For instance, we did not need to cotransfect an E3 ligase PIASy expression vector to detect hPPARα SUMOylation in the HuH-7 cell line, while it was critical in the Cos-7 cell line (15).

4. After 24 h of incubation with the transfection mix, further 24 h of incubation leads to a better expression level. These supplementary 24 h can also be a nice opportunity to assess the effect of different treatments on the regulation of PPAR SUMOylation. For that purpose, DMEM with lower concentrations of fetal calf serum such as 0.2% is recommended since natural PPAR ligands can be present. In addition, to improve treatment efficiency, 0.2% BSA in the medium can be used as efficient carrier for lipophilic PPAR ligands such as PPARα ligand GW7647.
5. Cells must be well resuspended while collecting the sample. To facilitate the cell pipeting, trim the extremity of a 1,000 μL pipette tip.
6. Sonication is used to shear DNA and to reduce viscosity, albeit any method can be used and depend on availability of equipment. Sonication with a tip or an ultrasonicating bath is however faster and leads to less sample waste. If a probe is used, clean it between each sample to avoid cross-contamination. Needles and syringes are cheaper but it takes longer time to shear the DNA especially with several tubes.
7. The protein amount recovery may vary between 100 and 600 μg. If a too low amount is used, it might be possible that the modified protein will not be detected. At the contrary, too much protein at the beginning can generate a lot of background. An amount of 150–300 μg of starting concentration should be optimal.
8. Washes are critical steps during which a portion of the beads may be lost. To avoid it, pinch the tip with a forceps and use it to remove the washing buffer. Furthermore, it might be helpful to leave 50 μL of buffer and avoid being in contact with the matrix during the first washing steps. The remaining washing buffer can be taken out at the very last wash before adding the elution buffer. Suction power should also be adjusted wisely to avoid the same problem.
9. We have seen a lot of background while we were performing our first SUMOylation assays. To avoid it, we adjusted the washing condition by adding NaCl 300 mM and Triton X-100 0.5% compared to the recommended conditions provided with the Ni-NTA beads.
10. At high concentration, imidazole competes with histidine tag to bind to the nickel ion. Therefore, it can also be successfully used to elute the His-tagged protein from the beads. Here, we

propose a protocol using both Laemmli buffer and imidazole 500 mM, which might enhance protein recovery. Alternatively, the 60 μL of imidazole + the 30 μL of Laemmli 3× can be replaced by 90 μL of Laemmli buffer 1×.

11. Western blotting can be performed following different protocols. However, incubation times and solutions (blocking, antibody, washes) should be modified regarding the use of a different antibody.
12. During the run in such device, gel might get warm preventing a proper migration. In that context, a cooling system might be used. However, this kind of material is quite expensive. Using an ice-cold buffer and performing the migration into a cold room should be considered as a cheaper alternative.
13. Roll a piece of pipettes onto the newly added layer each time to get rid of bubbles.
14. An example of a positive result is shown in Fig. 5. In this experiment, HuH-7 cells were transfected altogether with pSG5 negative control or with the pSG5-hPPARα expression vector and with pSG5 or pSG5-His_6-SUMO1 expression vector as indicated. Western blot from the Ni-NTA purified 6xHis tagged protein (see Fig. 5a) shows the enrichment of His_6-SUMO1-PPARα and confirms that PPARα is modified by SUMO-1 in HuH-7 cells. Western blots from crude lysate display the protein amount of His_6-SUMO-1 (see Fig. 5b), hPPARα (see Fig. 5c), and actin (see Fig. 5c) before the purification and are used as transfection loading control.
15. Usually, a band corresponding to the non-SUMOylated nuclear receptor might appear as shown for PPARα (15) or the estrogens receptor alpha (ERα) (32) even if they are not polyHistidine-tagged (see Fig. 5). As suggested by Sentis et al., nuclear receptors might be retained to the Nickel affinity column by their zinc finger domain, which is part of the DNA binding domain.
16. SUMO-2/3, but not SUMO-1, contain a ΨKxD/E consensus site and then are able to form some polySUMO chains (24). Therefore, the migration profile might be different from one shown in Fig. 5a and several bands corresponding to the protein modified with SUMO-2/3 should be observed.
17. From this technique, different questions relative to nuclear receptor SUMOylation might be addressed. One of the major issues is to localize the modified lysine. For that purpose, the systematic replacement of putative lysine acceptor site by an unSUMOylable arginine residue is commonly used. Then, the nuclear receptor wild type (WT) or the mutant protein must be exogenously expressed with the His_6-tagged SUMO constructs, purified with Ni-NTA beads and then analyzed by

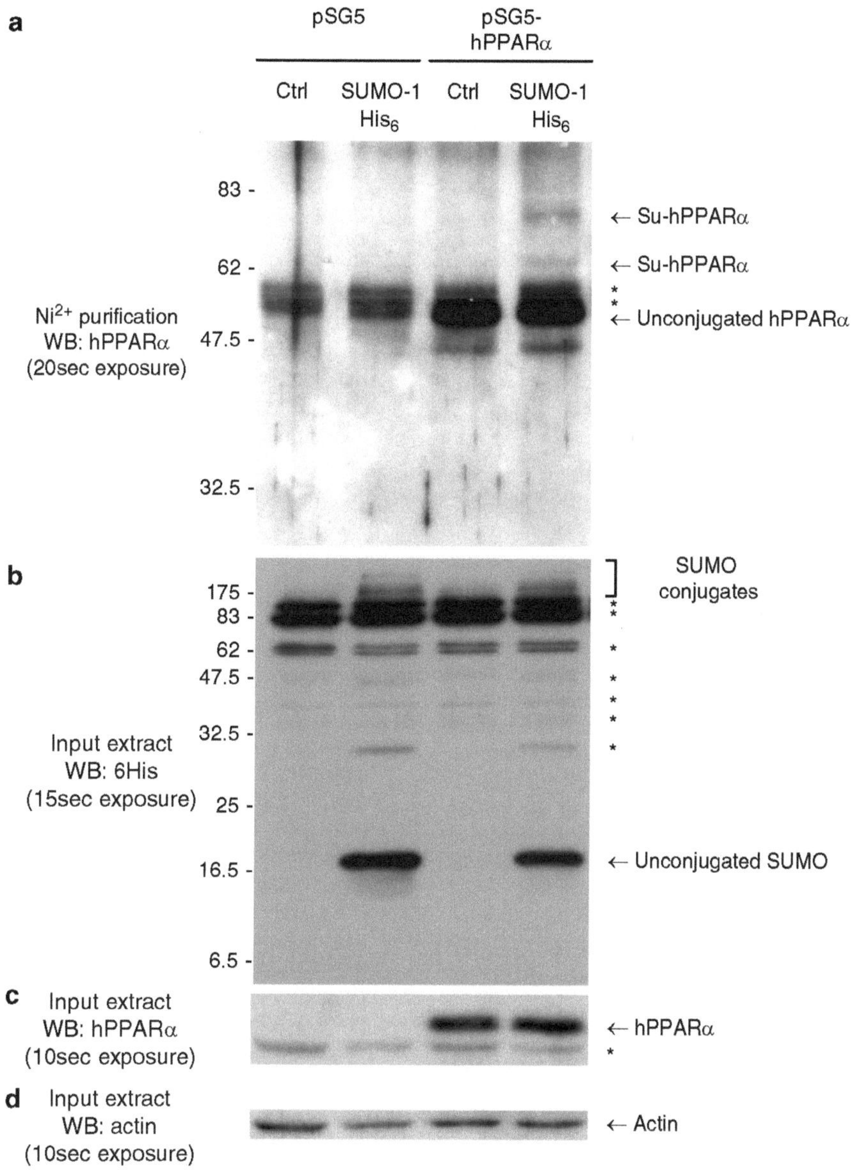

Fig. 5. A positive result example. HuH-7 cells were transfected as described with pSG5 or pSG5-hPPARα and with pSG5 or pSG5-His$_6$-SUMO-1. PPARα proteins were analyzed by western blotting after purification on Ni-NTA matrix (**a**) and in the crude lysates (**c**). As control, total His$_6$-tagged proteins (**b**) and actin proteins (**d**) were analyzed in the crude lysate as well. Molecular weights are indicated in kDa and asterisks designate nonspecific bands. Su-hPPARα: SUMO-modified PPARα.

western blotting. A band corresponding to the SUMOylated WT protein should appear as positive control while no band should be present at the same molecular weight with the mutant protein. However, a weaker signal can still persist even if the SUMOylation site is mutated (12, 13, 15, 35, 36). It has been suggested that such residual SUMOylation occurs on other alternative sites and might be influenced by the mutation itself (27).

References

1. Hochstrasser M (2009) Origin and function of ubiquitin-like proteins. Nature 458:422–429
2. Gareau JR, Lima CD (2010) The SUMO pathway: emerging mechanisms that shape specificity, conjugation and recognition. Nat Rev Mol Cell Biol 11:861–871
3. Geiss-Friedlander R, Melchior F (2007) Concepts in sumoylation: a decade on. Nat Rev Mol Cell Biol 8:947–956
4. Hay RT (2005) SUMO: a history of modification. Mol Cell 18:1–12
5. Kim JH, Baek SH (2009) Emerging roles of desumoylating enzymes. Biochim Biophys Acta 1792:155–162
6. Zhao J (2007) Sumoylation regulates diverse biological processes. Cell Mol Life Sci 64:3017–3033
7. Treuter E, Venteclef N (2010) Transcriptional control of metabolic and inflammatory pathways by nuclear receptor SUMOylation. Biochim Biophys Acta 1812(8):909–918
8. Pourcet B, Fuchart J, Staels B et al (2006) Selective PPAR modulators, dual and pan PPAR agonists: multimodal drugs for the treatment of type 2 diabetes and atherosclerosis. Expert Opin Emerg Drugs 11:379–401
9. Nolte RT, Wisely G, Westin S et al (1998) Ligand binding and co-activator assembly of the peroxisome proliferator-activated receptor-gamma. Nature 395:137–143
10. Xu HE, Lambert M, Montana V et al (1999) Molecular recognition of fatty acids by peroxisome proliferator-activated receptors. Mol Cell 3:397–403
11. Blanquart C, Barbier O, Fruchart J et al (2003) Peroxisome proliferator-activated receptors: regulation of transcriptional activities and roles in inflammation. J Steroid Biochem Mol Biol 85:267–273
12. Ghisletti S, Huang W, Ogawa S et al (2007) Parallel SUMOylation-dependent pathways mediate gene- and signal-specific transrepression by LXRs and PPARgamma. Mol Cell 25:57–70
13. Ohshima T, Koga H, Shimotohno K (2004) Transcriptional activity of peroxisome proliferator-activated receptor gamma is modulated by SUMO-1 modification. J Biol Chem 279:29551–29557
14. Pascual G, Fong A, Ogawa S et al (2005) A SUMOylation-dependent pathway mediates transrepression of inflammatory response genes by PPAR-gamma. Nature 437:759–763
15. Pourcet B, Pineda-Torra I, Derudas B et al (2010) SUMOylation of human peroxisome proliferator-activated receptor alpha inhibits its trans-activity through the recruitment of the nuclear corepressor NCoR. J Biol Chem 285:5983–5992
16. Blomster HA, Imanishi S, Siimes J et al (2010) *In vivo* identification of sumoylation sites by a signature tag and cysteine-targeted affinity purification. J Biol Chem 285:19324–19329
17. Bruderer R, Tatham M, Plechanovova A et al (2011) Purification and identification of endogenous polySUMO conjugates. EMBO Rep 12:142–148
18. Ganesan AK, Kho Y, Kim S et al (2007) Broad spectrum identification of SUMO substrates in melanoma cells. Proteomics 7:2216–2221
19. Golebiowski F, Matic I, Tatham M et al (2009) System-wide changes to SUMO modifications in response to heat shock. Sci Signal 2:ra24
20. Miller MJ, Barrett-Witt G, Hua Z et al (2010) Proteomic analyses identify a diverse array of nuclear processes affected by small ubiquitin-like modifier conjugation in Arabidopsis. Proc Natl Acad Sci USA 107:16512–16517
21. Choi SJ, Chung S, Rho E et al (2006) Negative modulation of RXRalpha transcriptional activity by small ubiquitin-related modifier (SUMO) modification and its reversal by SUMO-specific protease SUSP1. J Biol Chem 281: 30669–30677
22. Kaul S, Blackford J Jr, Cho S et al (2002) Ubc9 is a novel modulator of the induction properties of glucocorticoid receptors. J Biol Chem 277:12541–12549
23. Poukka H, Aarnisalo P, Karvonen U et al (1999) Ubc9 interacts with the androgen receptor and activates receptor-dependent transcription. J Biol Chem 274:19441–19446
24. Tatham MH, Jaffray E, Vaughan O et al (2001) Polymeric chains of SUMO-2 and SUMO-3 are conjugated to protein substrates by SAE1/SAE2 and Ubc9. J Biol Chem 276:35368–35374
25. Leuenberger N, Pradervand S, Wahli W (2009) Sumoylated PPARalpha mediates sex-specific gene repression and protects the liver from estrogen-induced toxicity in mice. J Clin Invest 119:3138–3148
26. Yamashita D, Yamaguchi T, Shimizu M et al (2004) The transactivating function of peroxisome proliferator-activated receptor gamma is negatively regulated by SUMO conjugation in the amino-terminal domain. Genes Cells 9:1017–1029
27. Tatham MH, Rodriguez M, Xirodimas D et al (2009) Detection of protein SUMOylation *in vivo*. Nat Protoc 4:1363–1371

28. Laemmli UK (1970) Cleavage of structural proteins during the assembly of the head of bacteriophage T4. Nature 227:680–685
29. Nakabayashi H, Taketa K, Miyano K et al (1982) Growth of human hepatoma cells lines with differentiated functions in chemically defined medium. Cancer Res 42:3858–3863
30. Staels B, Koenig W, Habib A et al (1998) Activation of human aortic smooth-muscle cells is inhibited by PPARalpha but not by PPARgamma activators. Nature 393:790–793
31. Degerny C, Monte D, Beaudoin C et al (2005) SUMO modification of the Ets-related transcription factor ERM inhibits its transcriptional activity. J Biol Chem 280:24330–24338
32. Sentis S, Le Romancer M, Bianchin C et al (2005) Sumoylation of the estrogen receptor alpha hinge region regulates its transcriptional activity. Mol Endocrinol 19:2671–2684
33. Hwang EJ, Lee J, Jeong J et al (2009) SUMOylation of RORalpha potentiates transcriptional activation function. Biochem Biophys Res Commun 378:513–517
34. Liu B, Shuai K (2008) Regulation of the sumoylation system in gene expression. Curr Opin Cell Biol 20:288–293
35. Poukka H, Karvonen U, Janne O et al (2000) Covalent modification of the androgen receptor by small ubiquitin-like modifier 1 (SUMO-1). Proc Natl Acad Sci USA 97:14145–14150
36. Tojo M, Matsuzaki K, Minami T et al (2002) The aryl hydrocarbon receptor nuclear transporter is modulated by the SUMO-1 conjugation system. J Biol Chem 277:46576–46585

Chapter 11

Analyzing Phosphorylation-Dependent Regulation of Subcellular Localization and Transcriptional Activity of Transcriptional Coactivator NT-PGC-1α

Ji Suk Chang and Thomas W. Gettys

Abstract

Peroxisome proliferator-activated receptor gamma coactivator 1 alpha (PGC-1α) is a nuclear transcriptional coactivator that regulates the genes involved in energy metabolism. Recent evidence has been provided that alternative splicing of PPARGC1A gene produces a functional but predominantly cytosolic isoform of PGC-1α (NT-PGC-1α). We have demonstrated that transcriptional coactivation capacity of NT-PGC-1α is directly correlated with its nuclear localization in a PKA phosphorylation-dependent manner. In this chapter, we describe quantitative imaging analysis methods that are developed to measure the relative fluorescence intensity of the protein of interest in the nucleus and cytoplasm in a single cell and the frequency distribution of nuclear/cytoplasmic intensity ratios in the population of cells, respectively. This chapter also describes transient cotransfection and dual-luciferase reporter gene assay that examine the ability of coactivators to activate the transcriptional activity of transcription factors.

Key words: Coactivator, PGC-1α, NT-PGC-1α, Nuclear receptor, PPARγ, Transcription

1. Introduction

Transcriptional activity of nuclear receptors and transcription factors is dynamically regulated in response to physiological stimuli by transcriptional coregulators such as coactivators and repressors. PGC-1α is one such protein that activates multiple nuclear receptors and transcription factors (1, 2) involved in global responses such as mitochondrial biogenesis (3–6) and tissue-specific responses such as adaptive thermogenesis (4), fatty acid oxidation (7, 8), gluconeogenesis (9–11), and muscle fiber type switching (12). Alternative splicing of PPARGC1A gene produces a functional but

Mostafa Z. Badr and Jihan Youssef (eds.), *Peroxisome Proliferator-Activated Receptors (PPARs): Methods and Protocols*, Methods in Molecular Biology, vol. 952, DOI 10.1007/978-1-62703-155-4_11, © Springer Science+Business Media New York 2013

predominantly cytosolic isoform of PGC-1α (NT-PGC-1α) (13). NT-PGC-1α retains the N-terminal transactivation and nuclear receptor interaction domains of PGC-1α but lacks the C-terminal domains containing nuclear localization signals. NT-PGC-1α enters the nucleus by passive diffusion but is actively translocated to the cytosol by CRM1-mediated nuclear export in basal conditions (14). There are several precedents that transcriptional activity of transcription factors/coregulators are negatively regulated by sequestering from the nucleus. Their nuclear translocation is often controlled by multiple extracellular signals (15–19). cAMP triggers nuclear accumulation of NT-PGC-1α that is directly phosphorylated by cAMP-activated PKA at residues S194, S241, and T256 (13, 14).

To better understand direct correlation between PKA phosphorylation-dependent nuclear localization of NT-PGC-1α and its enhanced coactivation capacity, we employed indirect immunofluorescence and luciferase reporter gene assay using HA-tagged wild type NT-PGC-1α, phosphorylation-resistant NT-PGC-1α-S194A/S241A/T256A, and phosphomimetic NT-PGC-1α-S194D/S241D/T256D. Quantitative microscopic image analysis is critical to precisely determine signaling-dependent changes in subcellular localization of the protein of interest. In this chapter, we describe a fluorescence quantification method using Image J software to quantify the relative fluorescence intensity of the protein in the nucleus and cytoplasm in the single cell. Mean value of the relative nuclear/cytoplasmic fluorescence intensity ratios represents average subcellular localization of the protein in single cells. The frequency distribution histogram of nuclear/cytoplasmic intensity ratios provides the overview of signaling-dependent responses in the population of cells.

This chapter also describes transient cotransfection and dual-luciferase reporter assay, where the activities of firefly and *Renilla* luciferases are measured sequentially in a single sample. Firefly luciferase is used as an experimental reporter and *Renilla* luciferase is used as an internal control to measure transfection efficiency. The luciferase reporter gene assay, where the luciferase reporter gene is driven by a promoter of interest, is commonly used for studying the transcriptional activity of transcription factors on their responsive promoters. This assay is also applied to investigate how transcriptional coregulators modulate transcription factors. The Gal4-DBD, fused to the protein of interest or the protein domain of interest, is a useful tool to analyze transactivation property or identify functional domain of transcription factors/coregulators. The Gal4-DBD fusion protein is recruited to the nucleus due to strong nuclear localization signals in Gal4-DBD and binds to a Gal4-responsive element in a luciferase reporter pGK-1 (20, 21).

2. Materials

2.1. CHO-K1 Cell Culture and Transfection

1. F-12K medium: F-12K supplemented with 1.26 g/L glucose and 2 mM glutamine.
2. F-12 + 10% FBS + 1% P/S: F-12K medium supplemented with 10% volume/volume fetal bovine serum and 50 μM penicillin and streptomycin.
3. 1× Phosphate-buffered saline (PBS), pH, 7.4: 155.17 mM NaCl, 2.97 mM Na_2HPO_4, 1.06 mM KH_2PO_4.
4. Trypsin-EDTA: 0.5 g/L of trypsin and 0.2 g/L of EDTA·4Na in Hanks' Balanced Salt Solution without $CaCl_2$, $MgCl_2·6H_2O$, and $MgSO_4·7H_2O$.
5. Opti-MEM I Medium.
6. FuGENE® 6 Transfection Reagent. Store at 4°C.
7. Hemocytometer.
8. Inverted microscope.
9. 37°C water bath.

2.2. Immunofluorescence

1. Poly-L-lysine-coated coverslips (12 mm diameter) and microscope slides (25 × 75 × 1 mm).
2. 1× PBS.
3. Paraformaldehyde: prepare fresh 4% (v/v) paraformaldehyde by diluting 16% paraformalde solution with 1× PBS.
4. Permeabilization solution: 0.5% (v/v) Triton X-100 in 1× PBS.
5. 5× Blocking solution: prepare 5% (v/v) normal goat serum (NGS) and 5% (v/v) bovine serum albumin, fraction V (BSA) in 1× PBS and filter it through a 0.22 μm filter unit, Durapore PVDF membrane (see Note 1).
6. HA-tag antibody.
7. Alexa Fluor 488-conjugated anti-rabbit immunoglobulin secondary antibody.
8. Vectashield mounting medium with DAPI (4, 6-diamino-2-phenylindole). Store at 4°C in the dark.
9. Nail polish.
10. Zeiss LSM510 confocal microscope.
11. Dibutyryl cAMP: prepare 100 mM dibutyryl cAMP in water.

2.3. Dual-Luciferase Reporter Assay

1. Dual-Luciferase® Reporter Assay Kit.
2. BRL49653: prepare 10 mM BRL49653 in DMSO.
3. 9-cis-RA: prepare 10 mM 9-cis-RA in DMSO.

4. Dibutyryl cAMP: prepare 100 mM dibutyryl cAMP in water.
5. White polystyrene 96-well plate.
6. Luminometer.

3. Methods

3.1. CHO-K1 Cell Culture and Transfection

1. CHO-K1 cells are grown in F-12K+10% FBS+1% P/S medium in 10 cm^2 culture dishes at 37°C until 80% confluent.
2. Wash the cells with 1× PBS.
3. Add ~1 mL of pre-warmed trypsin-EDTA to the cells drop by drop and incubate the dish at 37°C incubator for 2–3 min.
4. Stop trypsin digestion by adding 5 mL of F-12K+10% FBS+1% P/S medium.
5. Count the number of cells per mL using hemocytometer and seed 3×10^4 cells per well in 2 mL in 12-well cell culture plate 1 day before transfection. Next day, confluence is ~40% (see Note 2).
6. Pipet 6 μL of FuGENE 6 Reagent directly into 94 μL of Opti-MEM I medium per well without contacting the walls of the plastic tube, mix gently, and incubate for 5 min at room temperature (see Note 3).
7. Add 1 μg of DNA into the tube, mix gently, and incubate for 30 min at room temperature.
8. Add the FuGENE 6 reagent: DNA complex to the cells in a drop-wise manner. Incubate the cells at 37°C overnight.
9. Replace the medium with fresh F-12K+10% FBS+1% P/S medium.

3.2. Immunofluorescence

All the fixation and immunolabeling procedures were performed in a humid chamber at room temperature.

1. Seed 3×10^4 CHO-K1 cells on top of poly-L-lysine-coated coverslip that sits in a well of 12-well culture plate 1 day before transfection.
2. Replace old F-12K+10% FBS+1% P/S medium with F-12K+1% P/S medium 1 h before transfection (see Note 4).
3. Transfect CHO-K1 cells with wild type NT-PGC-1α-HA, NT-PGC-1α-S194A/S241A/T256A-HA, or NT-PGC-1α-S194D/S241D/T256D-HA using FuGENE 6 Transfection Reagent as described above.
4. 24 h after transfection, replace the medium with 1 mL of F-12K+1% P/S medium containing vehicle or 500 μM dibutyryl cAMP (dbcAMP) and incubate for 1 h.

5. Wash the cells with 1× PBS three times.
6. Add 1 mL of 4% paraformaldehyde solution per well and fix the cells for 10 min, followed by three washes for 5 min each with 1× PBS.
7. Permeabilize with 0.5% Triton X-100 Permeabilization solution for 5 min.
8. Incubate the fixed cells with 5× Blocking solution for 30 min.
9. Incubate the fixed cells for 1 h with rabbit anti-HA antibody (1:4,000, Abcam) diluted in 1× Blocking solution and wash three times for 10 min each with 1× PBS.
10. Incubate the fixed cells for 1 h in the dark with Alexa Fluor 488-conjugated anti-rabbit immunoglobulin secondary antibody (1:500) in 1× Blocking solution and wash three times for 10 min each with 1× PBS.
11. Put one drop of DAPI-containing Vectashield mounting medium on a microscope slide and carefully invert the coverslip into a drop of mounting medium. Gently press the coverslip to remove excess mounting medium. Seal the coverslip using nail polish when mounting medium is solidified. The sample can be viewed immediately after the nail polish is dry or be stored in the dark at 4°C.

3.3. Microscopy and Fluorescence Quantification

1. Acquire images using a Zeiss LSM 510 Meta confocal microscope (a Plan-Neofluar 40×/0.85 numerical aperture objective) coupled to a CCD camera. The same focal plane is used to obtain images of NT-PGC-1α (Alexa 488 channel) (Fig. 1a) and DAPI-stained nucleus (DAPI channel) in randomly selected positively transfected cells (150–200 cells).
2. To quantify the fluorescence signals in the nucleus and cytoplasm, the NIH Image J software is used. First, open the microscopic images obtained from the Alexa 488 and DAPI channels using Image J software and synchronize two windows. Manually outline the DAPI-stained nucleus using a freehand selection tool and measure the selected nuclear area and integrated density of nuclear fluorescence signals. Similarly, outline the entire cell and measure the selected area of the cell and integrated density of total fluorescence signals.
3. Measure the fluorescence background from the cells that do not express NT-PGC-1α. Average these background signals and subtract them from the specific fluorescence signals for each NT-PGC-1α-expressing cell.
4. The area of the cytoplasm and cytoplasmic fluorescence signals are obtained by subtracting total nuclear area and fluorescence from total cellular area and fluorescence, respectively.

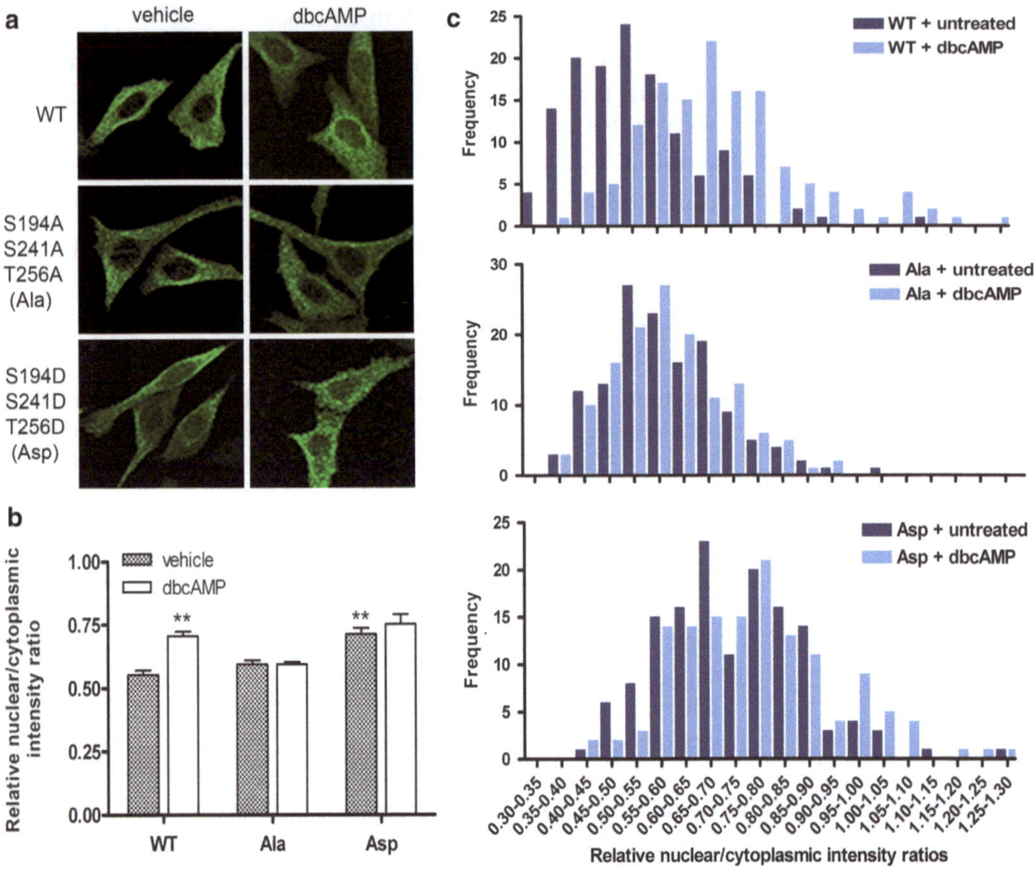

Fig. 1. PKA phosphorylation-dependent regulation of subcellular localization of NT-PGC-1α. (**a**) Localization of NT-PGC-1α-HA (WT), NT-PGC-1α-S194A/S241A/T256A-HA, and NT-PGC-1α-S194D/S241D/T256D-HA was analyzed in CHO-K1 cells that were treated with vehicle or 500 μM dibutyryl cAMP (dbcAMP) for 1 h after serum-starvation. (**b**) Quantification of the relative nuclear to cytoplasmic fluorescence intensity ratios of NT-PGC-1α wild type and mutants. The bar diagram of the relative ratios of nuclear to cytoplasmic fluorescence intensity is shown with the mean ± S.D. Significant deference by Student *t* test; $^{**}p < 0.01$. (**c**) Frequency distribution histogram of the relative nuclear/cytoplasmic intensity ratios in the population of cells. The number of cells was plotted over nuclear/cytoplasmic fluorescence intensity ratios. A significant shift to higher ratios is shown in the population of cells by dibutyryl cAMP treatment and PKA phosphorylation.

5. The relative nuclear/cytoplasmic (Nuc/Cyt) ratio of fluorescence intensity is determined by dividing the nuclear fluorescence intensity per unit area by the cytoplasmic fluorescence intensity per unit area (Fig. 1b). The nuclear/cytoplasmic intensity ratio of 1 represents even distribution of the protein between the nucleus and cytoplasm.
6. To determine overall responses of single cells to extracellular signals, create a frequency distribution histogram of the relative nuclear/cytoplasmic intensity ratios in the population of cells by plotting the number of cells vs. nuclear/cytoplasmic intensity ratios (Fig. 1c).

Table 1
Cotransfection of CHO-K1 cells for transcriptional coactivation assay using a $(PPRE)_3$-TK-luc reporter

Samples 1	1	2	3	4	5	6	7	8
$(PPRE)_3$-TK-luc	0.38	0.38	0.38	0.38	0.38	0.38	0.38	0.38
pRL-SV40	0.02	0.02	0.02	0.02	0.02	0.02	0.02	0.02
pSV sport-PPARγ	0.15	0.15	0.15	0.15	0.15	0.15	0.15	0.15
pSV sport-RXRα	0.15	0.15	0.15	0.15	0.15	0.15	0.15	0.15
pcDNA3.1	0.3	0.3						
pcDNA3.1-NT-PGC-1α			0.3	0.3				
pcDNA3.1-NT-PGC-1α-S194A/S241A/T256A					0.3	0.3		
pcDNA3.1-NT-PGC-1α-S194D/S241D/T256D							0.3	0.3
BRL49653+9-cis-RA	−	+	−	+	−	+	−	+
Total μg	1	1	1	1	1	1	1	1

3.4. Dual-Luciferase Reporter Assay Using a $(PPRE)_3$-TK-luc Reporter

1. Seed 3×10^4 CHO-K1 cells per well in 12-well plate 1 day before transfection.
2. Transfect CHO-K1 cells with multiple plasmids (Table 1) using FuGENE 6 Transfection Reagent and incubate the cells in F-12K+10% FBS+1% P/S media.
3. 24 h after transfection, treat the cells with vehicle or 1 μM BRL49653/1 μM 9-cis-RA for 6 h.
4. Wash the cells with 1× PBS.
5. Add 250 μL of 1× Passive Lysis Buffer (Promega Dual-Luciferase Reporter Assay Kit) to each well, place the 12-well culture plates on a rocking platform, and gently rock the plates for 15 min at room temperature.
6. Transfer the lysate to a microcentrifuge tube and centrifuge for 5 min. Transfer the supernatant to a new microcentrifuge tube.
7. Prepare Luciferase Assay Reagent II (LAR II) by dissolving the lyophilized Luciferase Assay Substrate in 10 mL of the Luciferase Assay Buffer II (see Note 5). LAR II should be prepared at ambient temperature just before use.
8. Prepare Stop & Glo Reagent (100 μL reagent per assay) by diluting 50× Stop & Glo Substrate to 1× with Stop & Glo Buffer. Stop & Glo Reagent should be prepared at ambient temperature just before use.

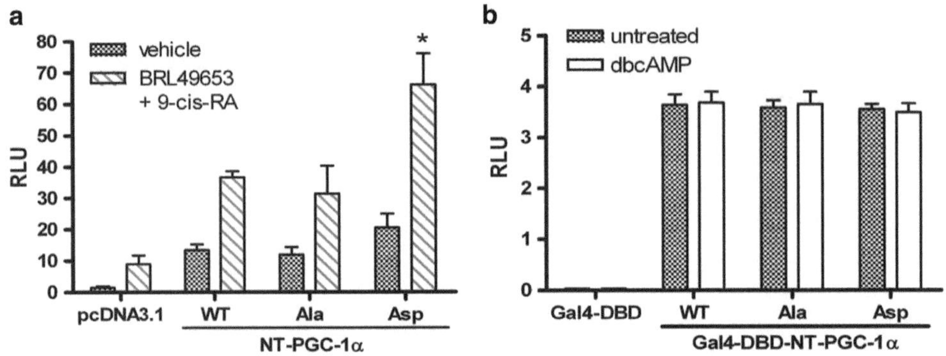

Fig. 2. Correlation of increased coactovation capacity of NT-PGC-1α with its nuclear concentration. (**a**) Transcriptional coactivation assay. pcDNA3.1, NT-PGC-1α (WT), NT-PGC-1α-S194A/S241A/T256A (Ala), or NT-PGC-1α-S194D/S241D/T256D (Asp) were cotransfected into CHO-K1 cells with a PPRE luciferase reporter [$(PPRE)_3$-TK-luc], RXRα, PPARγ, and a *Renilla* luciferase reporter. Twenty-four hour after transfection, cells were treated with vehicle or 1 μM BRL49653/1 μM 9-cis-RA for 6 h. The relative luciferase activity units were determined by dividing firefly luciferase activity measurements by *Renilla* luciferase activity measurements. Data represent mean ± S.D. of at least three independent experiments. Significant deference is determined by student *t* test; $*p<0.05$. Increased nuclear localization of NT-PGC-1α-S194D/S241D/T256D enhances its coactivation capacity. (**b**) Transcriptional activity assay. Gal4-DBD, Gal4-DBD-NT-PGC-1α (WT), Gal4-DBD-NT-PGC-1α-S194A/S241A/T256A (Ala), and Gal4-DBD-NT-PGC-1α-S194D/S241D/T256D (Asp) were cotransfected into CHO-K1 cells with a pGK-1 luciferase reporter containing Gal4 DNA binding sites and a *Renilla* luciferase reporter. 24 h after serum-starved transfection, cells were treated with vehicle or 500 μM dibutyryl cAMP (dbcAMP) for 1 h. Luciferase activity was determined as described above. Data represent mean ± S.D. of at least five independent experiments. All the Gal4-DBD fusion proteins are targeted to the nucleus. PKA phosphorylation does not increase the transcriptional activity itself of NT-PGC-1α.

9. To carry out luciferase reporter assay, predispense 100 μL of LAR II into each well of a 96-well white plate. Transfer 20 μL of the cleared lysate and mix by pipetting two or three times. Place the plate in the luminometer and initiate firefly luciferase activity measurement.
10. Remove the plate from the luminometer and add 100 μL of Stop & Glo Reagent into each well. Place the plate in the luminometer and initiate *Renilla* luciferase activity measurement.
11. To normalize luciferase activity for transfection efficiency, divide the firefly luciferase activity measurements by *Renilla* luciferase activity measurements (Fig. 2a).
12. Perform at least two more independent luciferase reporter assays for each sample.

3.5. Dual-Luciferase Reporter Assay Using a pGK-1 luc Reporter

1. Seed 3×10^4 CHO-K1 cells per well in 12-well plate 1 day before transfection.
2. Transfect CHO-K1 cells with multiple plasmids (Table 2) using FuGENE 6 Transfection Reagent and incubate the cells in F-12K+1% P/S medium for 24 h.
3. Treat the cells with vehicle or 500 μM dibutyryl cAMP for 1 h.
4. Follow steps 4–12 described in the Subheading 3.3 (Fig. 2b).

Table 2
Cotransfection of CHO-K1 cells for transcriptional activity assay using a pGK-luc reporter

Samples 2	1	2	3	4	5	6	7	8
pGK-luc	0.5	0.5	0.5	0.5	0.5	0.5	0.5	0.5
pRL-SV40	0.05	0.05	0.05	0.05	0.05	0.05	0.05	0.05
pCMV-DBD	0.1	0.1						
pCMV-DBD-NT-PGC-1α			0.1	0.1				
pCMV-DBD-NT-PGC-1α-S194A/S241A/T256A					0.1	0.1		
pCMV-DBD-NT-PGC-1α-S194D/S241D/T256D							0.1	0.1
Dibutyryl cAMP	−	+	−	+	−	+	−	+
Total μg	0.65	0.65	0.65	0.65	0.65	0.65	0.65	0.65

4. Notes

1. Filtration of 5× Blocking solution (5% NGS and 5% BSA) through a 0.22 μm filter significantly reduces tiny spot-like fluorescent backgrounds.
2. Transfection efficiency of CHO-K1 using FuGENE 6 is very low when >60% confluent. It is recommended to seed CHO-K1 cells at ~40% confluency to get ~50% transfection efficiency. Also, seeding the cells at low confluency allows single cells to nicely spread out without contacting each other 24 h after transfection.
3. The biological activity of undiluted FuGENE6 Reagent is significantly reduced by the contact with plastic surface. Thus it is critically important to avoid the contact of FuGENE 6 Reagent with plastic walls of the tube containing Opti-MEM I medium during the dilution step. Always add FuGENE 6 Reagent by pipetting directly into the Opti-MEM I medium.
4. In the presence of serum, CHO-K1 cells tend to be round up on a poly-L-lysine-coated or non-coated coverslip. Serum starvation allows a well-spread out morphology of CHO-K1 cells on a coverslip.
5. Repeated freeze/thawing cycles of the reconstituted LAR II reagent decrease assay performance. Thus it is highly recommended to dispense this reagent into 100 μL aliquots and store at −80°C. It is also recommended to thaw frozen aliquots of this reagent at room temperature.

Acknowledgments

We thank Anik Boudreau, Jeho Shin, Yagini Joshi, Chelsea Black, and Peter Huypens for their technical contributions, and David Burk and Courtney Cain for their bioimaging support of this project. This work was supported by NIH RO1 DK074772 (TWG), by a P&F award to JSC from the Pennington NORC (NIH 1P30 DK072476), and in part by NIH grant P20-RR021945 (TWG).

References

1. Finck BN, Kelly DP (2006) PGC-1 coactivators: inducible regulators of energy metabolism in health and disease. J Clin Invest 116:615–622
2. Handschin C, Spiegelman BM (2006) Peroxisome proliferator-activated receptor gamma coactivator 1 coactivators, energy homeostasis, and metabolism. Endocr Rev 27:728–735
3. Schreiber SN, Emter R, Hock MB, Knutti D, Cardenas J, Podvinec M, Oakeley EJ, Kralli A (2004) The estrogen-related receptor alpha (ERRalpha) functions in PPARgamma coactivator 1alpha (PGC-1alpha)-induced mitochondrial biogenesis. Proc Natl Acad Sci USA 101:6472–6477
4. Puigserver P, Wu Z, Park CW, Graves R, Wright M, Spiegelman BM (1998) A cold-inducible coactivator of nuclear receptors linked to adaptive thermogenesis. Cell 92:829–839
5. Wu Z, Puigserver P, Andersson U, Zhang C, Adelmant G, Mootha V, Troy A, Cinti S, Lowell B, Scarpulla RC, Spiegelman BM (1999) Mechanisms controlling mitochondrial biogenesis and respiration through the thermogenic coactivator PGC-1. Cell 98:115–124
6. Lehman JJ, Barger PM, Kovacs A, Saffitz JE, Medeiros DM, Kelly DP (2000) Peroxisome proliferator-activated receptor gamma coactivator-1 promotes cardiac mitochondrial biogenesis. J Clin Invest 106:847–856
7. Vega RB, Huss JM, Kelly DP (2000) The coactivator PGC-1 cooperates with peroxisome proliferator-activated receptor alpha in transcriptional control of nuclear genes encoding mitochondrial fatty acid oxidation enzymes. Mol Cell Biol 20:1868–1876
8. Wang YX, Lee CH, Tiep S, Yu RT, Ham J, Kang H, Evans RM (2003) Peroxisome-proliferator-activated receptor delta activates fat metabolism to prevent obesity. Cell 113:159–170
9. Herzig S, Long F, Jhala US, Hedrick S, Quinn R, Bauer A, Rudolph D, Schutz G, Yoon C, Puigserver P, Spiegelman B, Montminy M (2001) CREB regulates hepatic gluconeogenesis through the coactivator PGC-1. Nature 413:179–183
10. Yoon JC, Puigserver P, Chen G, Donovan J, Wu Z, Rhee J, Adelmant G, Stafford J, Kahn CR, Granner DK, Newgard CB, Spiegelman BM (2001) Control of hepatic gluconeogenesis through the transcriptional coactivator PGC-1. Nature 413:131–138
11. Rhee J, Inoue Y, Yoon JC, Puigserver P, Fan M, Gonzalez FJ, Spiegelman BM (2003) Regulation of hepatic fasting response by PPARgamma coactivator-1alpha (PGC-1): requirement for hepatocyte nuclear factor 4alpha in gluconeogenesis. Proc Natl Acad Sci USA 100:4012–4017
12. Lin J, Wu H, Tarr PT, Zhang CY, Wu Z, Boss O, Michael LF, Puigserver P, Isotani E, Olson EN, Lowell BB, Bassel-Duby R, Spiegelman BM (2002) Transcriptional co-activator PGC-1 alpha drives the formation of slow-twitch muscle fibres. Nature 418:797–801
13. Zhang Y, Huypens P, Adamson AW, Chang JS, Henagan TM, Lenard NR, Burk D, Klein J, Perwitz N, Shin J, Fasshauer M, Kralli A, Gettys TW (2009) Alternative mRNA splicing produces a novel biologically active short isoform of PGC-1{alpha}. J Biol Chem 284:32813–32826
14. Chang JS, Huypens P, Zhang Y, Black C, Kralli A, Gettys TW (2010) Regulation of NT-PGC-1alpha subcellular localization and function by protein kinase A-dependent modulation of nuclear export by CRM1. J Biol Chem 285:18039–18050
15. Biggs WH 3rd, Meisenhelder J, Hunter T, Cavenee WK, Arden KC (1999) Protein kinase

B/Akt-mediated phosphorylation promotes nuclear exclusion of the winged helix transcription factor FKHR1. Proc Natl Acad Sci USA 96:7421–7426

16. Brunet A, Kanai F, Stehn J, Xu J, Sarbassova D, Frangioni JV, Dalal SN, DeCaprio JA, Greenberg ME, Yaffe MB (2002) 14-3-3 transits to the nucleus and participates in dynamic nucleocytoplasmic transport. J Cell Biol 156:817–828
17. Hayden MS, Ghosh S (2004) Signaling to NF-kappaB. Genes Dev 18:2195–2224
18. Sasaki T, Kojima H, Kishimoto R, Ikeda A, Kunimoto H, Nakajima K (2006) Spatiotemporal regulation of c-Fos by ERK5 and the E3 ubiquitin ligase UBR1, and its biological role. Mol Cell 24:63–75
19. Roux P, Blanchard JM, Fernandez A, Lamb N, Jeanteur P, Piechaczyk M (1990) Nuclear localization of c-Fos, but not v-Fos proteins, is controlled by extracellular signals. Cell 63:341–351
20. Silver PA, Keegan LP, Ptashne M (1984) Amino terminus of the yeast GAL4 gene product is sufficient for nuclear localization. Proc Natl Acad Sci USA 81:5951–5955
21. Webb P, Lopez GN, Uht RM, Kushner PJ (1995) Tamoxifen activation of the estrogen receptor/AP-1 pathway: potential origin for the cell-specific estrogen-like effects of antiestrogens. Mol Endocrinol 9:443–456

Chapter 12

In Vivo Studies of PPAR-Chromatin Interactions: Chromatin Immunoprecipitation for Single-Locus and Genomewide Analyses

Till Adhikary and Rolf Müller

Abstract

The immunoprecipitation of protein complexes with associated DNA fragments, termed chromatin immunoprecipitation, allows for the analysis of the binding of endogenous factors. With the advent and widespread use of next generation high-throughput sequencing technologies, factor binding can be assessed in an unbiased and genomewide approach. The protocol outlined here yields sufficient material to subsequently carry out such analyses and discusses key steps in detail.

Key words: Chromatin immunoprecipitation, ChIP, Protocol, ChIP-Seq, Illumina, PPAR

1. Introduction

The chromatin immunoprecipitation (ChIP) assay, which was developed in the late 1970s and 1980s (1–4), renders the semi-quantitative analysis of the binding of endogenous factors to DNA possible. Its basic principle is the immunoprecipitation of protein complexes with associated DNA fragments; after washing and purification steps, the DNA is detected by PCR or high-throughput sequencing. The signal-to-noise ratio of the assay depends on a multitude of factors. Proteins that bind to DNA indirectly are notoriously more difficult to detect than histones or transcription factors which bind to DNA directly. The availability of suitable antibodies imposes further limits on the sensitivity and specificity of the assay. Most protocols make use of chemical cross-linkers to avoid the dissociation of bound proteins during shearing and washing steps, enabling the detection of nonhistone proteins.

Mostafa Z. Badr and Jihan Youssef (eds.), *Peroxisome Proliferator-Activated Receptors (PPARs): Methods and Protocols*, Methods in Molecular Biology, vol. 952, DOI 10.1007/978-1-62703-155-4_12, © Springer Science+Business Media New York 2013

Usually, the crosslinker is formaldehyde, which covalently connects amino and imino groups via a one-carbon spacer; a multitude of crosslinkers with longer spacers are commercially available for special applications. Crosslinks are reversed prior to purification of the DNA in order not to disturb subsequent analyses. Shearing of the DNA is achieved by means of sonication or enzymatic digestion. Enzymatic digestion may introduce a location bias because the cleavage sites are not uniformly distributed in a given DNA fragment. Here, we present a widely applicable protocol that we routinely use to detect histones, transcription factors, and cofactors following formaldehyde-mediated crosslinking and ultrasonic shearing of genomic human and murine chromatin from cell lines (adapted from personal communication, Caroline Bouchard). It can yield DNA which is of sufficient quality and quantity for massive parallel sequencing (ChIP-Seq) from a single IP. However, careful establishment, several controls, and special attention to key steps during the experiment and the analysis are required in order to generate reproducible and meaningful results.

2. Materials

Prepare all buffers with ultrapure water and analytical grade chemicals. Take care not to contaminate the reagents with DNA (no spoons in and no return of chemicals to stock vessels). Strictly follow all chemical hazard precautions and waste disposal regulations (special attention: formaldehyde, SDS, sodium deoxycholate). We do not add sodium azide to the buffers.

- Formaldehyde 37%.
- Glycin (2.5 M stock solution).
- Protease inhibitors (e.g., Sigma catalog no. P8340).
- PBS (ice-cold).
- Antibodies.
- Lysis buffer I (hypotonic lysis buffer).
  5 mM PIPES pH 8.0.
  85 mM KCl.
  0.5% (v/v) NP40.
  Add protease inhibitors prior to use.
- Lysis buffer II (RIPA buffer).
  10 mM Tris–HCl pH 7.5.
  150 mM NaCl.
  1% (v/v) NP40.

1% (w/v) sodium deoxycholate.
0.1% (w/v) SDS.
1 mM EDTA.
Add protease inhibitors prior to use.

- BSA (fraction V, 20 g/L, dissolved in TE pH 8.0).
- Sonicated salmon sperm DNA: stock at 10 g/L (stored at −20°C).
- Wash buffer I (mixed micelle buffer).
  20 mM Tris pH 8.1.
  150 mM NaCl.
  2 mM EDTA.
  0.1% (w/v) SDS.
  1% (v/v) Triton X-100.
  Store at 4°C.
- Wash buffer II (buffer 500).
  20 mM Tris pH 8.1.
  500 mM NaCl.
  2 mM EDTA.
  0.1% SDS.
  1% (v/v) Triton X-100.
  Store at 4°C.
- Wash buffer III (lithium chloride detergent buffer).
  10 mM Tris pH 8.1.
  250 mM LiCl.
  1% (v/v) NP40.
  1% (w/v) sodium deoxycholate.
  1 mM EDTA.
  Store at 4°C.
- Elution buffer (prepare fresh).
  1% (w/v) SDS.
  100 mM $NaHCO_3$.
- TE.
  10 mM Tris pH 7.4.
  1 mM EDTA.
- Reversion mix (per sample, prepare fresh).
  16 mL 5 M NaCl.
  16 mL 1 M Tris pH 6.8.
  8 mL 0.5 M EDTA.
  10 mg RNase A (10 g/L stock).
  20 µg proteinase K (20 g/L stock).
- PCR purification kit (Qiagen) and additional binding buffer PB (Qiagen).

3. Methods

1. Stimulate or treat the cells as appropriate (see Note 1).
2. Under a chemical fume hood, add formaldehyde to 1% final concentration directly to the medium, mix by gently swirling the dish or flask and incubate for 10 min at room temperature under the fume hood (see Note 2).
3. Add glycin to a final concentration of 125 mM in order to quench the remaining formaldehyde. Swirl gently and incubate for 5 min under the fume hood (see Note 3).
4. Under the fume hood, discard the supernatant and wash the cells twice with ice-cold PBS. Harvest cells with a rubber policeman and collect them in a 15 mL conical tube with a blue pipette tip (maximum number of 2.5×10^8 cells per tube). Pellet the cells by centrifugation at $1{,}200 \times g$ for 5 min at 4°C and aspirate the supernatant (see Note 4).
5. Gently but thoroughly resuspend the pellet in 1 mL of lysis buffer I per $\sim 2.5 \times 10^7$ cells with a blue tip. Incubate on ice for at least 20 min, then centrifuge at $1{,}200 \times g$ for 5 min at 4°C and aspirate the supernatant (see Note 5).
6. Resuspend the nuclei in 1 mL lysis buffer II per $\sim 2.5 \times 10^7$ nuclei and incubate on ice for at least 10 min. Ensuring the homogeneity of the suspension, prepare 1 mL aliquots in suitable tubes (see Note 6).
7. Sonicate the lysates. This step has to be established for every cell line and sonication device (see Note 7). The sample has to be cooled efficiently to prevent the reversal of crosslinks during sonication.
8. Transfer the samples to Eppendorf tubes and centrifuge for 15 min at maximum speed ($12{,}000\text{–}20{,}000 \times g$) in a tabletop centrifuge at 4°C. Transfer the supernatant(s) to new vessels, pooling identically treated samples. This is the soluble chromatin. Remove an aliquot of about 50 μL for the analysis of shearing efficiency. Store the chromatin stock at 4°C (see Note 8).
9. Add 20 μg of proteinase K and 20 μg of RNase A to the aliquot, mix, and add 100 μL of mineral oil. Incubate overnight at 65°C (see Note 9).
10. Add loading buffer to the sample and load a volume corresponding to 10 μL of the sample on a 1% agarose gel containing ethidium bromide. Run at a maximum voltage of 80 V. Estimate the size of the DNA fragments (see Note 10).
11. Wash an appropriate amount of protein A sepharose twice with 10 mL lysis buffer II in a 15 mL conical tube, centrifuge each time 5 min at $1{,}200 \times g$ at 4°C, and aspirate the supernatant.

Resuspend the beads in a volume of 10 mL lysis buffer II with 1 g/L BSA and 0.4 g/L sonicated salmon sperm DNA. Rotate overnight on a tumbler at 4°C to block the sepharose beads (see Note 11).

12. Centrifuge the blocked beads and remove a fraction of the supernatant to make a 50% slurry. Gently but thoroughly resuspend the beads. Incubate 10 μg unspecific antibodies (e.g., rabbit IgG from serum) per 100 μL of 50% bead slurry allocated for preclearing of the chromatin (see Note 12) in an Eppendorf tube. Prepare a sufficient volume for preclearing of all samples. Rotate at 4°C for at least 30 min. Centrifuge the suspension at the minimum possible speed in a tabletop centrifuge for 1 min. Aspirate the supernatant. Resuspend the beads in lysis buffer II and add 100 μL of 50% slurry per 1 mL of soluble chromatin. Rotate at 4°C for 30–60 min for preclearing of the chromatin, then centrifuge at 1,200 × *g* for 5 min at 4°C. Transfer the precleared chromatin supernatant to a fresh conical tube.
13. Distribute soluble chromatin to Eppendorf tubes. Per immunoprecipitation, an aliquot that corresponds to 1×10^6–1×10^7 cells should be used. Add antibody to each sample (see Note 13 for further information). Take aliquots of the soluble chromatin that correspond to 1% of the volume used per IP and store them at 4°C in Eppendorf tubes. These are the INPUT samples. Rotate the IP samples overnight at 4°C.
14. Add 50 μL of blocked bead slurry to each IP. Rotate for at least 1 h at 4°C (see Note 14).
15. Centrifuge the samples at the minimum possible speed in a tabletop centrifuge for 1 min. Aspirate the supernatant and wash the samples with 1 mL each of cold wash buffer I. Keep the samples on ice until step 17 is completed (see Note 15).
16. Centrifuge as in step 15, aspirate the supernatant, and wash with 1 mL each of cold wash buffer II (see Note 16).
17. Centrifuge as in step 15, aspirate the supernatant and wash with 1 mL each of cold wash buffer III. Repeat this step once (see Note 17).
18. Centrifuge as in step 15, aspirate the supernatant, and wash with 1 mL each of room temperature TE. Repeat this step once, not putting the samples on ice (see Note 18).
19. Aspirate the supernatant quantitatively. Add 200 μL of freshly prepared elution buffer to each sample and incubate for 15 min at room temperature with rotation or vigorous agitation (see Note 19).
20. Centrifuge at maximum speed in a tabletop centrifuge for 1 min. Carefully transfer the supernatants to new tubes and

label them appropriately. Again, add 200 µL of elution buffer each to the beads, and incubate for 15 min at room temperature with rotation or vigorous agitation (see Note 20).

21. Centrifuge at maximum speed in a tabletop centrifuge for 1 min. Carefully transfer the supernatants to the tubes from the previous step, thereby pooling them. Add 42 µL of freshly prepared reversion mix to each tube. Add 400 µL of elution buffer and 42 µL of reversion mix to the input samples from step 13. Briefly vortex and centrifuge samples. Incubate all samples at 65°C in a heating block overnight (see Note 21).
22. Purify the samples using the Qiagen PCR purification kit. 2.2 mL of binding buffer is used per sample. Elute twice with 50 µL of: elution buffer, incubating the columns with buffer for 5 min at room temperature before each centrifugation (see Note 22).
23. Use the samples as templates for PCR. Quantitative PCR is preferred (see Note 23). Primers for a genomic region where the factor of interest binds (positive control region) and primers for a non-bound region (negative control region) are needed.
24. Analyse the data and determine if the ChIP was specific (see Note 24). Estimate the signal-to-noise ratio.
25. If the samples are of appropriate quality, proceed to analyses of other genomic regions and/or deep sequencing for genome-wide analysis (see Note 25).

4. Notes

1. Most transcriptional events can take place rapidly upon stimulation, e.g., within 1 h when using PPAR ligands. Do a time course experiment with narrow intervals (10 or 15 min—the temporal resolution of the assay is roughly 10 min due to the fixation) to determine which conditions are to be used in later experiments. 1×10^6–1×10^7 cells per IP are needed. This depends largely on the quality of the antibody and on the abundance of the epitope in the sample. We found that 7.5×10^6 cells are sufficient for most ChIP-validated antibodies. Using 15 cm culture plates is most convenient. If the supply of cells is limited, determine the amount of cells needed to give a robust signal for the antibodies to be used before setting up large-scale experiments.
2. Do not contaminate the formaldehyde stock. Mix thoroughly prior to use. Formaldehyde is usually supplied as a 37%

solution and will form a gradient. It is best to store it in aliquots. Safety precautions apply.

3. Safety precautions apply.
4. Cell pellets can be snap frozen in liquid nitrogen and stored at −80°C. Thaw pellets on ice before continuing.
5. The first lysis step uses hypotonic conditions to disrupt the cellular membrane. Cytoplasmic components are removed. Efficiency of the lysis can be controlled under a microscope. It is critical that cell numbers and lysis conditions are reproducible for all steps of the protocol.
6. The number of nuclei per volume can be decreased if fragmentation of the chromatin is hard to achieve. We prepare the aliquots in 15 mL conical tubes because they can be easily attached to the sonication apparatus. Some machines provide receptacles for Eppendorf tubes.
7. Proper sonication of the chromatin is critical for the success of the assay. This step should be established with great care before doing any experiments. As a lead, we use 50–70 pulses of 1 s with 2 s of pause at 10–20% amplitude generated by a Branson Sonifier S250D equipped with a microtip, which is sufficient to shear the DNA of many cell lines. We installed clamps to align the sample tube with the tip and securely fasten it during sonication. For cooling, fill a large beaker with ice and add low-grade ethanol until the ice is afloat and immerse the sample tube immediately after starting the sonication device. After sonication is finished, remove the sample tube from the beaker immediately, or ice crystals will form which can deteriorate the quality of the sample. It is preferable to sonicate the samples as shortly and weakly as possible in order to preserve the epitopes. Some factors, like histones, can withstand considerable force, but indirectly bound proteins can be easily lost during the process. If using a tip, immerse it at about 60–80% of maximum depth, pointing at the bottom of the conical tube in a perfect right angle. This is to prevent foaming of the sample. If the sample is turned to foam, discard of it (epitopes are altered by oxidation and shearing efficiency decreases dramatically if the sample is not a liquid).
8. If the pellets are large and white, sonication was inefficient. If the pellets are small, transparent and darkly speckled, the number of pulses and/or the amplitude should be decreased in order to preserve epitopes.
9. Heating of the sample reverses the crosslinks and enables analysis of the DNA. Photometric quantitation of the chromatin is hampered by proteins crosslinked to the DNA. If the DNA content of the sample needs to be determined, purify it appropriately; however, we experienced that using chromatin with

equal numbers of nuclei and equal volume per IP yields more comparable input signals.

10. Use a sample buffer that does not contain dyes which run in the 500 bp range in order not to quench fluorescence from the chromatin sample in this region. Transferring small quantities of mineral oil into the gel chamber along with the sample is not harmful. Do not run the gel at high voltages because this will change the apparent size of the fragments. If using buffers other than the ones we specify, consider that high concentrations of SDS such as 1% will appear as a distinct bright band because it binds to ethidium bromide. Chromatin will always appear as a smear of heterogeneous fragments. The ethidium bromide runs in the opposite direction of the DNA, do not let the electrophoresis run for too long.

    A maximum size of the DNA fragments of 500–1,000 bp is desirable. For some cell lines and cell types, this cannot be achieved. The assay will work with somewhat larger fragments, but spatial resolution will decrease (the maximum achievable resolution for this assay is about 500 bp when using PCR-based analysis). In our experience, resolution will not decrease dramatically even with a maximum fragment size of 2,000 bp, probably because smaller fragments are immunoprecipitated preferentially. Keep in mind that stronger sonication can dramatically decrease the signals for weakly bound factors.

11. This step can be carried out in parallel with step 9. Blocked sepharose beads can be stored for months at 4°C. 50 μL of 50% slurry per IP and an amount of 100 μL per 1 mL of soluble chromatin for preclearing are needed. Prepare a sufficiently large amount of beads. Protein A and protein G sepharose display different binding properties. Choose as appropriate for the antibodies intended to use. We use protein A sepharose whenever possible because in our hands it binds better to antibodies derived from rabbit. It is possible to use a mix of A and G beads.

12. We aspire the supernatant from bead pellets with a vacuum source connected to standard syringe needles that have a diameter of 0.5 mm or less in order to limit the aspiration of beads. Precleared chromatin can be stored at 4°C (up to several weeks) or snap frozen in liquid nitrogen and stored at −80°C for longer periods.

13. For example, distribute 300 μL of soluble chromatin to each tube if the concentration of nuclei was 2.5×10^7/mL for the equivalent of 7.5×10^6 nuclei per IP. We found that 300 μL is a good minimal volume per IP to ensure proper mixing of the sample during incubation with the antibodies. We try to use polyclonal antibodies derived from rabbits because they have good binding properties compared to other species in

our hands. Use 4–5 μg per IP as a starting point. It might be necessary to titrate the optimum amount per IP for individual antibodies. If the concentration of the antibody is unknown, use 4–10 μL of serum as a starting point. As one of the negative controls of the assay, unspecific antibodies (e.g., rabbit IgG from serum, Sigma I5006) of the same isotype(s) as the antibodies used in the assay are needed. If using different isotypes of antibodies in one experiment, several of these unspecific IP samples are needed. This is to discriminate genomic regions which bind unspecifically to the sepharose or to the antibodies.

14. Make sure that the bead suspension is homogeneous before distributing it to the chromatin samples. We found a clipped-off yellow tip best for pipetting the slurry. Resuspend the beads after every single distribution to the samples.
15. The 1 min centrifugation step can be carried out at room temperature. When washing the beads, gently add the wash buffer, close the tube, then gently invert several times. When aspirating the supernatant, take special care not to remove any beads.
16. See Note 15.
17. See Note 15.
18. See Note 15.
19. The bead volume per sample should be equal. When quantitatively aspirating the supernatant, make sure that as little liquid phase as possible remains in the tube, e.g., by inserting the needle into the pellet with the opening facing the wall of the tube in order not to remove a substantial amount of beads. Remove residual beads from the needle before processing the next sample.
20. Do not transfer any beads along with the supernatant after the elution.
21. Do not transfer any beads along with the supernatant after the elution. Do not forget to process the input samples.
22. It is also possible to use phenol-chloroform extraction and ethanol precipitation or other kits to purify the DNA. However, we have found the specified kit to reproducibly yield high purities and sufficient amounts of DNA. Do not use columns with different lot numbers to purify samples from the same experiment since the DNA yield can still vary considerably between lots. Keep the buffers clean. Keep the bottles tightly closed whenever possible to prevent contamination and evaporation of volatile constituents. It is especially important not to contaminate the: elution buffer. We make aliquots and store them at −20°C. Also keep in mind that the maximum amount

of bound DNA is 10 μg per column. For this reason, do not increase the amount of input DNA to considerably more than 1% under the conditions specified here.

23. Standard PCR does not allow for the accurate comparison of samples with gross differences in template amounts since some reactions might have reached saturation at a given cycle number. Additionally, nonlinear fluorescence intensities hamper gel-based analyses. When performing quantitative PCR, we routinely use SYBR Green based assays because they are easy to establish and are less costly than Taqman-based approaches. For every experiment, perform PCR for positive and negative control genomic regions. Always include negative control IP samples and a no template control PCR.

24. If the amplification efficiency of the primers was not determined, a simplified approach equivalent to the Livak method for calculating the relative expression of transcripts (5) can be used. Keep in mind that at least three replicates for each sample need to be measured in order to calculate standard deviations. "Fold enrichment" ("fold IgG") values can lead to serious misinterpretation of data because a potentially strong signal is divided by a highly variable, weak signal. A "percent of input" approach gives more robust and reproducible results. One percent of input as a reference value is suitable because ChIP signals typically range from roughly 0.05 (background) to 20 (excellent anti-histone antibody) percent of input. When calculating error bars, take error propagation into account because the error range depends on the errors of both input sample and antibody sample replicates. When using a simple calculation employing Gaussian error propagation and the assumption that the PCR had 100% efficiency, the term (using input samples of 1%) for final values in percent is

$$2^{-(\mathrm{Ct(input)}-\mathrm{Ct(sample)})} \pm 2^{-(\mathrm{Ct(input)}-\mathrm{Ct(sample)})} \times \ln(2) \times \bullet \sqrt{\left(\mathrm{SD(input)}^2 + \mathrm{SD(sample)}^2\right)}$$

(Ct, cycle threshold; SD, standard deviation). The signal-to-noise ratio can be roughly estimated from the resulting bar graphs. Note that apparent differences in binding strength between samples precipitated with the same antibody can result from technical reasons (samples not treated equally during the assay) or from the accessibility of the antibody's epitope and do not necessarily mean that there is differential binding of the target protein. It is of paramount importance to robustly reproduce results several times in order to estimate signal-to-noise ratios and draw conclusions. In our experience, antibodies should reproducibly show at least a twofold enrichment over background for bound genomic regions and no enrichment in control regions to be considered valid for ChIP. Antibody specificity can be verified by specific knockout or knockdown of the target protein. To be considered for Illumina sequencing, the enrichment should be high. We successfully performed ChIP sequencing with a polyclonal antibody directed against PPARβ/δ (6) that displays an enrichment factor of about 20–30 at strongly bound genomic loci; we were able to detect peaks with low enrichment (e.g., twofold), albeit at the cost of higher false negative/false positive rates.

25. Ensure high purity of the samples if a genomewide analysis is to be carried out. All DNA contaminations will decrease the number of usable reads from the sequencing run. Use only clean reagents, pipettes, and plasticware. We found that autoclaving can introduce DNA contaminations. To measure total DNA content of the samples, use a fluorescent dye such as Pico Green. In our laboratory, this protocol results in a DNA concentration of about 1 ng/μL when eluting with a total volume of 100 μL for most antibodies. Antibodies against unmodified and modified histones can give higher yields. Using the manufacturer's protocol from Illumina for ChIP-Seq worked well for us. Illumina sequencing is the de facto standard for ChIP-Seq because the platform offers vastly more reads than other platforms commonly available at the time of writing. Data analysis should be carried out by an experienced bioinformatician.

References

1. Jackon V (1978) Studies on histone organization in the nucleosome using formaldehyde as a reversible cross-linking agent. Cell 15(3): 945 951
2. Gilmour DS, Lis JT (1984) Detecting protein-DNA interactions in vivo: distribution of RNA polymerase on specific bacterial genes. Proc Natl Acad Sci USA 81(14):4275–4279
3. Hebbes TR, Thorne AW, Crane-Robinson C (1988) A direct link between core histone acetylation and transcriptionally active chromatin. EMBO J 7(5):1395–1402
4. Solomon MJ, Larsen PL, Varshavsky A (1988) Mapping protein-DNA interactions in vivo with formaldehyde: evidence that histone H4 is retained on a highly transcribed gene. Cell 53(6):937–947
5. Livak KJ, Schmittgen TD (2001) Analysis of relative gene expression data using real-time quantitative PCR and the 2(–Delta Delta C(T)) Method. Methods 25(4):402–408
6. Adhikary T, Kaddatz K, Finkernagel F, Schönbauer A, Meissner W, Scharfe M, Jarek M, Blöcker H, Müller-Brüsselbach S, Müller R (2011) Genomewide analyses define different modes of transcriptional regulation by peroxisome proliferator-activated receptor-β/δ. PLoS One 6(1):e16344

Chapter 13

FISH Analysis Using *PPARγ*-Specific Probes for Detection of *PAX8-PPARγ* Translocation in Follicular Thyroid Neoplasms

Noor Akmal Syed Husain Sharifah, Zubaidah Zakaria, and Wai Kit Chia

Abstract

Fluorescence in situ hybridization (FISH) is increasingly gaining importance in clinical diagnostics settings. Due to the ability of the technique to detect chromosomal abnormalities in samples with low cellularity or containing a mixed population of cells even at a single-cell level, it has become more popular in cancer research and diagnosis. Here, we describe the FISH technique for detection of *PAX8-PPARγ* translocation in follicular thyroid neoplasms, and the optimal protocol for the detection of this fusion gene using in archival formalin-fixed paraffin-embedded (FFPE) thyroid tissue sections.

Key words: Fluorescence in situ hybridization (FISH), Formalin-fixed paraffin-embedded (FFPE) tissue, *PAX8-PPARγ* fusion gene, Follicular thyroid neoplasms

1. Introduction

The t(2;3)(q13;p25) translocation was first reported in follicular thyroid neoplasms by Kroll et al. (1). This translocation resulted in the fusion of the *PAX8* gene, which encodes the paired domain transcription factor, with the peroxisome proliferator-activated receptor (*PPARγ*) gene (1). Although the role of this fusion gene in the development and differentiation of follicular thyroid tumors is not understood, the identification of the *PAX8-PPARγ* fusion gene promises a new turning point in differentiating follicular thyroid tumors from other thyroid diseases (2).

Currently, only postoperative differentiation between follicular thyroid carcinoma and adenoma is possible based on the presence of capsular or vascular invasion. With the identification of this fusion gene paired with molecular cytogenetics technology, there is

Mostafa Z. Badr and Jihan Youssef (eds.), *Peroxisome Proliferator-Activated Receptors (PPARs): Methods and Protocols*, Methods in Molecular Biology, vol. 952, DOI 10.1007/978-1-62703-155-4_13, © Springer Science+Business Media New York 2013

a possibility to differentiate follicular thyroid carcinoma from its benign counterpart, even on a single-cell level (3).

With proven successes in the past, cytogenetics analysis is an invaluable method in the detection of chromosomal abnormalities in tumor samples. Even though this technique is still considered to be the "gold-standard" in tumor cytogenetics (4), there are still setbacks in karyotyping to obtain good and analyzable metaphase spreads through short-term culture in primary cancer specimens (5). Furthermore, the lack in availability of fresh material, low mitotic index, and/or percentage of neoplastic cells, the cytogenetics complexity, and the time-consuming nature of analysis has further restricted the use of this technology in studies involving solid tumors (4).

Therefore, fluorescence in situ hybridization (FISH) plays an important role as a contingency technique to identify specific numeric or structural chromosome abnormalities in cancer research (5). First introduced through a successful experiment on human chromosome in 1986 (6), FISH is gaining increasing clinical application in routine cytogenetic diagnostics (7). This is a well-established technique that makes it possible to obtain "cytogenetic information" from interphase nuclei and particularly valuable in the study of archival material, such as formalin-fixed paraffin-embedded tissue (FFPE) specimens (3).

FISH is particularly advantageous for samples of low cellularity or containing a mixed population of cells (8). It is the only tool used to investigate specific numerical chromosomal aberrations, chromosomal translocations, amplification of oncogenes, or deletion of tumor suppressor genes in archival samples on a single-cell level (3).

However, the success of probe hybridization is dependent on the optimal preparation of samples (8) and it is technically demanding as it is difficult to obtain good quality FISH signals reproducibly (9). A thorough understanding of the tissue of interest and of slide preparation techniques is the first step to obtaining high quality results (8). Herewith, we describe a protocol to detect the *PAX8-PPAR*γ fusion gene based on our recent study using the FISH technique (5).

2. Materials

2.1. Chemicals

1. 10× SkipDewax (Insitus Biotechnologies, Albuquerque, NM).
2. 0.1% *w/v* Poly-l-lysine.
3. Pepsin.
4. Bacterial artificial chromosomes (BACs)—RP11-57D6, RP11-402P11, and RP11-434I13 (Open Biosystems, USA).
5. Nick Translation Labeling Kit (Vysis, USA).
6. SpectrumOrange- and SpectrumGreen-dUTP (Vysis, USA).

7. Hybridization Buffer (Vysis, USA).
8. Sodium acetate.
9. Formamide.
10. 100% Ethanol.
11. QuickSpin Columns Sephadex G-50 (Roche Diagnostics, Sweden).

2.2. Reagents to be Prepared

1. 1× *SkipDewax.*

   Add 45 mL purified H_2O and 5 mL 10× SkipDewax Deparaffinize Solution to a glass Coplin jar and mix thoroughly. Prepare fresh before use.
2. 1.5 mg/mL *Pepsin.*

   Dilute 75 mg pepsin in 50 mL 0.2 N hydrochloric acid in a Coplin jar and mix well.
3. *Denaturation buffer.*

   Add 35 mL formamide, 5 mL 20× SSC (pH 5.3), and 10 mL purified H_2O to a glass Coplin jar and mix thoroughly. Verify that the pH is 7.0–7.5 by measuring the pH at ambient temperature. Between use, store covered at 4°C. Discard after 7 days.
4. *Ethanol series.*

   Dilute 100% ethanol v/v with purified H_2O to prepare the wash solutions. Between uses, store covered at ambient temperature. Discard stock solutions after 6 months.
5. 20× *SSC.*

   Mix thoroughly 66 g 20× SSC in 200 mL purified H_2O. Adjust to pH 5.3 at ambient temperature using concentrated HCl and fill to quantity sufficient (QS) to final volume of 250 mL. Store at ambient temperature. Discard stock solution after 6 months or sooner if solution appears cloudy or contaminated.
6. *Wash Buffer I* (0.4× *SSC/0.3% NP-40*).

   Mix thoroughly 20 mL of 20× SSC with 950 mL purified H_2O. Add 3 mL nonyl phenoxypolyethoxylethanol-40 (NP-40). Mix thoroughly until NP-40 is dissolved. Adjust pH to 7.0–7.5 with NaOH. Add purified H_2O to bring final volume to 1 L. Store the solution at ambient temperature. Discard stock solution after 6 months or sooner if solution appears cloudy or contaminated.
7. *Wash Buffer II* (2× *SSC/0.1% NP-40*).

   Add 100 mL of 20× SSC to 850 mL purified H_2O. Add 1 mL NP-40. Adjust pH to 7.0–7.5 with NaOH. Add purified H_2O to bring final volume to 1 L. Store the solution at ambient temperature. Discard stock solution after 6 months or sooner if solution appears cloudy or contaminated.

2.3. Consumables

1. Rubber cement.
2. Clear nail polish.
3. Pipettor tips.
4. Coverslips (22 × 22 mm).
5. Microcentrifuge tubes (0.2 and 1.5 mL).

2.4. Equipments

1. Microtome.
2. Float bath.
3. Ovens (60 and 37°C).
4. Shaker incubator (37°C).
5. Waterbaths (37 and 75°C).
6. HyBrite/ThermoBrite or self-prepared humidified hybridization chamber (see Note 1).
7. Thermal cycler.

3. Methods

3.1. Slide Preparation on Mounted Formalin-Fixed Paraffin-Embedded Tissue

3.1.1. Preparing Tissue Sections

1. Mount 3–5 μm sections of the tissue on poly-L-lysine-coated microscopy slides.
2. Dry the slides at 60°C overnight.
3. Cut a corresponding tissue section adjacent and stain with hematoxylin and eosin (H&E) for histological evaluation.
4. Identify appropriate areas of tumor cells on the H&E slides and evaluate the slide for tissue tumor composition and quality of fixation.

3.1.2. Deparaffinization and Pepsin Digestion

1. Remove paraffin from the tissue sections by immersing the slides into 50 mL of SkipDewax 1× preheated at 80°C for 30 min.
2. Rinse the slides by immersing into a Coplin jar containing distilled water for 30 s, and then air-dry the slides.
3. Digest the tissue by immersing into a preheated Coplin jar containing pepsin solution (1.5 mg/mL) in a 37°C waterbath for 20–60 min (see Note 2).
4. Agitate the Coplin jar every 5–10 min to ensure an even digestion of the mounted tissue.
5. Rinse in 1× PBS in a Coplin jar for 2 min and air-dry the slides at room temperature.

3.1.3. DAPI Evaluation for Digestion Adequacy

1. Add 10 μL of 4′,6-diamino-2-phenylindole (DAPI) counterstain and place a 22 × 22 mm coverslip onto the tissue section.
2. Check for adequacy of digestion using the epifluorescence microscope (see Note 3).

3. Remove the coverslip and dehydrate the mounted tissue sections in an ethanol series (70, 95, and 100%) for 1 min each.
4. Air-dry the slides at room temperature until hybridization.

3.2. Probe Preparation and Labeling

While there are FISH probes designed specifically for clinical diagnostic purposes that are available commercially (e.g., Vysis, Poseidon, etc.), other FISH probes, if not readily available in the market, can be prepared separately for specific detection of chromosomal abnormalities (deletion, amplification, aneusomy, inversion, or translocation). In this section, a protocol for probe labeling which was previously described by (10) and which have been subsequently modified Chia et al. (5) to detect the *PAX8-PPAR*γ translocation in follicular thyroid cancer using DNA extracted from specific BAC clones for the *PAX8* and *PPAR*γ regions will be described in detail.

3.2.1. Probe Labeling

Perform the nick translation labeling method using the Vysis Nick Translation Labeling Kit. The kit labels 2 μg of BAC DNA in each reaction tube.

1. Prepare three microcentrifuge tubes for each of the BAC DNAs (RP11-57D6, RP11-402P11, and RP11-434I13). Two BAC clones, RP11-402P11 and RP11-57D6, are labeled with *SpectrumOrange*-dUTP while RP11-434I13 is labeled with *SpectrumGreen*-dUTP.
2. Mix the reagents provided in the kit in each tube according to Table 1.
3. Mix each tube briefly by tapping the bottom of the tubes and briefly centrifuge to collect the mixture at the bottom of the tube.

Table 1
Reagent mixture for nick translation labeling

Reagent	Volume/reaction (μL)
Nuclease-free water	(17.5–*X*)
Extracted BAC DNA (2 μg)	*X*
SpectrumGreen or SpectrumRed dUTP (0.2 mM)	2.5
dTTP (0.1 mM)	5
dNTP mix	10
10× Nick translation buffer	5
Nick translation enzyme	10
Total	50

X volume of 2 μg extracted BAC DNA

4. Incubate the mixtures at 15°C for 60 min in a thermal cycler.
5. Perform agarose gel electrophoresis for each probe mixture to determine the size of the labeled probes (see Note 4).
6. Deactivate the nick translation enzyme by heating the tubes at 70°C for 10 min when the desired size of labeled DNA is acquired.

3.2.2. Probe Filtration and Precipitation

1. Transfer each of the labeled probes separately into three Sephadex G-50 QuickSpin Columns (see Note 5).
2. Centrifuge the columns at 900×*g* for 5 min.
3. The eluted probes (about 50 μL of each probe) are mixed in a 1.5 mL microcentrifuge tube.
4. Add the following to the probe mixture: 20 μL Cot-1 DNA, 15 μL sodium acetate 3 M, and 375 μL cold absolute ethanol.
5. Mix the probe mixture by inverting the tube and keep in −80°C overnight.
6. Centrifuge the tube at 13,000 rpm is 15,400g. at 4°C for 30 min to pellet the probe.
7. Discard supernatant and resuspend the pellet with 500 μL of 70% ethanol.
8. Repeat step 6 and 7.
9. Drain the supernatant and air-dry the probe pellet in a dark room for 10–15 min.
10. Reconstitute the probe by adding 10 μL of hybridization buffer.
11. Store the probe at −20°C until denaturation and hybridization.

3.3. Denaturation and Hybridization

Denaturation conditions for FISH are similar to those required for purified DNA, except that they are performed in the presence of formamide, a chaotropic agent, which lowers the melting point of duplexed nucleic acids (8).

3.3.1. Denaturation of Target DNA

1. Preheat 2× SSC and denaturation solution in a 75°C waterbath.
2. Place the slides from Subheading 3.1.3 into the preheated 2× SSC and incubate at 75°C for 30 min.
3. Remove the Coplin jar containing 2× SSC and the slides from the waterbath and cool down to 37°C for 20 min.
4. Transfer the slides into a Coplin jar containing 0.1× SSC for 1 min at room temperature (RT).
5. Transfer the slides into the preheated denaturation solution and incubate at 75°C for 1 min.
6. Transfer the slides into a Coplin jar containing chilled 0.1× SSC for 1 min.

7. Transfer the slides into a Coplin jar containing chilled 2× SSC for 1 min.
8. Dehydrate the slides in an ethanol series (70, 95, and 100%) for 1 min each.
9. Air-dry the slides at RT.

3.3.2. Denaturation of Fluorescence-Labeled Probes

1. Heat the tube containing the probe mixture from Subheading 3.2.2 in a 75°C waterbath for 5 min.
2. Immediately transfer the tube onto ice for 30 s.
3. Transfer and incubate the tube in a 37°C waterbath for 30 min.

3.3.3. Hybridization

1. Add 10 μL denatured probes from Subheading 3.3.2 onto the denatured slides from Subheading 3.3.1.
2. Place a 22 × 22 mm coverslip over the probe and allow it to spread evenly under the coverslip.
3. Seal the coverslip by applying rubber cement around the periphery of the coverslip.
4. Place the slides in a prewarmed humidified chamber (see Note 1).
5. Incubate the slide at 37°C overnight.

3.4. Post hybridization Wash

1. Perform post hybridization wash in a dark room.
2. Remove the rubber cement seal and coverslip using a pair of forceps.
3. Immerse slides in preheated wash buffer I (0.4 × SSC/0.3% NP-40) at 75°C for 2 min.
4. Transfer the slides into wash buffer II (2 × SSC/0.1% NP-40) for 30 s at RT.
5. Remove and air-dry the slides in the dark room.
6. Apply 10 μL DAPI counterstain to the hybridized area.
7. Place a 22 × 22 mm coverslip over the probe and allow it to spread evenly under the coverslip.
8. Seal the coverslip by applying clear nail polish around the periphery of the coverslip.
9. Store the hybridized slide in the dark at −20°C.

3.5. Quantification of Hybridization Signals Using Epifluorescence Microscope

1. Identify regions for FISH analysis on H&E-stained sections for each case.
2. Analyze the hybridization signals on each nucleus using an epifluorescence microscope. Score at least 150–200 nuclei from identified tumor areas within the slide where possible (see Note 6).
3. Analyze and document the slide using the Applied Spectral Imaging System (Fig. 1).

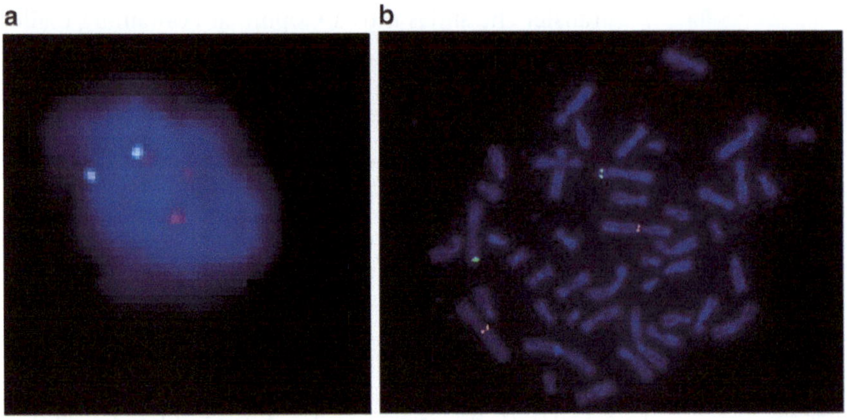

Fig. 1. Interphase FISH analysis in the detection of *PAX8-PPARγ* translocation. (**a**) Two *orange*, one *green*, and one fused *orange–green* (*yellow*) signal pattern observed in a nucleus positive for the *PAX8-PPARγ* translocation. (**b**) *Orange* signals on the *PAX8* region of locus 2q14 and *green* signals on the *PPARγ* region of locus 3p25 observed in a normal metaphase spread (magnification ×1,000).

4. Notes

1. Prepare the humidified hybridization chamber in the laboratory by wetting paper towels with water in a plastic container. Elevate the slides by putting the slides on two adjacently placed serological pipettes on the wet paper towels.
2. Optimization of enzymatic nucleus digestion is critical to ensure success of assay. Digestion time depends on the type of tissue used. In our laboratory, digestion time of ~30–40 min is required for breast tissues. For thyroid tissues, an optimal digestion of 60 min is required.
3. DAPI evaluation is a self-developed technique to assess the adequacy of pepsin digestion (Fig. 2, Chia et al., unpublished data). If tissue sections are underdigested, redigest by repeating step 3–5 in Subheading 3.1.2 for 5–30 min, depending on the extent of underdigestion. Reevaluation after redigestion is optional. Autofluorescence will become problematic and underestimation of signal copy number may result if tissue sections are underdigested. If tissue sections are overdigested, discard the slide and replace with new tissue sections. Overdigestion of tissue will decrease signal intensity and destroy nuclear morphology.
4. The size of the DNA smear on agarose gel electrophoresis should be in the range of 300–3,000 base pairs (bp). If the size of the smears is >3,000 bp, the reaction tube should be re-incubated at 15°C for another 15 min, followed by agarose gel electrophoresis. If the DNA smear is <300 bp, the probe labeling step should be repeated from the first step of Subheading 3.2.1

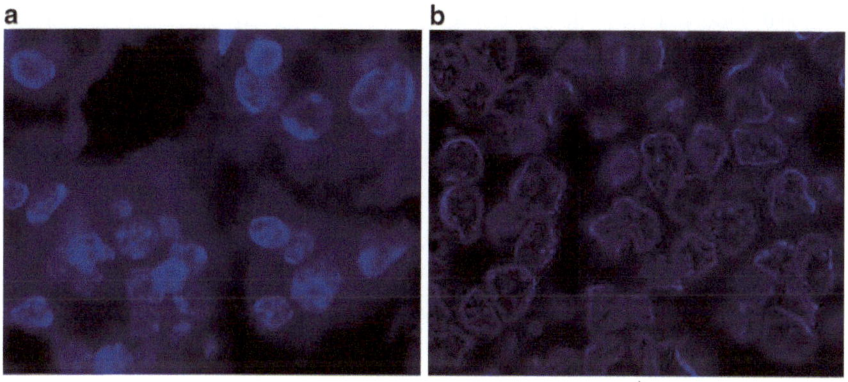

Fig. 2. DAPI Evaluation. (**a**) Underdigested nuclei showed hazy background with "cloud-like-structure" obstructing the nuclei. Fluorescence probe will not be able to penetrate into underdigested nuclei for hybridization. (**b**) Overdigested nuclei showed "hollow" nuclei which are not suitable for hybridization. Overdigested sections are to be discarded and to be replaced with a new tissue section (magnification ×200).

by reducing the incubation time at 15°C for the particular probe. Probe fragments >3,000 bp will produce low fluorescence intensity, while probe fragments <300 bp will result in nonspecific hybridizations.

5. Unlabeled fluorescence molecules will be filtered by using Sephadex G-50 QuickSpin Columns. Unlabeled fluorescence increases fluorescence background and reduces signal intensity if not filtered.

6. It is important for the person who conducts the FISH analysis to become familiar with the size, shape, and fluorescent intensity of each of the DNA probe signals in order to accurately evaluate the hybridization results. The signal should be bright, distinct, and easily enumerated. The background should appear dark or black and free of fluorescing particles and the probe should not cross-hybridize to other loci (Fig. 1).

References

1. Kroll TG, Sarraf P, Pecciarini L et al (2000) *PAX8-PPARγ1* fusion oncogene in human thyroid carcinoma. Science 289:1357–1360
2. Lui WO, Foukakis T, Liden J et al (2004) Expression profiling reveals a distinct transcription signature in follicular thyroid carcinomas with a PAX8-PPARγ fusion oncogene. Oncogene 24:1467–1476
3. Liehr T, Manvelyan M (2009) Characterization of archived formalin-fixed/paraffin-embedded or cryofixed tissue, including nucleus extraction. In: Liehr T (ed) Fluorescence in situ hybridization (FISH)—application guide. Springer, Berlin, Heidelberg, Germany, pp 147–156
4. Ventura RA, Martin-Subero J, Jones M et al (2006) FISH analysis for the detection of lymphoma-associated chromosomal abnormalities in routine paraffin-embedded tissue. J Mol Diagn 8:141–151
5. Chia WK, Sharifah N, Reena R et al (2010) Fluorescence in situ hybridization analysis using PAX8- and PPARG-specific probes reveals the presence of PAX8-PPARG translocation and 3p25 aneusomy in follicular thyroid neoplasms. Cancer Genet Cytogenet 196:7–13
6. Pinkel D, Straume TA, Gray J (1986) Cytogenetic analysis using quantitative, high-sensitivity, fluorescence hybridization. Proc Natl Acad Sci USA 83:2934–2938

7. Liehr T (2010) Fluorescence in situ hybridization (FISH)—quality issues in molecular cytogenetics. In: Kristoffersson U, Schmidtke J, Cassiman J-J (eds) Quality issues in clinical genetic services. Springer Science+Business Media, New York, pp 315–320
8. van Stedum S, King W (2002) Basic FISH technique and troubleshooting. In: Fan Y-S (ed) Molecular cytogenetics: protocols and applications, vol 204. Humana Press, Totowa, pp 51–63
9. Chin S-F, Daigo Y, Huang H et al (2003) A simple and reliable pretreatment protocol facilitates fluorescent in situ hybridisation on tissue microarrays of paraffin wax embedded tumour samples. Mol Pathol 56:275–279
10. Dwight T, Thoppe SR, Foukakis T, Lui WO, Wallin G, Hoog A, Frisk T, Larsson C, Zedenius J (2003) Involvement of the PAX8/peroxisome proliferator-activated receptor gamma rearrangement in follicular thyroid tumors. J Clin Endocrinol Metab 88:4440–4445

Chapter 14

Immunohistochemical Techniques to Identify and Localize Proteins of Interest in Paraffin Embedded Tissue Sections

Carolyn M. Komar and Mary Jane Long

Abstract

Proteins carry out cellular functions. Identifying proteins within tissues, which are characteristically comprised of various cell types, is critical to understanding how the tissue functions. Being able to assess protein expression in tissues is also essential to gaining insight into how tissues change under different physiological conditions, in pathological states, in response to treatments, etc. Immunohistochemistry exploits antibody-antigen associations to identify specific proteins within tissues. This is a very powerful technique as it allows for studying tissues intact, preserving cellular relationships, and tissue structure. Here, we discuss the process of using an antibody specific for PPARγ to identify ovarian cells that express this transcription factor, and how its expression changes during the ovarian cycle.

Key words: Peroxisome proliferator-activated receptor γ (PPARγ), Ovary, Immunohistochemistry, Paraffin, Tissue sections, Antibody, Immunostaining

1. Introduction

Histology is the study of the microscopic structure of tissues in relation to their function. Tissues are comprised of cells, and most tissues contain multiple cell types. Knowledge of how varied cellular components of a tissue work together is critical for understanding normal function and assessing pathological conditions.

Cellular function is carried out by proteins. The proteins expressed by a cell represent the "cellular workforce," and regulation and modification of protein expression controls cellular

Mostafa Z. Badr and Jihan Youssef (eds.), *Peroxisome Proliferator-Activated Receptors (PPARs): Methods and Protocols*, Methods in Molecular Biology, vol. 952, DOI 10.1007/978-1-62703-155-4_14, © Springer Science+Business Media New York 2013

function. Immunohistochemistry is a powerful technique because it enables investigation of protein expression and localization in tissue, preserving cellular relationships and tissue structure.

Immunohistochemical techniques utilize antibodies to identify specific proteins of interest within a tissue. One immunohistochemical method, immunofluorescence, uses a fluorescently tagged antibody against a specific antigen to identify cellular expression and location of the antigen when the tissue is observed with a fluorescent microscope. A second method uses antibodies complexed with an enzyme that when incubated with substrate, generates a color reaction to identify where the antigen is located (Fig. 1). Tissues processed with this latter method can be observed with a light microscope. Labeling using this approach is permanent because unlike fluorescence, the color does not fade.

Both processes of immunolocalizing protein in tissues are able to identify low quantities of proteins. The sensitivity of the latter method is greatly enhanced by exploiting the affinity of avidin/strepavidin for biotin, to generate an immunocomplex that amplifies the enzymatic color reaction. Avidin is a large protein found in egg whites; strepavidin is its bacterial counterpart isolated from *Streptomyces aridimii.* Biotin is a small, organic molecule that belongs to the group of B-complex, water soluble vitamins. Avidin/strepavidin has four binding sites for biotin and the non-covalent interaction between the two is essentially irreversible. Antibodies can be tagged with biotin, and when mixed with a complex of biotinylated-horseradish peroxidase (HRP)/ avidin, increases the amount of enzyme in the area of the protein of interest (Fig. 1). The resultant robust color that develops following the addition of substrate allows detection of even small amounts of protein.

Analysis of tissues using immunohistochemistry identifies cellular patterns of protein expression, and how expression may change under physiological or pathological situations, in response to treatment(s), etc. The procedure described below details the immunolocalization of an antigen using a biotinylated primary antibody and the biotinylated HRP/avidin complex. The antigen, or protein of interest, is peroxisome proliferator-activated receptor γ (PPARγ). The tissue, representing one made up of many different cell types, is the ovary. Figure 2 illustrate the usefulness of this technique to identify protein expression patterns in tissues and how that expression changes under normal physiological conditions, in this case, during the female reproductive cycle.

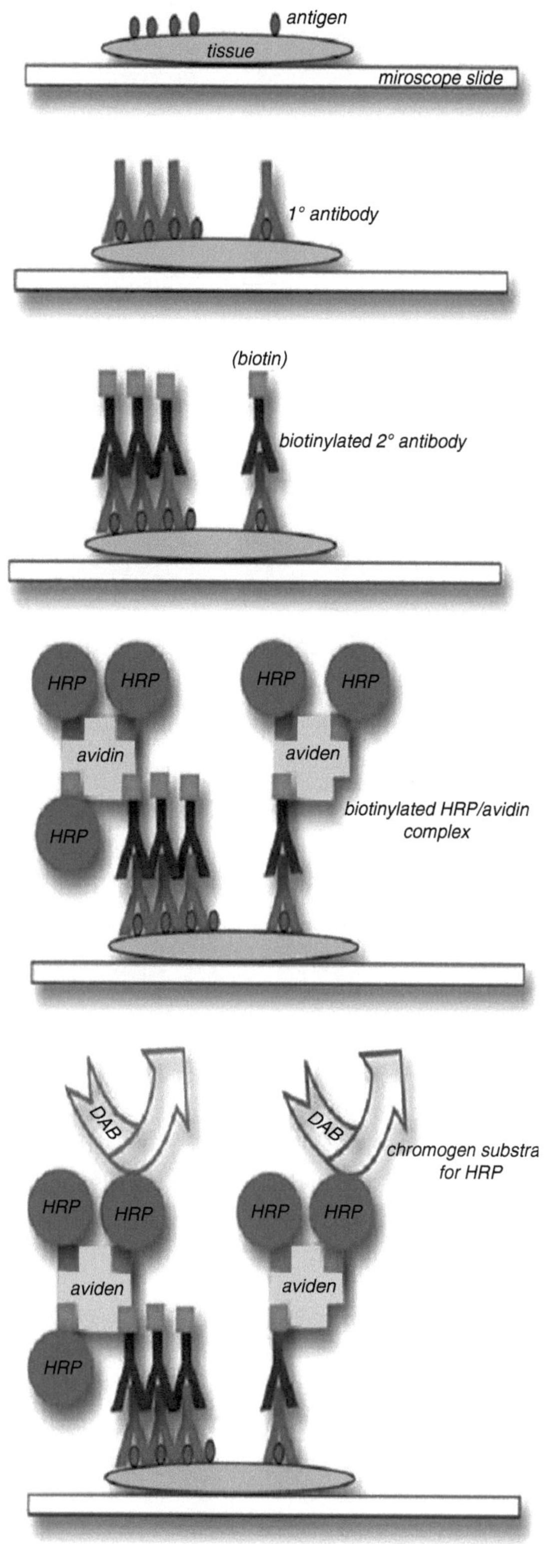

Fig. 1. Schematic of the immunohistochemical labeling process. See text for details. 1°: primary; 2°: secondary; DAB: 3, 3, diaminobenzidine tetrahydrochloride; HRP: horseradish peroxidase; avidin or its bacterial counterpart, strepavidin.

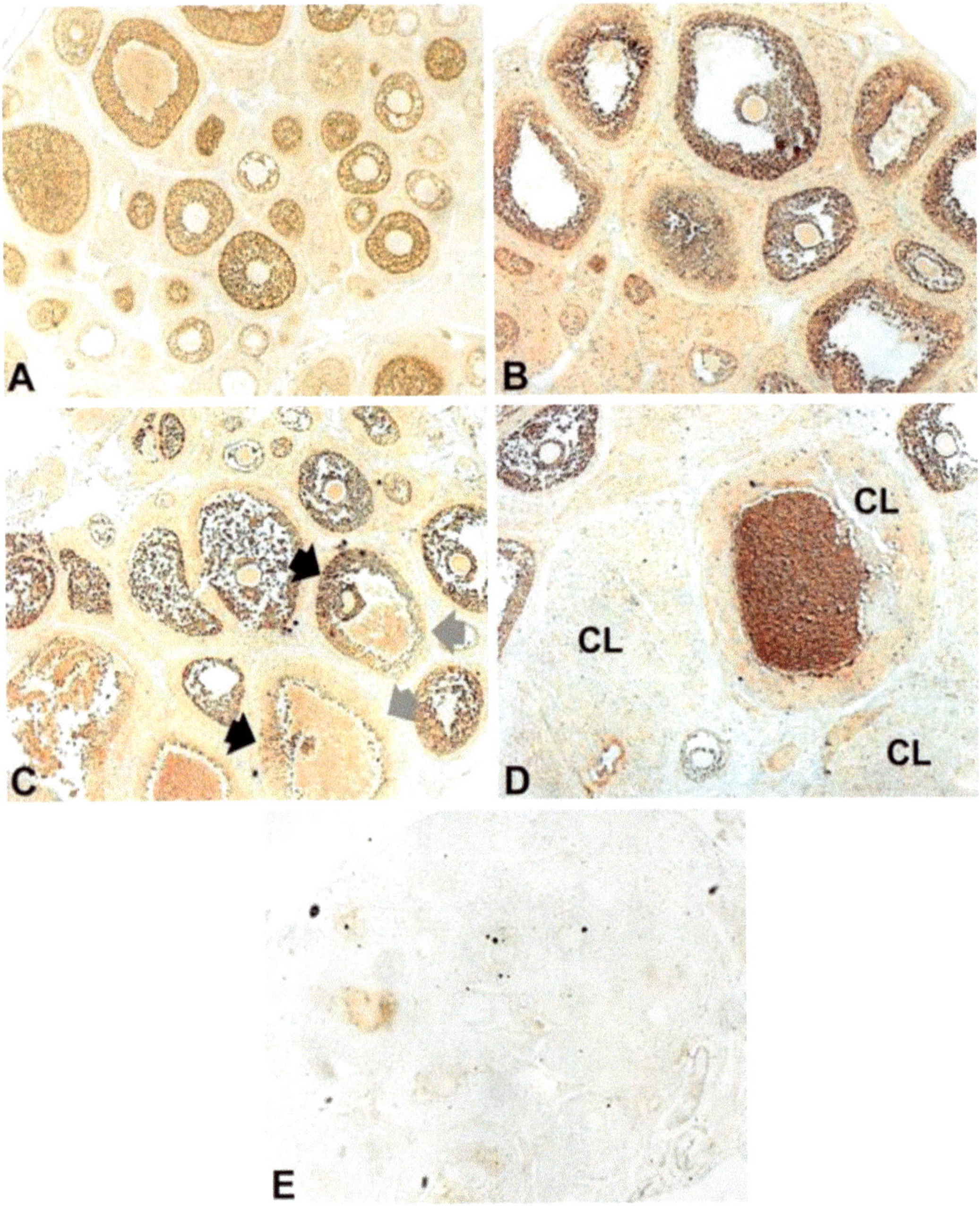

Fig. 2. Immunolocalization of PPARγ in ovarian tissue sections collected from rats 0 h (**a**, **e**) and 48 h (**b**) after administration of pregnant-mares' serum gonadotropin to stimulate follicular development, and 4 h (**c**) and 24 h (**d**) after treatment with human chorionic gonadotropin to induce ovulation and luteinization. Protein corresponding to PPARγ was identified by the *dark* (*brown*) reaction product. The tissue section pictured in (**e**) was treated with normal goat serum in place of the anti-PPARγ antibody. *Arrows* indicate granulosa cells expressing high levels of PPARγ (*black*) vs. those that no longer express PPARγ (*gray*) within the same follicle. *CL* corpus luteum. Magnification: 100×. Originally published in (1).

2. Materials

2.1. Buffers

1. Phosphate buffered saline (PBS): 10× stock solution, pH 7.2. For 1 L of 10× PBS, add 76 g sodium chloride ($NaCl_2$), 11.2 g sodium phosphate (NaH_2PO_4), 14 g disodium hydrogen phosphate (Na_2HPO_4) to ~900 mL water (see Note 1). Mix and adjust pH with HCl or NaOH as needed. Bring up to 1 L with water. Store at room temperature.
2. 1 M Tris–HCl buffer, pH 8.5: Add 157.6 g Tris–HCl to ~900 mL of water. Mix and adjust pH as above. Store at room temperature.

2.2. Immunohistochemical Components

1. Blocking serum, biotinylated secondary antibody (goat anti-mouse IgG), and biotinylated HRP/avidin were purchased from Vector Laboratories (Burlingame, CA).
2. The primary antibody against PPARγ, a mouse monoclonal IgG, was obtained from Santa Cruz Biotechnology (E8; Santa Cruz, CA).
3. Chromogens for use as substrate for peroxidase—3,3, diaminobenzidine or 3, amino-9-ethylcarbazole (AEC), were purchased from Zymed (now InVitrogen, Carlsbad, CA) (see Note 2).
4. The aqueous mounting media for use with AEC was obtained from Vector Laboratories.

2.3. Non-disposable Processing Supplies

Slides can be processed using Coplin jars, Wheaton staining dishes, or any staining dish with appropriate slide rack (see Note 3).

3. Methods

3.1. Slide Preparation and Rehydration (see Note 4)

1. Deparaffinize tissues by placing slides in xylenes (see Note 5) for 5 min, 2× (see Note 6). Blot slides on an absorbent pad before the next step.
2. Rehydrate tissue by processing slides though decreasing concentrations of ethanol: 100, 95, 70, and 50 for 10–15 s each. Blot slides on absorbent paper.
3. Place slides in water.

3.2. Antigen Retrieval (see Note 7)

1. Place slides in Coplin jars filled with 0.01 M Tris–HCl (dilute 1 M stock solution of Tris–HCl with water).
2. Cover jars loosely with cap or Saran wrap.
3. Heat in a microwave on the highest setting (750 W) for 5 min.

4. Check the level of solution in Coplin jars and refill with 0.01 M Tris–HCl to fully cover slides.
5. Heat slides for another 5 min at the same setting and then remove from microwave and allow to cool to room temperature (15 min) (see Note 8).

3.3. Quench Endogenous Peroxidase Activity

1. Dilute hydrogen peroxide to 1.5% with methanol (see Note 9).
2. Place slides in 1.5% H_2O_2 for 20 min at room temperature.
3. Wash slides 2× in 1× PBS (dilute 10× PBS with water) for 5 min.

3.4. Blocking and Immunolabeling (see Note 10)

1. For immunolabeling using the E8 PPARγ antibody, use normal goat serum (Vector Laboratories) diluted according to the manufacturer's instructions using 1× PBS.
2. Remove slides from the last PBS wash one at a time.
3. Using a KimWipe, wipe excess buffer off the back of each slide and the front while carefully avoiding the tissue (see Note 11).
4. Place an equal volume of diluted serum on each slide (see Note 12).
5. Carefully place slides in a humid chamber (see Note 13) at 37°C for 10 min (see Note 14).
6. After a 10 min incubation, blot the excess serum solution from each slide by tipping the slide and touching its edge to an absorbent pad.
7. Add an equal volume of primary antibody solution (1:100 (see Note 15); diluted in 1× PBS) to each designated "treated" slide, or normal serum of an equivalent dilution (1:100 in 1× PBS) to slides serving as negative controls. Again, ensure coverage of tissue, but do not add solutions in excess.
8. Return slides to humid chamber and incubate at 4°C overnight (see Note 16).
9. The following day, blot off remaining solution and place slides in Coplan jars or slide holders.
10. Wash 2× in 1× PBS for 5 min each.

3.5. Chromogen for Protein Localization

1. Remove slides from the last PBS wash and wipe with a KimWipe as described above.
2. Add an equal volume of biotinylated secondary antibody solution (in this case, goat anti-mouse IgG diluted 1:100 (see Note 17) in 1× PBS) to each slide.
3. Incubate for 10 min at room temperature in a humidified chamber.
4. Blot off excess solution and place slides in 1× PBS.

5. Wash slides 2× in 1× PBS for 5 min each.
6. Prepare the solution of biotinylated HRP/avidin according to manufacturer's directions.
7. Remove slides from PBS and wipe with a KimWipe as described above.
8. Add an equal volume of the biotinylated HRP/avidin solution to each slide.
9. Incubate slides at room temperature for 30 min in a humid chamber.
10. Wash slides 2× in 1× PBS for 5 min. During this time, prepare the chromogen solution (DAB; AEC) according to manufacturer's directions.
11. Remove slides from PBS wiping with a KimWipe as described above.
12. Add an equal volume of chromogen solution to each slide and incubate at room temperature for 3–5 min (see Note 18). If using DAB, keep solution in the dark and incubate slides in the dark.
13. To stop the peroxidase/chromogen color reaction, place slides in water (see Note 19).
14. If AEC was used, remove slides one at a time from water and mount with an aqueous mounting medium (Fig. 3).

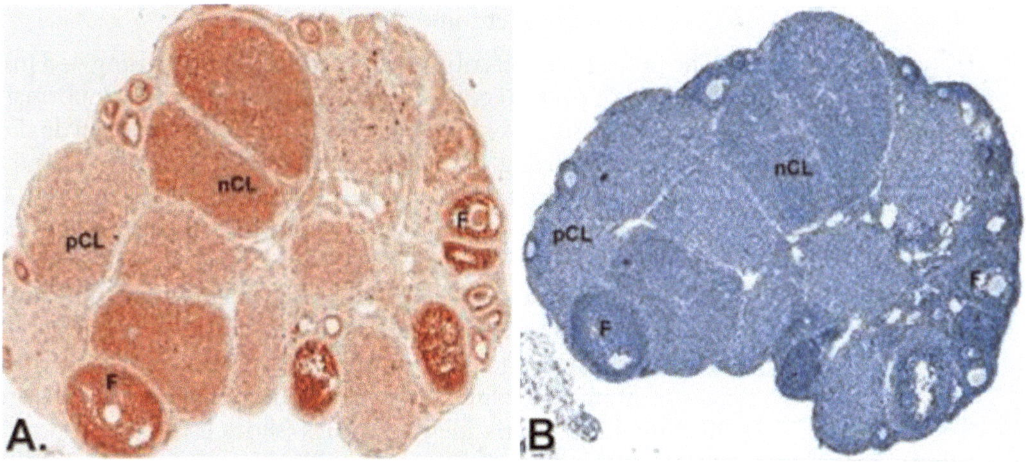

Fig. 3. Immunolocalization of the transcriptional co-activator with PDZ-binding motif (TAZ) in the ovary. Ovaries collected from mature, cycling, female mice were fixed in 4% paraformaldehyde and embedded in paraffin. Tissue sections (8 μm) were processed with an anti-TAZ antibody ((**a**); IMGENEX; 1:500), or normal rabbit serum (**b**) as a negative control. The latter tissue was counterstained with hematoxylin to assist visualization of the tissue. TAZ is denoted by the *dark* (*red*) reaction product (AEC used as chromogen). Magnification: ×20. *F* follicle; *nCL* corpus luteum from the current reproductive cycle; *pCL* corpus luteum from the previous reproductive cycle.

15. If DAB was used, proceed to dehydrate tissues by passing them through increasing concentrations of ethanol for 10–15 s each: 50, 70, 95, and 100%.
    (a) Blot slides on an absorbent pad and place in xylenes.
    (b) Remove slides from xylenes one at a time and mount with permount (Fig. 2).

4. Notes

1. Ultrapure water (18 MΩ at room temperature) or distilled water should be used for preparation and dilution of all solutions.
2. Waste should be properly handled according to specific disposal regulations.
3. It is important that the jar/dish used is filled with enough of each solution to ensure complete coverage of the tissue on the slide. To ensure complete coverage of tissue on slides, use enough reagent to cover 2/3 of the slide or the entire glass portion up to the frosted marking area. It is recommended that no more than 15 slides be processed at a time, especially until one becomes proficient with the procedure. Tissues must not dry during the labeling process as this will greatly increase the background. This is particularly relevant when slides are individually hand-processed: blocking; addition of each antibody; addition of chromogen.
4. The procedures described below were optimized using 4–5 μm sections of paraffin embedded tissue fixed with 4% paraformaldehyde. Tissues fixed in Bouin's solution or formaldehyde did not yield as strong a reaction, or any reaction, respectively, in our hands with this particular primary antibody. Once the paraffin is removed from the tissue sections, the sections can be difficult to identify on slides. Prior to starting the procedure, a PAP pen can be used to circle tissue sections on the slide. The PAP pen will leave behind a "waxy" substance that will both identify the location of the tissue and act as a barrier to concentrate the solutions on the surface of the tissue. Before beginning, properly identify tissues/slides using a lead pencil to mark the frosted area of the slide. Tissues treated with the primary antibody against the protein of interest (in this case, PPARγ) may be listed as "treated" or identified by the primary antibody used. Tissues serving as negative controls will not be treated with the primary antibody, but rather nonimmune or pre-immune serum from the same species used to generate the primary antibody. If this serum is not available, normal serum from the species used to generate the secondary antibody can be used in place of the primary antibody.

5. Waste should be properly handled according to specific disposal regulations. Hemo-D is an effective, less toxic alternative to xylenes.
6. To decrease the occurrence of air bubbles when slides are placed in a solution, dunk slides once or twice before leaving the slides in the solution for the desired time.
7. During fixation, protein cross-linking can occur that prevents antibody access to antigen in tissue. Antigen retrieval is a procedure that can enhance immunohistochemical staining. This step can be omitted to determine if antigen retrieval is necessary for optimum immunostaining of your tissue. If antigen retrieval is skipped, remove slides from water and place in 1× PBS for 3 min before proceeding to treatment with 1.5% hydrogen peroxide.
8. An alternative solution that may enhance the effects of microwaving is a 0.01 M citrate buffer (pH = 6.0) *in lieu* of the 0.01 M Tris–HCl buffer. There are other options for antigen retrieval that may be identified empirically for the tissue and antibodies of interest. Other options include boiling slides in citrate or Tris buffer; treating tissue with 0.1% trypsin for 1 h at 37°C.
9. The percentage of hydrogen peroxide may be adjusted according to the endogenous peroxidase activity of the tissue being analyzed. If there is high background or nonspecific staining, one way to address this is by increasing the percentage of hydrogen peroxide.
10. Normal (pre-immune) serum from the species used to generate the secondary antibody is used to block potential interactive sites in the tissue to decrease nonspecific binding.
11. Drying the slides in this manner will help prevent "bleeding" of the blocking serum from the surface of the tissue and will also prevent drying of the tissue.
12. Make sure enough serum solution is added to ensure complete coverage of the tissue, but not so much that it is at risk of running off the slide. A guide is to use between 250 and 500 μL/slide, depending on the size/amount of the tissue to be covered. The volume used should generate a hydrostatic bubble over the tissue sections.
13. A humid chamber can be constructed from any shallow box/pan with a lid, or foil/Saran wrap cover. For example, serological pipettes (5 or 10 mL size) can be taped to the bottom of a 9×13 in. Pyrex dish along its length. Cut strips of Whatman filter paper to cover the bottom of the dish or place paper towels in the bottom of the dish and set in between the pipettes. Soak the paper with 1× PBS. To create a humid chamber, cover tightly with foil or Saran wrap.

14. Do not place slides in the humid chamber close enough so that they touch. If the slides touch, there is the risk that the solution may run off one slide and onto another, or onto the pipettes suspending them. This may result in the tissues drying and increasing background, or unequal exposure of tissues to solutions which will lead to uneven immunolabeling.
15. At this point it is important not to have slides touching in the humid chamber for the reasons listed above, but more importantly, to prevent contamination of slides. Since not all slides will receive the primary antibody solution, as those serving as negative controls will get normal serum, it will prevent cross contamination.
16. This step can also be conducted for 1 h at 37°C.
17. The dilution of primary and secondary antibody that gives the most optimal immunostaining will need to be derived empirically for each tissue and antibody of interest. Initial experiments using a series of dilutions will determine optimum concentrations. A good initial dilution series is: 1:10; 1:100, 1:1,000, 1:10,000. A second experiment using a more focused dilution series is often necessary to ensure optimum labeling, and to avoid unnecessarily wasting antibody. For example, if 1:1,000 yielded the best results in the initial experiment, the second experiment might test: 1:250, 1:500, 1:1,000, 1:1,500, 1:2,000.
18. The length of time tissues are exposed to chromogen solution can be monitored by periodically looking at slides briefly under a light microscope. As the color develops, the reaction should be stopped before the background becomes too dark.
19. At this point in the protocol, slides can be counterstained with hematoxylin. However, depending on the reaction, counterstaining may be omitted to better illustrate the immunolabeling.

Acknowledgments

CMK is indebted to C. A. Smith, Research Associate, College of Veterinary Medicine, Cornell University, who taught her the "tricks of the trade." The critical reading of this chapter by Bobbi Hoover and Dr. Bonny Dickenson is also gratefully acknowledged. The authors also acknowledge the technical assistance of Karen Ayers for constructing Fig. 1.

Reference

1. Banerjee J, Komar CM (2006) Effects of luteinizing hormone on peroxisome proliferator-activated receptor γ in the rat ovary before and after the gonadotropin surge. Reproduction 131:93–101

Chapter 15

Determination of PPAR Expression by Western Blot

Aziz Guellich and Catherine Coirault

Abstract

Peroxisome proliferator-activated receptors (PPARs) are key nuclear receptors which mediate the regulation of the transcription of many genes, especially those involved in lipid catabolism and inflammation. In mammals, three types of PPARs named PPARα, PPARβ (also called PPARδ), and PPARγ have been identified. Up and down-regulations of one or more of these receptors are encountered during development, physiological adaptations, and in several diseases. Therefore, the ability to accurately measure PPAR expression in cells and tissues is an important aspect of PPAR research. This can be done by Western blot with specific antibodies. In this chapter, we describe the use of our techniques 1,2 to measure PPARs expression, describe the adequate way to extract proteins, the best antibodies currently available and discuss potential misleading results in the absence of appropriate controls.

Key words: Electrophoresis, PPARs, Immunoblot

1. Introduction

Western blot is an analytical technique which consists of transferring proteins from a gel to a nitrocellulose or nylon membrane following electrophoresis separation. It allows the detection and the relative quantification of the expression of a specific protein in tissue or cell homogenates. Western blot requires specific primary antibodies directed against the desired protein of interest and a detection system which usually involves horseradish peroxidase-conjugated secondary antibodies and chemiluminescence.

Mammals have three PPARs, they are encoded by three different genes and in general, they differ in their tissue distribution pattern and they have specific and distinct function (1–2). PPARs share 60–80% homology in their ligand- and DNA-binding domains (3). They have also close molecular weights which lead to a "superposed" migration pattern when subjected to electrophoretic separation.

Mostafa Z. Badr and Jihan Youssef (eds.), *Peroxisome Proliferator-Activated Receptors (PPARs): Methods and Protocols*, Methods in Molecular Biology, vol. 952, DOI 10.1007/978-1-62703-155-4_15, © Springer Science+Business Media New York 2013

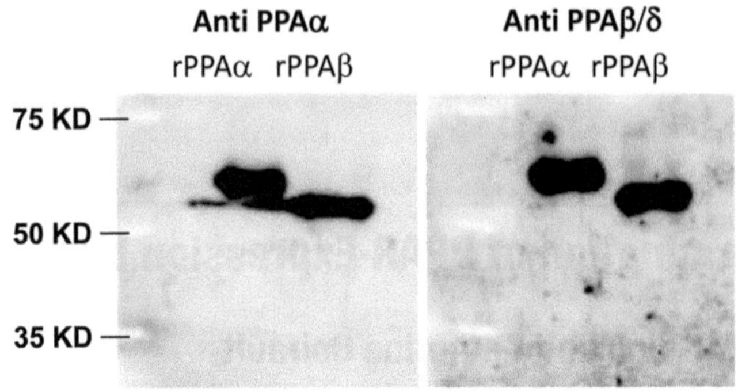

Fig. 1. Example of inter-cross-reactivity between two commercial anti-PPARα and anti-PPARδ, 200 ng of recombinant PPARα or PPARβ/δ were loaded and detected with anti-PPARα and anti-PPARβ/δ. Both antibodies are not specific and so not suitable for western blot use.

Given these considerations, only antibodies able to discriminate reliably between theses subtypes could be used to carry out western blot for PPARs.

PPARs are nuclear receptors; they are so naturally more abundant in the nucleus than in the cytoplasm. In some cases, using nuclear extracts may help to overcome lower sensitivity issues and improve the detected signal.

Once the band of a PPAR appears at the expected molecular weight, a strict quality control must be done to assure the specificity of each antibody. This is extremely crucial, failure to have the adequate controls such recombinant proteins, protein extracts from tissues known to express high level of a specific PPAR (liver for PPARα, muscle for PPARδ, fat tissue for PPARγ), and neutralization experiments may lead to confusing results. Example of a western blot image showing a considerable cross-reactivity of two commercial anti-PPARα and anti-PPARδ antibodies is shown in Fig. 1. Such issues should be formally assessed before considering the use of valuable samples. Information such as the peptide (when it's not a full length protein) used to raise the antibody against provides important elements to predict some issues. In silico analysis may give an idea on possible overlapping by comparing and reviewing the sequence of antigen and the isotypes. Species-to-species and batch-to-batch variabilities must also be taken in consideration. The same antibody characterization must be done each time a new species or new batch of antibody is used.

2. Materials

2.1. Equipment

1. Tissue homogenizer.
2. Centrifuge.
3. Lab microwave oven.

4. Nanodrop (ND-1000 spectrophotometer, NanoDrop Products, Wilmington, DE) or standard spectrophotometer for dosage of proteins.
5. Complete electrophoresis system, with glass plates, spacers, and 10 teeth comb (0.75 mm). In our lab, we use Mini-Protean II (Bio-Rad).
6. Power supply with constant current and/or voltage.
7. Complete transfer system (Mini Trans-Blot cell) with tank transfer, electrode module, two cassettes, and four fiber pads (sponges).
8. Detection system.
9. Agitator.

2.2. Supplies

1. 1.5, 15, and 50 mL tubes.
2. 10, 100, and 1,000 μL pipettes.
3. 5 and 10 mL pipettes.
4. Cell scrapers.
5. Spectrophotometer cuvettes.
6. 25 μL syringe, type Microliter (Hamilton).
7. Filter paper.
8. Nitrocellulose membrane with 0.45 μm pore sizes (Schleicher & Schuell).
9. Heat sealable plastic bag.
10. Plastic box.
11. Flat tweezers.
12. Magnetic stir.

2.3. Reagents

1. Distillated water (ddH_2O) (see Note 1).
2. NP-40.
3. Sodium deoxycholate ($C_{24}H_{39}NaO_4$).
4. Sodium orthovanadate (Na_3VO_4).
5. Ethanol 70%.
6. Absolute methanol.
7. Isopropanol.
8. Filtered sodium dodecyl sulfate (SDS) 10% (w/v).
9. Tris (Trizma-base): powder form and prepared solutions (Tris–HCl 1.5 M-pH 8.8, Tris–HCl 1 M-pH 7.4 and 0.5 M-pH 6.8) (see Note 2).
10. Acrylamide/bis 30% (see Note 3).
11. Ammonium persulfate (APS) 10% (w/v) in ddH_2O (see Note 4).
12. *N, N, N,N*-tetramethyl-ethylenediamine (TEMED) (Sigma). Store at 4°C.

13. Protein molecular-weight standards.
14. Tetraacetic acid (TCA) 0.3%.
15. Bromophenol blue.
16. Glycerol.
17. Beta-Mercaptoethanol.
18. Dithiothretiol (DTT) 1 M.
19. Ponceau red.
20. Tween.
21. Non fat dry milk.

2.4. Kits, Recombinant Proteins, and Antibodies

1. NE-PER nuclear and cytoplasmic extraction kit (Pierce).
2. Protease inhibitor cocktail (Roche).
3. BCA protein assay kit (Pierce) in the case of dosing proteins with a spectrophotometer.
4. Recombinant PPARs (Santa Cruz).
5. Primary anti-PPARs.
6. Secondary antibodies.
7. ECL plus (Amersham).

2.5. Stock Solutions, Gels, and Buffers

Filter all buffers before use. For Tris solutions, systematically check pH (see Note 5).

1. 1.5 M Tris–HCl, pH 8.8. Dissolve 90.86 g of Tris base (MW = 121.14) in 500 mL of ddH_2O. Adjust pH to 8.8 using HCl. Store at 4°C.
2. 1 M Tris–HCl, pH 7.4. Dissolve 60.58 g of Tris base (MW = 121.14) in 500 mL of ddH_2O. Adjust the pH to 7.4 using HCl. Store at 4°C.
3. 0.5 M Tris–HCl, pH 6.8. Dissolve 30.29 g of Tris base in 500 mL of ddH_2O. Adjust pH to 6.8 using HCl. Store at 4°C.
4. SDS 10% (w/v). Dissolve 50 g of SDS in 500 mL of ddH_2O. Store at room temperature.
5. TCA 0.3%. Dissolve 3 g of TCA in 1 L of ddH_2O.
6. Tris buffered saline (TBS 10×). 1.37 M NaCl, 0.1 M Tris, pH 7.4. Weight 30 g Tris base and 80 g NaCl in 1,000 mL of ddH_2O. Adjust pH to 7.4 using HCl.
7. TBS containing 0.05% Tween (TBS-T). To make TBS-T, add 900 mL of ddH_2O to 100 mL of TBS (10×), then add 1 mL of tween. Adjust pH to 7.4 and mix well.
8. Resolving gel solution. In our studies we use either 9 or 12% SDS polyacrylamide gels.

9. Prepare the 12% SDS polyacrylamide gel by mixing 2.5 mL of Tris–HCl 1.5 M, pH 8.8 with 100 μL of SDS, 4 mL of acrylamide/Bis solution and 3.35 mL of ddH2O. Add 50 μL of APS and 12.5 μL of TEMED.
10. Prepare the 9% SDS polyacrylamide gel by mixing 2.5 mL Tris–HCl 1.5 M, pH 8.8 with 100 μL of SDS, 3 mL of acrylamide mixture and 4.35 mL of ddH$_2$O. Add 50 μL of APS and 12.5 μL of TEMED.
11. Allow space for stacking gel and gently overlay with water or isobutanol (see Note 6).
12. Stacking gel solution. Prepare the stacking gel by mixing 1.25 mL of Tris–HCl 1.5 M pH 8.8 with 50 μL of SDS, 750 μL of acrylamide/Bis solution and 3.05 mL of ddH$_2$O. Add 50 μL of APS and 7.5 μL of TEMED. Insert a 10-well gel comb immediately without introducing air bubbles.
13. SDS extraction buffer: 1% SDS, sodium vanadate 1 mM, Tris–HCl 10 mM, protease cocktail inhibitor (see Note 7).
14. RIPA buffer: Tris–HCl 50 mM, NaCl 150 mM, NP-40 1%, SDS 0.1%, sodium deoxycholate 0.25%, protease cocktail inhibitor (see Note 8).
15. Laemmli buffer (2×): Tris 62.5 mM, 2% SDS, 20% Glycerol, 0.2% (w/v) bromophenol Blue, 5% β-mercaptoethanol (w/v), 0.1% (v/v) 1 M DTT (see Note 9).
16. Electrophoresis running buffer: 0.25 M Tris, 192 mM glycine, 0.1% SDS (see Note 10).
17. Protein transfer buffer: 0.25 M Tris, 192 mM glycine, 0.01% SDS, 15% methanol (see Note 11).

2.6. Protein Extraction

2.6.1. Total Protein Extract from Cells

1. Add 0.2 mL of RIPA buffer to the cells in the plate.
2. Scrap off adherent cells with a scraper at 4°C.
3. Transfer the resulting lysate to a microcentrifuge tube.
4. Incubate 40 min on ice.
5. Centrifuge cell lysate at 10,000 × *g* for 10 min at 4°C.
6. Transfer the resulted supernatant which corresponds to total proteins to a new tube. Keep at −80°C.

This buffer can be used also for tissues.

2.6.2. Total Protein Extract from Muscles

1. Take a piece of muscle biopsies of approximately 20 mg.
2. Homogenize in boiling SDS buffer using an electric homogenizer (Turex).
3. Centrifuge at 12,000 × *g* for 5 min at 15°C.
4. Collect the supernatant which corresponds to total protein. Keep at −20°C.

2.6.3. Nuclear-Enriched Fraction

We use a commercial kit (NE-PER nuclear and cytoplasmic extraction reagents, Pierce) for nuclear-enriched protein fraction. Follow the manufacturer's instructions. Keep at −80°C.

2.6.4. Protein Concentration

Protein concentration is determined using NanoDrop® (ND-1000 spectrophotometer, NanoDrop Products, Wilmington, DE) or a kit "BCA protein assay kit" (Pierce). Store the samples at −20°C for future use.

3. Methods

1. Clean glass plates with distilled water followed by ethanol 70% and by distilled water once again.
2. Assemble the glass-plate sandwich of the electrophoresis apparatus.
3. Clip the sandwich to the casting stand.
4. Pour distilled water in the sandwich and make sure there is no leak.
5. Pour out the water and carefully eliminate the rest using absorbent paper.
6. Mark the glass in order to leave enough space for stacking gel.
7. Once everything is ready, add APS 10% and TEMED to the resolving gel solution. Mix well.
8. Using a 5 mL pipette, add the resolving gel solution to the sandwich by applying its tip slightly against one of the spacers until the solution reaches the limit line. If you are running two gels, make sure to pour the same quantity of the gel solution in both sides.
9. Using 1 mL pipette and an edge cut tip, gently add 1 mL of isopropanol against the edge of one of the spacers (see Note 12). Make sure that the gel surface is flat by putting the eye axe at the same level as the top limit of the gel. Allow the gel solution to polymerize 40 min at room temperature (see Note 13).

3.1. Stacking Gel

1. Pour off isopropanol and rinse with distilled water.
2. Wick off residual liquid using absorbent paper.
3. Quickly add APS and TEMED to the stacking gel solution as indicated in the table.
4. Add the staking gel using 1 mL pipette. Avoid air bubbles, they can easily be pushed by the comb and deform the bottom of the well.

5. Insert a 0.75-mm comb into the stacking gel solution.
6. Add stacking gel solution in the corners of the comb and along it.
7. Allow the stacking gel solution to polymerize for 30–45 min at room temperature.

3.2. Sample and Load the Gel

1. Dilute a portion of the protein sample to be analyzed 1:1 (v/v) with 2× Laemmli buffer.
2. Heat immediately the samples for 5 min at 95°C.
3. Centrifuge the samples to remove insoluble debris which could produce streaks in the protein lanes (see Note 14).

3.3. Loading Samples and Electrophoresis

1. Remove the comb genteelly, a kind of resistant may be felt at the beginning. Don't pull out strongly.
2. Rinse the wells with electrophoresis buffer to remove unpolymerized gel and debris.
3. Reassemble the gel sandwich.
4. Fill the lower buffer chamber with ~250 mL of electrophoresis buffer.
5. Place the sandwich in the tank and fill up the upper chamber until the sample wells are filled with buffer.
6. Using a 25 μL Hamilton syringe, load the first well with molecular-weight standards, the second with a positive control if any and then proceed to protein sample(s) (10 μg/10 μL). Rinse the syringe with ddH_2O between uses. Load empty wells with 1× Laemmli buffer. Equal volume should be loaded in all the wells.
7. Connect the electrophoresis system to power supply and run for 30 min at 90 V until the samples enter the separating gel, then increase to 120 V for 2 h.
8. Stop power.
9. Discard the buffer and take out the gel sandwich.
10. Remember the orientation of you gels (see Note 15).
11. Disassemble the sandwich in transfer buffer and slide a plastic wedge between the glass plates in order to separate them and expose the gel.
12. Carefully cut off the stacking gel, peel off gently the resolving gel and place it in the transfer cassette.

3.4. Transfer

At least, half an hour before the end of electrophoresis, start preparing what needed for transfer.

1. Cut two pieces of Whatman paper and a piece of nitrocellulose membrane to an appropriate size, and soak them in transfer buffer (see Note 16).

2. Pour 500 mL of transfer buffer in a plastic or plixy glass container, and for each gel, wet four pieces of Whatman papers and two sponges for 10 min. Place a cassette in another container filled with 500 mL of cold transfer buffer with its black side oriented to the bottom.

In the following order:

3. Place one sponge.
4. Two pieces of Whatman paper.
5. The gel.
6. The membrane,—mark its orientation by cutting an edge for example.
7. Cover the membrane by a second piece of Whatman.
8. Gently roll out bubbles between the gel and the membrane.
9. Add the second sponge.
10. Clamp the sandwich in to the black part by sliding over the white clip. It is important to keep the sandwich cassette submerged in transfer buffer during all the transfer procedure.
11. Slide the sandwiches in the electrode module by respecting the code colors: the black side of the sandwich facing the black side of the module.
12. Assemble the electrode module and place it in the buffer tank.
13. Place the frozen ice block in the transfer tank.
14. Put a magnetic stirring bar in the tank and fill up with buffer.
15. Connect the transfer system to the power supply.
16. Place tank on a magnetic stirrer, turn on the agitator and the power supply for 2 h in the cold room.

3.5. Control of the Quality of Transfer and Blotting

To check the quality of the transfer and ensure that the same amount of protein is loaded we highly advice to stain the membranes with Ponceau red

1. Dismantle each sandwich and take out the membrane avoiding touching where the proteins are. Handle the corner only with flat forceps.
2. In a plastic box, add an adequate quantity of Ponceau red to cover all the membrane.
3. Place the membrane in the plastic box and stain for 30 s.
4. Detain the membrane by transferring it to other plastic box filled with ddH_2O (several changes are needed).
5. Put the membrane in a plastic pocket and scan it. Take a picture and rinse off the Ponceau red with 0.3% TCA.
6. Mark the bands of standard molecular weight using a pen or pencil.

7. Saturate the membrane in order to block unspecific sites by incubating it in 5%-milk in TBS-T (w/v), pH = 7.4 for 1 h under agitation.
8. Incubate for 1 h.
9. Incubate with the appropriate anti-PPAR.
10. Prepare an adequate volume of the first antibody in 5% milk—TBST. Add it and seal the blots in heat sealable plastic bags.
11. Incubate at 4°C overnight under agitation.
12. Wash the blot with two quick changes of TBST, then 2 for 5 min each followed by a fifth change of 15 min under agitation.
13. Prepare the adequate HRP conjugated second antibody diluted in 5% milk- TBST for 1 h 30 min at room temperature with agitation.
14. Incubate the membrane with the prepared second antibody for 1 h 30 min.
15. Wash the blot with two quick changes of TBST, then two for 5 min each followed by a fifth change of 15 min under agitation.

3.6. Detection

1. Put the gel on a flat surface.
2. Add ECL-Plus detection solution. Ensure that the entire membrane is covered by detection solution. For a small blot 1 mL is largely sufficient.
3. Incubate for 5 min at room temperature in the dark.
4. Wick off excess detection fluid and place the membrane inside a plastic bag.
5. Roll out bubbles.
6. Put the membrane in the FujiFilmX imager machine (Laser 3000, Fujifilm), adjust the focus, scan on fluorescence to detect the ladder bands and then on chemiluminescence. Adapt the exposition time in order to see clearly the bands without arriving to saturation.
7. For normalization. Cut the membrane while stained with Ponceau red and incubate it as described previously for PPARs with the appropriate antibody (anti-actin or anti-calsequestrin or other).

4. Notes

1. All solutions are prepared using ultrapure water (purify deionized water with a sensitivity of 18.2 MΩ cm at 25°C).
2. Tris can be dissolved faster if the water is warmed to about 37°C. However, care should be taken to bring the solution to room temperature before adjusting pH.

3. We use commercially available acrylamide:Bis solution (BioRad) containing 29:1 acrylamide:Bis. The solution can be stored at 4°C for 1 month.
4. Prepare and store aliquots at −20°C to avoid repeated freeze/thaw cycles.
5. pH of Tris solution may change, especially during the first days following its preparation. If you see any impurity, discard the buffer and don't hesitate to prepare a new one. That will save you a lot of time.
6. Overlaying the gel with water or isobutanol prevents contact with atmospheric oxygen (which inhibits acrylamide polymerization) and helps to level the resolving gel solution.
7. Prepare the SDS extraction buffer by mixing 1 mL of 10% SDS, 100 μL of sodium vanadate 100 mM, 200 μL of Tris–HCl 1 M, 1 mL of protease cocktail inhibitor and 7.7 mL of ddH_2O. Protease cocktail inhibitor (Roche) should be prepared according to manufacture instructions (1 tablet/2 mL of ddH_2O).
8. Prepare RIPA buffer by mixing 500 μL of Tris–HCl 1 M, 300 μL of NaCl 5 M, 100 μL of NP-40, 100 μL of SDS, 250 μL of sodium deoxycholate, and 200 μL of protease cocktail inhibitor.
9. Prepare Laemmli buffer (2×) by mixing 125 μL of Tris 0.5 M, 200 μL of SDS, 200 μL of Glycerol, 2 mg of bromophenol Blue, 50 μL of β-mercaptoethanol, 1 μL of 1 M DTT, and 424 μL of ddH_2O.
10. Prepare 10× native buffer (0.25 M Tris, 192 mM glycine). Weight 30.3 g Tris and 144 g glycine, mix, and make it to 1 L with ddH_2O. Dilute 100 mL of 10X native buffer to 990 mL with ddH_2O and add 10 mL of 10% SDS. Care should be taken to add SDS solution last, since it makes bubbles.
11. Dilute 100 mL of 10× native buffer to 990 mL with ddH_2O. Add 150 mL methanol and 1 mL of 10% SDS. Place at 4°C (or in cold room).
12. Overlay the resolving gel with isobutanol prevents contact with atmospheric oxygen (which inhibits acrylamide polymerization) in addition to helping to level the resolving gel solution.
13. In some cases, especially when working in a hot area, the polymerization process may be fast and addition of isopropanol may result in a curved line.
14. Centrifuging the samples prior to the run helps remove non soluble debris, which could produce streaks in the protein lanes (revealed when stained with Coomassie Blue).
15. Cut a tiny wedge from the bottom left side of the marker lane and the main membrane for orientation purposes.
16. One can also use PVDF membrane. In this case an additional step consisting of its activation in methanol is necessary.

References

1. Djouadi F, Lecarpentier Y, Hebert JL, Charron P, Bastin J, Coirault C (2009) A potential link between peroxisome proliferator-activated receptor signalling and the pathogenesis of arrhythmogenic right ventricular cardiomyopathy. Cardiovasc Res 84:83–90
2. Salvi N, Guellich A, Michelet P, Demoule A, Le Guen M et al (2010) Upregulation of PPARbeta/delta is associated with structural and functional changes in the type I diabetes rat diaphragm. PLoS One 5:e11494
3. Arck P, Toth B, Pestka A, Jeschke U (2010) Nuclear receptors of the peroxisome proliferator-activated receptor (PPAR) family in gestational diabetes: from animal models to clinical trials. Biol Reprod 83:168–176

Chapter 16

Fluorescence Resonance Energy Transfer Techniques to Study Ligand-Mediated Interactions of PPARs with Coregulators

Nico Mitro, Cristina Godio, and Maurizio Crestani

Abstract

The capacity to induce the association of peroxisome proliferator-activated receptors (PPARs) with different transcriptional coregulators is determined by the peculiar 3D-structure that the receptors adopt when bound with a specific ligand. The fluorescence resonance energy transfer assay is a technique widely used to evaluate coregulator recruitment to nuclear receptors induced by ligands. With this assay it is possible to quantitatively determine the interaction and the affinity of coregulators with PPARs when these receptors are complexed with ligands. Here, we describe the use of this technique to assess the preferential interaction and the affinity of PPARγ with coregulators as a function of the chemical structure of the bound ligand.

Key words: Nuclear receptors, Peroxisome proliferator-activated receptors, Coactivators, Corepressors, Protein–protein interaction, Fluorescence resonance energy transfer

1. Introduction

Nuclear receptors activate the transcription of target genes by interacting with a large group of proteins termed transcriptional coregulators (1–3). The chemical structure of a given ligand imparts the spatial modification of the receptor conformation (4) and dictates the preferential interaction with different coregulators (5). This determines the final transcriptional output elicited by the ligand as a function of the relative abundance of a specific coregulator in different tissues and cell types. It is, therefore, critical to study the preferential interaction between nuclear receptors and coregulators in the presence of different ligands to

Mostafa Z. Badr and Jihan Youssef (eds.), *Peroxisome Proliferator-Activated Receptors (PPARs): Methods and Protocols*, Methods in Molecular Biology, vol. 952, DOI 10.1007/978-1-62703-155-4_16, © Springer Science+Business Media New York 2013

understand the physiology of nuclear receptors and the pharmacology of synthetic ligands. This could be achieved by using typical protein–protein interaction techniques like coimmunoprecipitation with specific antibodies, pull-down assays by affinity chromatography with tagged proteins followed by western blot analysis. Nonetheless, although informative, these techniques are not quantitative and are not suitable to determine the affinity of nuclear receptors with coregulators in the presence of specific ligands. Fluorescence resonance energy transfer (FRET) assay has become a useful technique to test the selective recruitment of coregulators to nuclear receptors complexed with a ligand (6). Typically, the technique is based on the ligand-induced recruitment of coregulator biotinylated peptide containing the LXXLL motif that represents the interface for the physical contact between the coregulator and the nuclear receptor (7). This interaction can be detected with a Europium-labeled antibody against the tagged nuclear receptor, which emits fluorescence at 615 nm when excited with a light beam at 340 nm. Allophycocyanin–streptavidin emits fluorescence at 665 nm when excited with light at 615 nm. This happens only when the ligand induces a receptor conformational change resulting in the recruitment of the biotinylated peptide that brings the Eu-labeled antibody in close proximity of allophycocyanin–streptavidin (see Fig. 1). FRET assays allow to test in a quick and quantitative manner which coregulators are preferentially recruited and can be predictive of the behavior of a specific ligand in different cell types and tissues depending on the relative abundance of coregulators. The technology can be easily adapted to high throughput format allowing the execution of screening campaigns of small molecule libraries also in industrial environments. Here we describe the use of FRET to study the capacity of ligands of the nuclear receptor PPARγ to preferentially recruit distinct coregulators depending on the peculiar chemical structure of the specific ligand complexed with the ligand-binding pocket of the receptor.

In particular, in this chapter we will present an experimental protocol that has been used in our laboratory to test ligand-induced coregulator recruitment on the nuclear receptor PPARγ (4, 5). This protocol has been broken down into three different steps as follows:

1. Expression and purification of PPARγ ligand-binding domain (LBD).
2. Coactivator recruitment assay to test the ability of new PPARγ ligands to recruit coregulators.
3. Coregulator interaction assay to determine the preferential coregulator recruitment by a specific or newly identified PPARγ ligand.

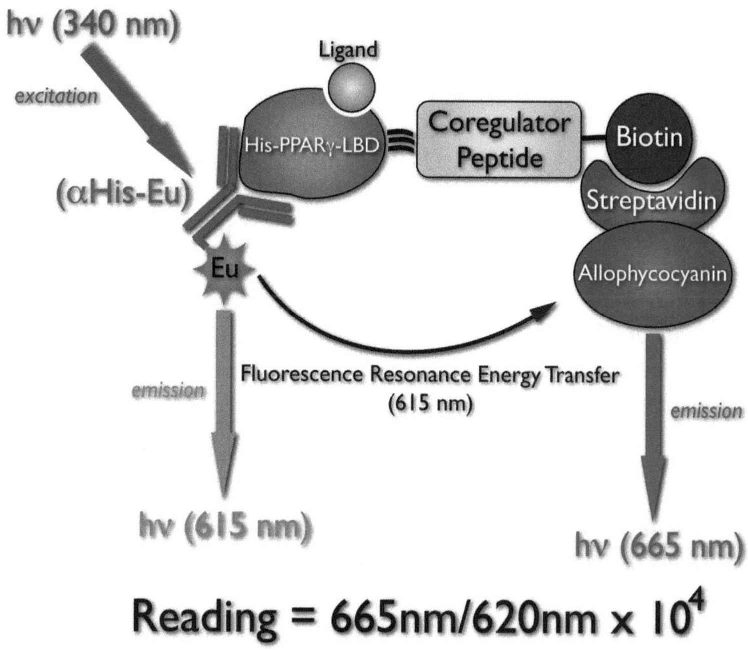

Fig. 1. Cartoon showing the principle of the coregulator recruitment assay by means of FRET technology. Excitation of Eu-labeled anti-histidine antibody (αHis tag-Eu) with 340 nm UV light causes emission at 615 nm. When the tested ligand is bound to the LBD of PPARγ it induces the conformational change required for the recruitment of the biotinylated coregulator peptide associated to the allophycocyanin-labeled streptavidin, which in turn emits light at 665 nm when excited at 615 nm by the light emitted by Eu.

2. Materials

2.1. Materials and Solutions for Expression and Purification of Histidine-Tagged PPARγ LBD

1. *E. coli* strain BL21 DE3.
2. *Luria Bertani (LB) broth with kanamycin*: 10 g tryptone/L, 5 g yeast extract/l, 10 g NaCl/l, pH 7. Dispense 200 mL in 2 L flasks. Autoclave 20 min at 1 atm to sterilize the medium, cool down and add kanamycin to a final concentration of 30 μg/mL.
3. 0.1 mM isopropyl-β-d-thiogalactopyranoside (IPTG).
4. Ni^{2+}-nitrilotriacetic acid column for the HPLC method or resin for the in-batch purification.
5. *Buffer A*: 20 mM Tris, 150 mM NaCl, 10% glycerol, 1 mM Tris 2-carboxyethylphosphine HCl, pH 8.
6. *Buffer B:* 20 mM Tris, 10% glycerol, 1 mM Tris 2 carboxyethylphosphine HCl, pH 8.
7. *Buffer C*: 20 mM Tris, 20 mM NaCl, 10% glycerol, 1 mM Tris 2-carboxyethylphosphine HCl, pH 8.

8. Shaking incubator with temperature control in the range between 10 and 42°C.
9. Plate reader for homogeneous time-resolved fluorescence detection in the 384- or 1,536-well format (e.g., EnVision 2101 Multilabel Reader, PerkinElmer).

2.2. Materials and Solutions for FRET Assay to Test the Ability of PPARγ Ligands to Recruit Coregulators

1. *FRET buffer:* 50 mM KCl, 50 mM Tris pH 7.5, 0.1 mg/mL fatty acid-free bovine serum albumin, 1 mM dithiothreitol (see Note 1).
2. Biotinylated coactivator peptide (length: around 20 amino acids) (see Note 2).
3. Europium (Eu)-labeled anti-histidine antibody.
4. Allophycocyanin-labeled streptavidin.

2.3. Materials and Solutions for FRET Assay to Determine the Preferential Coregulator Recruitment by a Specific or Newly Identified PPARγ Ligand

1. *FRET buffer:* 50 mM KCl, 50 mM Tris pH 7.5, 0.1 mg/mL fatty acid-free bovine serum albumin, 1 mM dithiothreitol (see Note 1).
2. Biotinylated coactivator peptide (length: around 20 amino acids) (see Note 2).
3. Europium-labeled anti-histidine antibody.
4. Allophycocyanin-labeled streptavidin.

3. Methods

3.1. His-Tagged PPARγ-LBD Protein Expression and Purification

The LBD of PPARγ is expressed in *E. coli* as N-terminal histidine-tagged protein using a pET28 vector and subsequently purified. The yield is typically 3–4 mg for PPARγ and around 1–2 mg for PPARα of recombinant protein/L of culture.

3.1.1. Experimental Protocol

1. Freshly transform *E. coli* cells (strain BL21 DE3) according to a standard transformation protocol described in the manufactured instructions and grow them in 1 L of LB medium containing kanamycin at 37°C to 0.6 OD_{600}.
2. Induce the culture with 0.1 mM isopropyl-β-d-thiogalactopyranoside (IPTG) and further incubate at 18°C for 20 h to maximize the expression of the recombinant-tagged protein. *All the next steps should be performed at 4°C or in a cold room.*
3. Harvest the cells by centrifuging the culture at 5,000 × *g* and resuspend cells in 20 mL/L culture of Buffer A (20 mM Tris, 150 mM NaCl, 10% glycerol, 1 mM Tris 2-carboxyethylphosphine HCl, pH 8) in the presence of protease inhibitors (e.g., Complete Mini EDTA-free; Roche Applied Science).

4. Sonicate the cells (5 min total: 30 s on-off pulse) (see Note 3) and isolate the soluble fraction by centrifugation (35,000 × *g* for 45 min).
5. Load the supernatant onto an Ni^{2+}-nitrilotriacetic acid column (GE Healthcare) and elute with a gradient of imidazole 0–300 mM in Buffer A with an HPLC system (Jasco, Lecco, Italy). This step could be also carried out by using an in-batch Ni^{2+} resin.
6. Collect the fractions containing the protein and quantitate with a standard method for protein analysis (e.g., Bradford assay).
7. Analyze the recombinant protein on 12% SDS-PAGE to check the purity of the preparation.
8. Dialyze the protein over Buffer C (20 mM Tris, 20 mM NaCl, 10% glycerol, 1 mM Tris 2-carboxyethylphosphine HCl, pH 8) to remove imidazole.
9. Determine the identity of the histidine-tagged protein on the basis of the molecular weight and whenever possible by sequencing the tryptic peptides obtained by liquid chromatography-electrospray ionization mass spectrometry and tandem mass spectrometry (LC-MS/MS), respectively.
10. Load the protein onto a Q-Sepharose HP column (GE Healthcare) and elute it with a gradient of 0–500 mM NaCl in Buffer B (20 mM Tris, 10% glycerol, 1 mM Tris 2-carboxyethylphosphine HCl, pH 8).
11. Dialyze the protein over buffer B and keep it frozen at −80°C in aliquots at a concentration of 1 mg/mL (see Note 4).

3.2. FRET Assay to Test the Ability of PPARγ Ligands to Recruit Coregulators

This protocol allows testing the ability of a ligand to induce the conformational changes required to favor the recruitment of transcriptional coregulators. The following protocol has been successfully used to test PPARγ ligands. However, this protocol could be adapted to study other nuclear receptors by adjusting the experimental conditions to obtain the best signal to noise ratio (see Note 5).

3.2.1. Experimental Protocol

The following protocol for FRET assays to evaluate ligand-induced coactivator recruitment is optimized in 384-well plates in a final volume of 10 μL. The protocol could be adapted to 96- or 1,536-well plates by scaling up or down respectively the volume of the reagents.

1. Prepare the compound plate(s) with 12 point dilution curves of each ligand to be tested. Ligands usually are diluted in dimethyl sulfoxide (DMSO) since it is less volatile than other solvents such as ethanol or methanol (see Note 6).

2. Prepare a mix containing the previously purified histidine-PPARγ-LBD (20–50 nM), the biotinylated coactivator peptide (500 nM), the Eu-labeled anti-histidine antibody (1 nM), and the allophycocyanin-labeled streptavidin (100 nM) in FRET buffer.
3. Aliquot 9 μL of the previously prepared mix (see Subheading 3.2, step 2) in black flat-bottom 384-well plates.
4. Aliquot 1 μL of the ligand to be tested (see Subheading 3.2, step 1) and incubate for 1 h at room temperature avoiding direct light exposure (see Note 7).
5. Measure time-resolved fluorescence in a plate reader by setting the excitation at 340 nm, and reading the emission at 615 nm for europium and at 665 nm for allophycocyanin fluorescence's.
6. Express the results as the ratio of allophycocyanin fluorescence ($\lambda = 665$ nm) to europium fluorescence ($\lambda = 615$ nm) multiplied by 10,000. The final results should be represented by this ratio minus the background (vehicle ratio) of at least three independent experiments (see Fig. 2).
7. Calculate the EC_{50} and K_d values of the ligands using commercially available software such as GraphPad Prism (GraphPad Software, San Diego, CA).

3.3. FRET Assay to Determine the Preferential Coregulator Recruitment by a Specific or Newly Identified PPARγ Ligand

This protocol is a variation of the standard FRET assay described in Subheading 3.2 and allows understanding which coregulators are preferentially recruited by the nuclear receptor in the presence of a given ligand. Finally, it also makes it possible to evaluate whether a ligand is a full agonist, a partial agonist or an antagonist depending on the ability of the ligand to favor the recruitment of coactivators and/or corepressors.

The following protocol has been successfully used to test PPARγ ligands and could be used for other nuclear receptors by adjusting the experimental conditions to obtain the best signal to noise ratio (see Note 5).

The protocol has been optimized to evaluate the recruitment of different coactivators induced by a specific ligand in 384-well plates in a final volume of 10 μL. The protocol could be adapted to 96- and 1,536-well plates by scaling up or down respectively the volume of the reagents.

1. Prepare a mix of 8 ng (see Note 5) of recombinant PPARγ-LBD, 0.8 ng of europium-labeled anti-His antibody and 86 ng of allophycocyanin-labeled streptavidin in a FRET buffer.
2. Add the ligand to be tested at a concentration ensuring the saturation of the receptor (see Note 8).

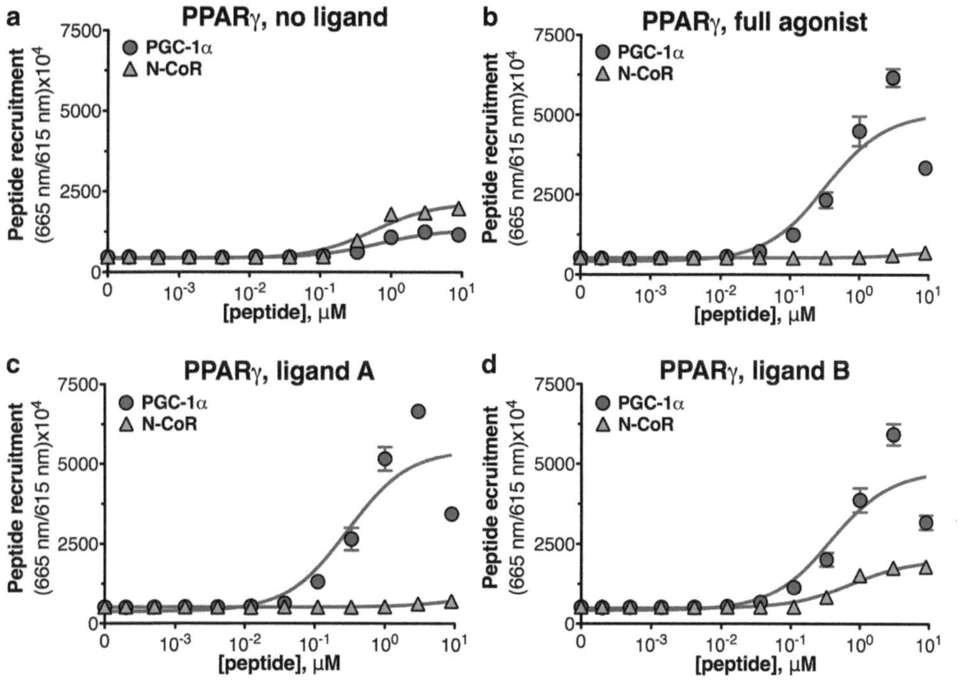

Fig. 2. Example of differential recruitment of coactivators/corepressors induced by different PPARγ ligands. FRET assays were run in 384-well format to test the capacity of two chemically distinct ligands to recruit the coactivator PPARγ coactivator-1α (PGC-1α) or the corepressor N-CoR to PPARγ. (**a**) The assay was run in the absence of ligand to show that in the inactive state the receptor is associated to the corepressor N-CoR and to low level of the coactivator PGC-1α. (**b**) The assay was run with a full agonist used as a reference compound able to recruit the coactivator PGC-1α and to displace the corepressor N-CoR. (**c**) The experimental ligand A showed recruitment of PGC-1α but not of N-CoR, indicating that this ligand is a full agonist of PPARγ. (**d**) The experimental ligand B showed recruitment of both PGC-1α and of N-CoR indicating that this ligand is a partial agonist of PPARγ. These results are consistent with cell-based transcription assays with the luciferase reporter, which had shown that the efficacy of ligand A was the same of that of the full agonist used in (**b**), whereas the efficacy of ligand B was lower than that of the full agonist used in (**b**).

3. Prepare a 12-point dose response curve of the biotinylated peptides starting at 9 μM as the highest concentration. Usually peptides are diluted in water (see Note 9).
4. Incubate the reactions for 1 h at room temperature avoiding direct light exposure.
5. Measure time-resolved fluorescence in a plate reader by setting the excitation at 340 nm, and reading the emission at 615 nm for europium and 665 nm for allophycocyanin fluorescence's.
6. Express the results as the ratio of allophycocyanin fluorescence (λ=665 nm) to europium fluorescence (λ=615 nm) multiplied by 10,000 (see Fig. 1). The final results should be represented by this ratio minus the background (vehicle ratio) of at least three independent experiments.
7. Calculate the EC_{50} and K_d values of the coregulator using commercially available software.

4. Notes

1. Albumin and dithiothreitol must be added right before the assay. It is possible to prepare a 10× stock solution of KCl and Tris pH 7.5 and complete the 1× FRET buffer with albumin and dithiothreitol before the assay.
2. Nowadays some companies develop custom coactivator peptides directly conjugated with fluorophores avoiding the use of biotinylated peptide and the allophycocyanin-labeled streptavidin. In order to maximize the signal to noise ratio it is fundamental to consider the close proximity between the donor fluorophore (Eu) and the acceptor fluorophore (allophycocyanin) to produce the best FRET signal. In fact, the signal is inversely related to the sixth power of distance between the two fluorophores, therefore by directly labeling the coregulator peptide, rather than using the biotin-streptavidin arm, the signal intensity should increase significantly.
3. The optimal intensity and the total duration of the sonication should be determined in pilot tests as they depend on the brand, model, and efficiency of the sonicator.
4. All the solutions buffer A, B, and C for the histidine-tagged PPARγ-LBD protein expression and purification should be freshly prepared and kept at 4°C since Tris 2-carboxyethylphosphine HCl is not stable for long time. Moreover to obtain a better purification and yield of the purified protein we strongly recommend to freshly prepare the imidazole gradient for protein elution.
5. To obtain the best signal to noise ratio it is important to optimize the following points in preliminary experiments:
   (a) Amount of tagged protein depending on the well plate format chosen to run the assay (for example if 8 ng of protein well plate are needed for the assay in 384 format, in 96-well plate the amount of tagged protein should be increased proportionally; however, we strongly recommend to run preliminary tests to fine tune the optimal amount of recombinant-tagged receptor).
   (b) Choose the best interacting coactivator and its amount depending on the nuclear receptor under investigation. When testing new ligands it is important to determine the best interacting coactivator with the nuclear receptor under investigation by using a known agonist as a positive control.
   (c) Optimize the amount of the allophycocyanin-labeled streptavidin and the Europium-labeled anti-histidine antibody respectively, since they are responsible of the final

fluorescence ratio 665/615 nm. For example if the amount of the biotinylated peptide coactivator has to be increased the amount of APC should increase as well to get better signal when the interaction occurs. Accordingly, the amount of the Europium-labeled anti-histidine antibody should be adjusted when changing the protein quantity. Finally, it is important to consider also the nature of the nuclear receptor to be tested. For example: the assay set up for another nuclear receptor may be different from PPARs and needs all the described optimizations.

6. The compound plate should contain the ligand ten times more concentrated than the desired final concentration.
7. Include at least four points of just vehicle to be subtracted as a background.
8. Prepare as many mixes as the ligands to be tested. To evaluate the basal interaction with coactivators and corepressors it is important to include also a mix with the unliganded nuclear receptor (i.e., vehicle alone).
9. The master plate should contain the coregulator ten times more concentrated than the desired final concentration.

Acknowledgement

This work was supported by grants from the European Commission (LSHM-CT2006-037498), the Giovanni Armenise-Harvard Career Development Award and the Cariplo Foundation (2008.2511 and 2009.2727). We are grateful to Miss Elda Desiderio Pinto for invaluable administrative support.

References

1. Hermanson O, Glass C, Rosenfeld M (2002) Nuclear receptor coregulators: multiple modes of modification. Trends Endocrinol Metab 13:55–60
2. Nettles KW, Greene GL (2005) Ligand control of coregulator recruitment to nuclear receptors. Ann Review Physiol 67:309–333
3. Rosenfeld MG, Lunyak VV, Glass CK (2006) Sensors and signals: a coactivator/corepressor/epigenetic code for integrating signal-dependent programs of transcriptional response. Genes Dev 20:1405–1428
4. Pochetti G, Godio C, Mitro N et al (2007) Insights into the mechanism of partial agonism: crystal structures of the peroxisome proliferator-activated receptor gamma ligand-binding domainin the complex with two enatiomeric ligands. J Biol Chem 282:17314–17324
5. Pochetti G, Mitro N, Lavecchia A et al (2010) Structural insight into peroxisome proliferator-activated receptor gamma binding of two ureidofibrate-like enantiomers by molecular dynamics, cofactor interaction analysis, and site-directed mutagenesis. J Med Chem 53:4354–4366
6. Liu J, Knappenberger K, Kack H et al (2003) A homogeneous in vitro functional assay for estrogen receptors: coactivator recruitment. Mol Endocrinol 17:346–355
7. Darimont BD, Wagner R, Apriletti J et al (1998) Structure and specificity of nuclear receptor-coactivator interactions. Genes Dev 12: 3343–3356

Chapter 17

Estimation of the PPARα Agonism of Fibrates by a Combined MM-Docking Approach

Fabio Lannutti, Alessandro Marrone, and Nazzareno Re

Abstract

Fibrates are peroxisome proliferator-activated alpha receptor (PPARα) activators derived from fibric acid and are the most clinically used therapeutics in the treatment of hypertriglyceridemia. Recently, we reported a computational approach for the investigation of the binding properties of fibrates, characterized by similar carboxylic heads but differing in the size and orientation of the hydrophobic portion. This procedure is based on a combination of standard docking and molecular mechanics approaches to better describe the adaptation of the protein target to the bound ligand. The application of our approach to a set of 23 fibrates and the use of an effective regression procedure, allowed the development of predictive models of the PPARα agonism. The obtained models are characterized by good performances realizing a fair trade-off between accuracy and computational costs. The best model is more specialized in the ranking of fibrate agonists whose binding is mainly controlled by steric rather than by electronic modulation. Here, we describe in details the application of this computational procedure for the prediction of PPARα agonism of fibrate ligands.

Key words: PPAR, Fibrates, Docking, Binding energy, Linear regression

1. Introduction

Among the most effective molecules acting as peroxisome proliferator-activated alpha receptor (PPARα) activators, fibric acid derivatives are clinically used to treat hypertriglyceridemia (1–3). The pathogenic potential attributed to the latter condition, explains the immense interest in the design of new fibrates with improved potency and selectivity toward the activation of PPARα. The PPARα agonism of fibrates is due to their ability to induce an activated conformation of the receptor by binding to the so-called ligand binding domain (LBD). Agonist binding promotes and stabilizes a target conformation that sustains the activation of PPARα. The investigation of the binding properties of known PPARα modulators is thus

Mostafa Z. Badr and Jihan Youssef (eds.), *Peroxisome Proliferator-Activated Receptors (PPARs): Methods and Protocols*, Methods in Molecular Biology, vol. 952, DOI 10.1007/978-1-62703-155-4_17, © Springer Science+Business Media New York 2013

essential to assess the molecular basis of fibrate agonism and to support the design of new agonists. To this purpose, a detailed knowledge of how a potent fibrate such as GW590735 interacts with the PPARα activated conformation can then be used to model the binding properties of other fibrate analogs by means of target-directed rational approaches. Several alternative computational approaches have been proposed to improve the accuracy in the investigation of the binding properties of potentially active ligands, balancing both the accuracy and practicability of computational efforts of binding free energy estimations. Docking approaches have recently been used to investigate the binding properties of either phthalate monoesters or newly designed fibrates as potential ligands of PPARγ or PPARα, respectively (4, 5). Recently, we developed a linear interaction energy with continuum electrostatics (LIECE) approach through the combination of docking and molecular mechanics (MM) calculations to estimate the binding geometry and the molecular interactions of a series of 23 fibrates with PPARα (6). Two alternative regression approaches were tested and compared for the generation of training subsets from the 23 considered fibrates, namely the user-defined training subset (UDTS) and the automatically generated training subset (AGTS) procedures. The best predictions were obtained with AGTS, which was more effective in filtering out outliers and led to more specialized models (6). In this contribution we describe in details how to use our computational approaches to estimate the PPARα agonism of fibrate ligands.

2. Materials

The following programs are needed to carry out the proposed computational procedure:

1. Maestro (7) and Pymol (8)—graphical user interfaces to be used in molecular models, preparation and analysis.
2. MacroModel (9)—Molecular modeling program to be used in local minimization of ligand structures, local minimization of docking complexes, and solvation docking calculations (*vide infra*).
3. Glide (10, 11)—Docking and scoring program.

3. Methods

3.1. Docking Studies

1. Download the crystal structure of a complex between the PPARα binding subunit and GW590735 (1) from the protein data bank (PDB) archives (entry code 2P54) and employ it as the target in docking calculations (see Note 1).

2. Graphically inspect and analyze the co-crystallized complex to find possible structural mismatches using Maestro and/or Pymol (see Note 2).
3. Prepare the protein for the docking studies using the protein preparation module implemented in Maestro. Remove the solvent molecules in the crystal structure add hydrogen atoms. Relax the structure by activating the "Hydrogens only" option and using the Optimized Potential for Liquid Simulations (OPLS2001) force field (12, 13) at a gradient threshold of 0.30 kJ/mol/Å (see Note 3).
4. Build the 3D structures of ligands in Maestro and subsequently minimize them by using the MacroModel program (9). Perform the local minimization with the all-atoms OPLS2001 force field (12, 13) through the truncated-Newton conjugate gradient (TNCG) algorithm by applying the gradient threshold criteria at 0.05 kJ/mol/Å, a maximum of 500 iterations, and simulating the water environment with the generalized-Born/solvation area (GB/SA) polarizable continuum method implemented in the program (see Note 4).
5. Carry out the docking calculations with Glide (10, 11) in two steps (see Note 5):
   (a) Calculate the target-ligand interaction fields by performing grid calculations on the 2P54 complex obtained after protein preparation. Select the co-crystallized ligand (1) to center and fix the dimensions of interaction grids. No constrains must be applied.
   (b) Dock flexibly and score all compounds using the Glide standard-precision (SP) mode (11). Keep at default values the docking set-up parameters. Upon completion of each docking calculation, save 30 poses per ligand. Rank the best-docked structures using a model energy score (E_{model}) derived from a combination of GlideScore (G_{score}, a modified and extended version of the empirically based ChemScore function), Coulombic, and van der Waals energies, and the strain energy of the ligands (11). Dock and score again the top ranked SP poses with the Glide extra-precision (XP) mode (11); the best ten XP poses are finally selected as final docking solutions.
6. Analyze the docking solutions (≤10 per each ligand) by the comparison with reference experimental pose of 1. Select from docking solutions the binding pose most similar to 1 in both orientation and position of carboxylate head (see Note 6).

3.2. Molecular Mechanics

1. Perform the local minimization of selected receptor-agonist complex (see Subheading 3.1, step 6), in gas phase using the MacroModel program to take into account part of the ligand-induced fit effect of the receptor structure. Select the Amber

force field with default treatment of charges and non-bonded interactions cut-offs. Without applying any constraints, set a maximum of 5,000 steps of Low-memory Broyden-Fletcher-Goldfarb-Shanno (LBFGS) algorithm and convergence limit of 0.05 kJ/mol/Å (see Note 7).

2. Inspect and analyze the effect of the MM relaxation on both the bound agonist pose and the protein conformation (see Note 8).
3. Estimate the protein–ligand binding energies as the energy difference between the optimized complexes and the separated receptor and ligand structures calculated in the presence of a water environment (GB/SA continuum solvation) within the same Amber force field employed in the complex optimizations (see Note 9).

3.3. Estimation of PPARα Agonism

1. Breakdown the binding energies from the MM calculations (ΔE_{ff}) obtained at step 3 of Subheading 3.2 into their single energy contributions, ΔE_{els}, ΔE_{VdW}, ΔE_{HB}, and $\Delta\Delta G_{solv}$. Each term can be obtained through the general scheme: $\Delta E_x = E_x^{complex} - E_x^{protein} - E_x^{ligand}$ (see Note 10).
2. Estimate the PPARα agonism of each ligand by employing the predictive model developed by Lannutti et al. (6) (see Note 11)

   $pEC_{50} = b_0 + b_{VdW}\Delta E_{VdW} + b_{solv}\Delta\Delta G_{solv}$ with $b_0 = 12.0, b_{VdW} = -0.0281$ and $b_{solv} = -0.0297$.

4. Notes

1. The entry 2P54 was employed as prototype of PPARα fibrate bound complex in the development of predictive models of agonism so that the use of this X-ray source is essential.
2. We have revealed (6) an atom type mismatch affecting the imidazole nitrogen atoms of the His 440. This mismatch must be fixed before going on to the next step. At this purpose, you can either use Maestro or a text editor to exchange the atom type labels of the nitrogens 1875 (NE2→ND1) and 1872 (ND1→NE2).
3. Check the consistency of H-bond patterns involving ligand 1, in particular at carboxylate head. This step must be used to feedback the correction of the His 440 mismatch performed in the previous step.
4. The fibrate ligands must be considered in their respective anionic forms (carboxylates). Consider the presence of any other ionizable group. The predictive models have been developed (6) on a set of 23 anionic fibrates, this means that when

a different total charge for the ligand is concerned, the estimation of the PPARα agonism could result much less accurate.

5. The use of version 3.5 of Glide (employed in the development of the computational approach (6)) is preferred. The increase in the number of poses per ligand in SP calculations can be considered if no acceptable pose is retrieved by 30 docking solutions. To easily filter the SP solutions, use the Pose Viewer tool by leaving the reference pose of 1 included in the Maestro workspace. Undisplaying the target atoms could be also helpful in pose filtering. The Pose Viewer tool can also be used to highlight the major differences in the target-ligand interactions affecting the docking solutions, as well as to characterize the target residues involved in ligand binding.
6. Due to the importance of carboxylate head binding in receptor activation, a more detailed analysis of this moiety binding modes could be useful. For this purpose, the pose selected as final docking solution per each ligand can be superimposed to the reference pose of 1 for the calculation of the in-place root mean square deviations (RMSD) between the carboxylate head portions.
7. The use of the Multi Minimization module is preferred at this step whenever more than one ligand is considered. In this case, the non-conformers option must be specified during the input preparation. Warning! The output structures must be a local minimum so that minimization should be started up again if convergence criteria were not fully satisfied.
8. This step can include either simple graphical inspections of ligand poses after relaxation or RMSD calculations extended to the whole bound complexes to unveil the effects of local minimization on target structure. See Lannutti et al. (6) for a more detailed description.
9. Starting from the bound complexes minimized at step 1, generate three groups of entries including complexes, proteins, and ligands in the same order. This step can be easily performed by using the "split-by-molecule" option of Project Table menu. The energy calculation on multiple structures inputs cannot be performed straightforwardly from the Maestro interface, but can be performed as follows:
   (a) Prepare an input of current energy calculations for each group and write the corresponding *.com* files in the working directory.
   (b) Export each group in corresponding multiple *.mae* structure files in the working directory.
   (c) Edit each of the three (complexes, proteins, and ligands) *.com* files by replacing all lines after the BDCO line with the following combination of keywords.

CRMS	0	0	0	0	0.0	0.5	0.0	0.0
BGIN	0	0	0	0	0.0	0.0	0.0	0.0
READ	0	0	0	0	0.0	0.0	0.0	0.0
ELST	0	0	0	0	0.0	0.0	0.0	0.0
WRIT	0	0	0	0	0.0	0.0	0.0	0.0
END	0	0	0	0	0.0	0.0	0.0	0.0

(d) Run Macromodel from terminal: <terminal-prompt>*bmin* "job-name."

(e) Import the output structures from the corresponding *–out.maegz* files.

After steps (a)–(e) the E_{VdW} and ΔG_{solv} for each structure is automatically reported in the project table and can be easily exported as *.csv* files for further processing.

10. The single energy contributions collected in *.csv* files (see Subheading 3.3, step 1) can be read and processed with Excel or other spreadsheets to obtain the corresponding energy variations, in particular ΔE_{VdW} and $\Delta\Delta G_{solv}$. Consider the sign of ΔE_{VdW} and $\Delta\Delta G_{solv}$ because the former is expected to be negative (attractive) whereas the second is expected to be positive (non attractive) and in the order of the $\Delta G_{solv}{}^{lig}$.
11. The predictive model was developed assuming the energy contributions expressed in kJ/mol, so please convert data to these units before input into the equation. The $p\mathrm{EC}_{50}$ values calculated by the proposed model can be further reported in μM units by multiplication for 1×10^6.

Acknowledgments

We thank the Ministry of Education, Universities and Research (MIUR) for financial support.

References

1. Schoonjans K, Staels BM, Auwerx J (1996) Role of the peroxisome proliferator activated receptor (PPAR) in mediating the effects of fibrates and fatty acids on gene expression. J Lipid Res 37:907–925
2. Staels B, Auwerx J (1997) Role of PPAR in the pharmacological regulation of lipoprotein metabolism by fibrates and thiazolidinediones. Curr Pharm Design 3:1–14
3. Fruchart JC, Duriez P, Staels B (1999) Molecular mechanism of action of the fibrates. J Soc Biol 193:67–75
4. Kaya T, Mohr SC, Waxman DJ, Vajda S (2006) Computational screening of phthalate monoesters for binding to PPARgamma. Chem Res Toxicol 19:999–1009
5. Li X, Zou H, Wu A, Ye Y, Shen J (2007) Structure-based drug design of a novel family

of chalcones as PPAR_ agonists: virtual screening, synthesis, and biological activities in vitro. Acta Pharmacol Sin 28:2040–2052

6. Lannutti F, Marrone A, Re N (2011) Prediction of the PPARα agonism of fibrates by combined MM-docking approaches. J Mol Graph Model 29:865–875
7. Maestro (2008) Version 8.5, Schrodinger, New York
8. DeLano WL (2002) The PyMOL molecular graphics system. DeLano Scientific, Palo Alto. http://www.pymol.org
9. MacroModel (2005) Version 9.0, Schrodinger, New York
10. Glide (2005) Version 3.5, Schrodinger, New York
11. Friesner RA, Murphy RB, Repasky MP, Frye LL, Greenwood JR, Halgren TA, Sanschagrin PC, Mainz DT (2006) Extra precision Glide: docking and scoring incorporating a model of hydrophobic enclosure for protein-ligand complexes. J Med Chem 49:6177–6196
12. Jorgensen WL, Maxwell DS, Tirado-Rives J (1996) Development and testing of the OPLS all-atom force field on conformational energetics and properties of organic liquids. J Am Chem Soc 118:11225–11236
13. Kaminski GA, Friesner RA, Tirado-Rives J, Jorgensen WL (2001) Evaluation and reparametrization of the OPLS-AA force field for proteins via comparison with accurate quantum chemical calculations on peptides. J Phys Chem B 105:6474–6487

Chapter 18

Combined Biophysical and Cell-Based Approaches for the Assessment of Ligand Binding to PPARγ

Silvia Zorrilla and Dolores Pérez-Sala

Abstract

Transcription factors of the peroxisome proliferator-activated receptor (PPAR) family are ligand-activated receptors that play key roles in lipid metabolism and inflammation. The γ isoform (PPARγ) is involved in adipocyte differentiation, insulin sensitization, and vascular pathophysiology, including inflammation and atherosclerosis, for which it is considered an important drug target. PPARγ ligands display varied structures and include fatty acids, electrophilic lipids, and certain drugs. These agonists promote conformational changes allowing interaction of PPARγ with coactivators and hence transcriptional regulation. Here we present a panoply of methods to study PPARγ interactions with ligands and activation in vitro and in cells. The first method is based on the competition of the fluorescent dye 1-anilinonaphthalene-8-sulfonic acid (ANS) with PPARγ ligands for the ligand binding pocket, allowing detection and quantification of ligand binding to PPARγ. This method is specific for PPARγ while ANS displays negligible interaction with other nuclear receptors such as PPARα and retinoid X receptor α (RXRα). The ANS competition assay has been validated through comparison of the affinities determined for well-known PPARγ ligands by this method with those reported in the literature. We also describe here gel-based competition assays that show limited performance with non-covalently bound ligands. In addition, we present a fluorescence anisotropy assay to analyze PPARγ activation by ligands in vitro through their capacity of eliciting PPARγ interaction with a fluorescently labeled peptide derived from one of its coactivators (SRC-1). Finally, we show cell-based assays to investigate PPARγ activation by interaction with its ligands. We believe that combined approaches using ANS, fluorescent coactivator peptides, and in-cell assays to monitor PPARγ binding and interactions may provide valuable strategies for the identification and characterization of PPARγ ligands.

Key words: PPAR activity assay, Fluorescence, PPAR ligand, 15d-PGJ_2, Fluorescence anisotropy assay, ANS

1. Introduction

Peroxisome proliferator-activated receptors (PPARs) play important roles in physiology and pathophysiology because they are involved in the regulation of lipid metabolism, inflammation, and

Mostafa Z. Badr and Jihan Youssef (eds.), *Peroxisome Proliferator-Activated Receptors (PPARs): Methods and Protocols*, Methods in Molecular Biology, vol. 952, DOI 10.1007/978-1-62703-155-4_18, © Springer Science+Business Media New York 2013

glucose homeostasis among other functions (1). Therefore, PPARs constitute important targets for therapeutic intervention. Indeed, although several drugs exist that activate PPAR, they are not exempt of adverse effects and some of them have been withdrawn from the market. In this context there is an interest in the search for new agonists potentially useful in clinical practice (2, 3). PPARs are ligand-activated transcription factors. Like several other nuclear receptors, including retinoic acid or vitamin D receptors, PPARs associate with retinoid X receptor (RXR) to form heterodimers which constitute the transcriptionally active species whose interaction with the DNA operators is modulated by coregulator proteins (4). There are three PPAR subtypes (α, β or δ and γ), which display ligand selectivity and different functions. PPARγ has been identified as an important negative regulator of inflammation, induces adipocyte differentiation and plays important roles in vascular biology. Several oxidized metabolites of fatty acids displaying electrophilic moieties, including 15-deoxy-$\Delta^{12,14}$-prostaglandin J_2 (15d-PGJ_2), oxo-eicosatetraenoic acids, hydroxyoctadecadienoic acids, and nitrated fatty acids, bind PPARγ with high affinity (5–7). In addition, exogenous compounds like the thiazolidinediones, which have been used as antidiabetic agents, interact with and activate PPARγ (1, 8).

PPAR proteins structure comprise a transcriptional activation domain located in the N-terminus followed by a twin zinc finger DNA binding domain and a highly conserved C-terminal domain connected through a hinge region (4). The ligand binding domain (PPARγ–LBD) is placed in the C-terminal region of the protein, through which interaction with coregulator proteins and functional heterodimerization with RXR also take place. The ligand binding pocket is a wide cavity which, as detailed above, accommodates ligands with diverse structure. These can be classified into agonists, antagonists, or partial agonists, depending on their ability to induce conformational changes on PPAR leading to the recruitment of coregulator proteins. Moreover, it has been shown that two fatty acid molecules may fit simultaneously in the PPARγ ligand binding domain and that some ligands bind covalently to a cysteine residue present in this cavity (6). These different binding modes need to be taken into account when developing and interpreting the results of studies aimed at the identification and characterization of PPARγ ligands.

The assessment of the interactions of PPARγ with its ligands is crucial not only to get insight into the mechanisms underlying the biological function of this key nuclear receptor but also to develop therapeutic strategies. Competition with radiolabeled ligands, followed by size exclusion chromatography is currently the most extended approach to quantify PPARγ interaction with ligands (9–11). This method is time-consuming, since it involves steps of separation of the complex and the free ligand and presents the drawback of involving radiolabeled compounds. On the other hand, the interaction of PPARγ with its ligands has also been addressed by crystallographic approaches, which are not accessible to all laboratories (6). Fluorescence spectroscopy methods are among the most widely used techniques for the detection and measurement of biomolecular interactions due to their high sensitivity, low sample consumption, and reproducibility (12). Another salient advantage of fluorescence – based procedures is that they allow measuring interactions under true equilibrium conditions with no need of additional steps to separate the free ligand from the complexes. Besides, fluorescence methods have proven very useful for the development of systematic screening assays, since most of these measurements are compatible with the use of automated plate-readers. Competition with dyes that exhibit a fluorescence emission strongly dependent on the nature of the surrounding environment, as it is the case of ANS, is a simple method towards determination of the affinity of protein-ligand binding (13, 14). A remarkable advantage of ANS binding assays is their overall low coast, since the fluorophore is inexpensive and the instrumentation is relatively simple. The nature of the ligand binding site of PPARγ comprising both hydrophobic and hydrophilic regions enables this nuclear receptor to bind ANS. Furthermore, ANS binds to PPARγ with moderate affinity being displaced by its specific ligands (15), providing a reliable means to determine the affinity of the ligands for the protein. Therefore, the ANS competition assay presents important advantages when compared with the above mentioned radiometric and crystallographic approaches, and it is also more straightforward than fluorescence anisotropy-based procedures devised to study ligand binding to PPARγ (16, 17). Gel-based assays, such as the one presented here, constitute a good complement to the ANS binding method. These assays rely on the capacity of PPARγ ligands to occupy the ligand binding pocket, therefore precluding access of a general cysteine reagent such as iodoacetamide or maleimide to the cysteine residue located in the ligand binding pocket (cysteine 285), and provide a useful strategy for the detection of compounds covalently binding to PPARγ.

Determination of PPARγ affinity for its ligands is crucial but not sufficient to fully characterize the functional implications of ligand binding to this nuclear receptor. Thus, to understand the physiological role of a compound interacting with PPARγ, additional studies are required to discern whether the interaction with the receptor is productive in terms of coregulator recruitment and transcriptional regulation. Here we propose using two kinds of approaches to complement the ligand binding study by competition with ANS and gel analysis, namely fluorescence anisotropy titrations of a SRC-1 peptide and PPRE reporter assays. Fluorescence anisotropy is a parameter that allows studying interactions in which the complexes are significantly larger in size than a fluorescent ligand involved in the complex (18), as it is the case of a fluorescently labeled peptide derived from the SRC-1 coactivator sequence and its complex with PPARγ. The anisotropy assay proposed allows selecting from the compounds that display significant interaction with the receptor those that, in addition, elicit the interaction with coregulators required for recognition of the DNA operators. This approach has already been used to study the interaction of other nuclear receptors with their coactivators in the presence of ligands (19, 20). For the ligands of PPARγ that induce the conformational changes leading to coactivator recruitment, the ability to activate transcriptional regulation can then be studied in cells by the PPRE reporter assay.

The global approach we propose here comprising a number of biophysical, biochemical, and cell biology methods may be useful to thoroughly characterize already known or new PPARγ ligands. Thus, by using this strategy, new ligands can be identified and their affinity measured (ANS competition and gel-based assays) and the functional consequences of binding can be addressed (SRC-1 fluorescence anisotropy and PPRE reporter assays). In this way new insight into the function of the emerging therapeutic target PPARγ can be obtained.

2. Materials

2.1. ANS Binding Assays

1. PPARγ-LBD: The ligand binding domain of human PPARγ (aa 193–475) was expressed as N-terminal His-tagged protein from a pET28a vector kindly gifted by Krister Bamberg (AstraZeneca, Sweden). The protein was purified by affinity chromatography following a previously described protocol (21).
2. ANS was from Fluka Chemie.
3. Rosiglitazone was from Alexis Biochemicals, 15-deoxy-$\Delta^{12,14}$-prostaglandin J_2, 9,10- dihydro-15-deoxy -$\Delta^{12,14}$-prostaglandin J_2 were from Cayman Chemical. Stock solutions were prepared in DMSO.
4. 3×3 mm quartz cuvettes for fluorescence (from Starna or Hellma).
5. Spectrophotometer: Cary 3E (Kontron Instruments).
6. Quartz cuvettes 1 cm path length for absorbance measurements (from Starna or Hellma).
7. Spectrofluorometer: PC1 photon counting steady-state ISS spectrofluorometer (Champaign, IL, USA).
8. Buffer for the assay: 20 mM Tris–HCl, 150 mM NaCl, 2 mM ethylenediaminetetraacetic acid (EDTA), 10% glycerol, and 1 mM tris(2-carboxyethyl)phosphine (TCEP), pH 7.5 (buffer A).

2.2. Gel-Based Competition Assays

1. PPAR agonists: Rosiglitazone from Alexis Biochemicals, 15-deoxy-$\Delta^{12,14}$-prostaglandin J_2, 9,10-dihydro-15-deoxy-$\Delta^{12,14}$-prostaglandin J_2 and prostaglandin A_1 were from Cayman Chemical. Stock solutions of agonists were prepared in DMSO at concentrations of 100 μM, except for Rosiglitazone that was prepared at 10 μM.
2. Recombinant PPARγ-LBD purified as previously described (21) was at a concentration of 9.5 μM.
3. Biotinylated iodoacetamide (Sigma) stock solution was prepared fresh in water at 2 mM and it was used immediately afterwards.
4. Incubation buffer: 20 mM Tris–HCl pH 7.5, 150 mM NaCl, 5% glycerol, and 0.1 mM DTT.
5. Reagents for gel electrophoresis were from Bio-Rad. Immobilon-P membranes from Millipore.
6. Horseradish peroxidase-conjugated streptavidin from GE Healthcare.
7. Wash buffer: Tween-Tris Buffered Saline (T-TBS) consists of 20 mM Tris–HCl pH 7.5, 500 mM NaCl, 0.05% Tween 20.
8. ECL detection reagents and Hyperfilm were from GE Healthcare.
9. Anti-PPARγ antibody can be obtained from Millipore and HRP-conjugated goat anti-rabbit immunoglobulins from Dako.

2.3. Anisotropy Assays

1. Fluorescein-labeled peptide SRC-1/NR2 (Fl–^{686}RHKILHRLLQEGS^{698}) corresponding to the NR box 2 binding motif of SRC-1 from the Universidad Pompeu Fabra (Barcelona, Spain).
2. Purified PPARγ-LBD (see above).
3. Rosiglitazone was from Alexis Biochemicals, 15-deoxy-$\Delta^{12,14}$-prostaglandin J_2 was from Cayman Chemical. Stock solutions were prepared in DMSO.
4. 3×3 mm quartz cuvettes for fluorescence measurements (from Hellma or Starna).
5. Spectrofluorometer equipped with polarizers: PC1 photon counting steady-state ISS spectrofluorometer (Champaign, IL, USA).
6. Buffer for the assay: Buffer A (see above).

2.4. Cell-Based PPRE Assays

1. Cells: Human mesangial cells, which express all three PPAR isoforms, can be obtained by primary culture as previously described (22) or purchased from a commercial source like Lonza (Walkersville, MD, USA).

2. Cells are maintained in RPMI1640 medium supplemented with 10% Fetal bovine serum as previously described and passaged by trypsinization with trypsin–EDTA from GIBCO.
3. Reporter plasmids containing the PPAR response elements from PPAR target genes fused with the coding sequence of Frirefly luciferase can be obtained commercially or through standard cloning procedures.
4. A reporter plasmid with a basal promoter fused to the renilla reniformis luciferase coding sequence, like the pRL-CMV plasmid can be obtained from Promega.
5. PPAR agonists to be used as positive controls, like 15d-PGJ$_2$ or rosiglitazone (see above).
6. Transfection reagents can be obtained from Invitrogen. In this protocol, Lipofectamine 2000 is used. Optimem medium from Gibco is used to prepare the transfection mixtures.
7. For determination of luciferase activities, the Dual luciferase reporter assay kit from Promega was used.

3. Methods

3.1. ANS binding Titrations and Competition Assays

The analysis of the interaction of PPARγ with its ligands by competition with ANS is carried out in two steps. First, a titration of PPARγ-LBD with the fluorophore is performed in the absence of ligand and then, the same titration is repeated in the presence of a fixed amount of the ligand. The binding isotherms are then globally analyzed. A schematic description of the assay can be found in Fig. 1.

(a) *Reference isotherm: ANS binding titration in the absence of ligand.*

1. Switch on the spectrophotometer and allow it to equilibrate for 30–45 min before taking an absorbance spectrum.
2. Prepare a stock solution of ANS in buffer A. Measure its absorption in the spectrophotometer and calculate its concentration ($\varepsilon = 5{,}000$ M/cm at 350 nm). Protect the solution from direct light.
3. Switch on the spectrofluorometer and allow the excitation beam to equilibrate for 30–45 min before starting the measurements. Set the temperature control at 20°C.
4. Thaw one aliquot of purified PPARγ-LBD and keep it on ice.
5. Prepare a 160 μL sample containing 100 μM ANS and 340 nM PPARγ-LBD in buffer A (solution (a)).

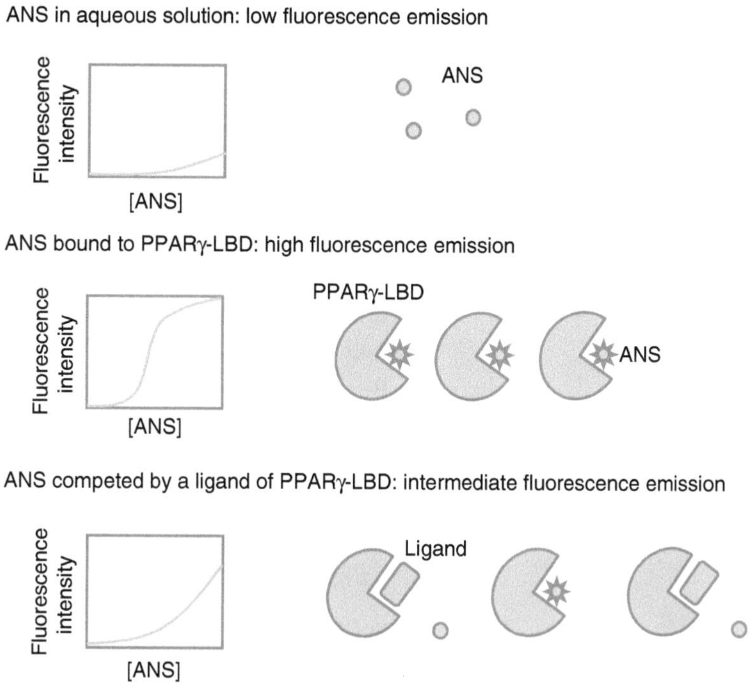

Fig. 1. ANS binding titrations and competition assays. ANS emits very weak fluorescence in aqueous solution. Upon binding to the ligand binding pocket within PPARγ-LBD the fluorescence of this dye is enhanced, providing a means of measuring the affinity of the interaction. Competition of the specific ligands of the nuclear receptor with ANS for the ligand binding site provokes a decrease of the emission of the dye. Analysis of the level of competition allows determining the affinity of the ligand for PPARγ-LBD.

6. Prepare 650–950 μL of a solution containing 340 nM PPARγ-LBD in the same buffer without ANS (solution (b)).
7. Divide solution (b) into 10–15 eppendorf tubes, put 60 μL in each tube.
8. Add 90 μL of solution (a) into the first tube containing solution (b) and mix by pipetting up and down. Withdraw 90 μL of this tube, put them into the second tube containing solution (b) and gently mix. Repeat the procedure with the remaining tubes containing solution (b).
9. Prepare a blank sample containing only ANS at 100 μM in buffer A and perform serial dilutions with buffer as in steps 7–8.
10. Drop 60 μL of solution (a) into a 3 × 3 mm cuvette for fluorescence and record a spectrum by exciting at 375 nm and with the emission set from 430 to 520 nm. Select the appropriate slits. Take the emission at 465 nm.

11. Measure the solutions prepared as described in step 8 as for solution (a) in step 10 and take the emission at 465 nm.
12. Measure the blank samples prepared as described in step 9 in the same way as stated in step 10 and take the emission at 465 nm.
13. Subtract the emission values obtained for the blanks from those of the samples with the same ANS concentration.
14. Plot the blank corrected emission intensity at 465 nm as a function of the total ANS concentration in the solution to build an isotherm.
15. Repeat the entire procedure at least three times.
16. Calculate an average isotherm (±SD) from the three independent titrations.
17. Analyze this isotherm using BIOEQS software (23, 24) and obtain the affinity of PPARγ-LBD for ANS. Consider a single PPARγ-LBD/ANS complex of stoichiometry 1:1.

(b) *Competition isotherm: ANS binding titration in the presence of ligand.*

1. Repeat steps 5–15 including the ligand to be tested, for example rosiglitazone, at 2–3 different concentrations, for example 10, 5, and 1 μM.
2. Plot the emission values as a function of the ANS concentration to build the isotherms.
3. Analyze globally the isotherms using the software BIOEQS (23, 24), assuming two different complexes PPARγ-LBD/ANS and PPARγ-LBD/ligand and determine the affinity of PPARγ-LBD for the ligand.

3.2. Gel-Based Competition Assays

1. Pre-incubation mixtures were prepared with 1.1 μL of PPAR-LBD stock, 1 μL of ligand or DMSO, and 7.9 μL of dilution buffer.
2. Incubate for 1 h at room temperature.
3. Add 1 μL of biotinylated iodoacetamide stock and incubate for 30 min at room temperature.
4. Add 2.7 μL of 5× Laemmli buffer and incubate for 5 min at 95°C.
5. Separate 8 μL of each sample in 15% polyacrylamide gels.
6. Gels are transferred to Immobilon-P membranes on a semidry transblot system from Bio-Rad set at 22v for 50 min using a three buffer system according to the instructions of the manufacturer.
7. Block the membrane by incubation with 2% defatted milk for 1 h at room temperature with shaking.

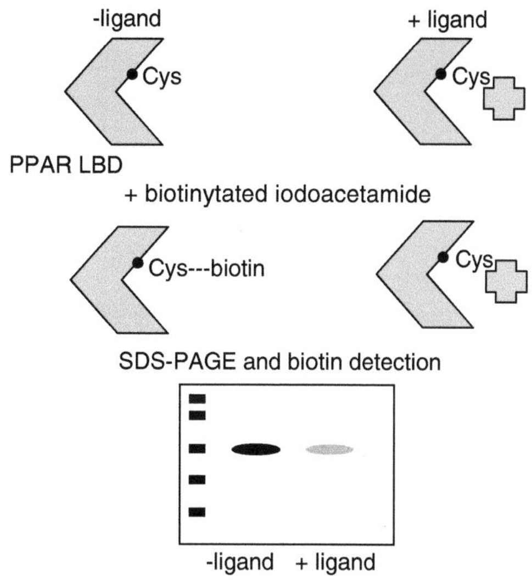

Fig. 2. Scheme of the gel-based competition assays. Incubation of the PPAR ligand binding domain in the presence of agonists may shield cysteine 285 from modification by a labeled cysteine reagent like biotinylated iodoacetamide. The biotin signal incorporated into the LBD in the absence or presence of ligands is later visualized after gel electrophoresis, blot, and detection with HRP-streptavidin. Signals can be semiquantitated to provide an estimation of the extent of competition.

8. Incubate for 30 min with horseradish peroxidase-conjugated streptavidin diluted 1:1000 in 1% BSA prepared in T-TBS.
9. Wash three times for 5 min with T-TBS.
10. For detection, membranes are incubated with ECL reagents for 1 min as specified in the manufacturer instructions. Signal is obtained by exposing membranes to Hyperfilm from GE Healthcare for various time periods to ensure non-saturated signals.
11. Gel-imaging software (like the freely available Scion Image from Scion Corporation, Inc.) should be used for signal quantitation.
12. To ensure even loading and transfer of the protein, the membranes may be incubated with an anti-PPARγ antibody followed by an HRP-conjugated secondary antibody and ECL detection according to standard Western blot procedures. A summary of this procedure is depicted in Fig. 2.

3.3. Fluorescence Anisotropy-Based Ligand Activation Assays

The ability of the ligands to activate PPARγ is assessed by comparison of the fluorescence anisotropy-based isotherm of binding of PPARγ-LBD to a peptide containing the binding sequence of the coactivator SRC-1 in the presence of the ligand with that obtained in its absence. A description of the assay can be found in Fig. 3.

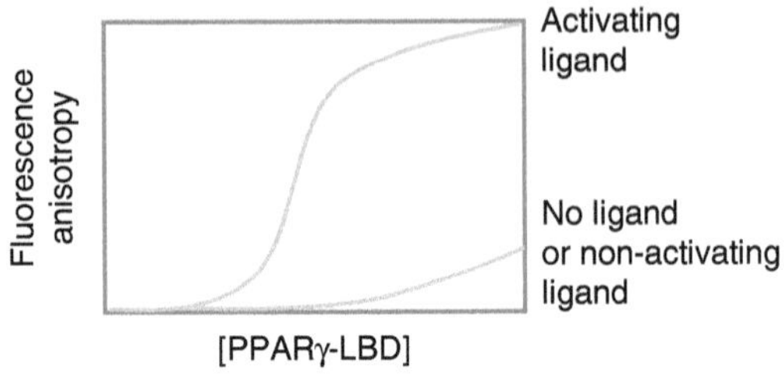

Fig. 3. Fluorescence anisotropy-based ligand activation assays. The interaction of the fluorescently labeled SRC-1 peptide with PPARγ-LBD provokes an increase in the fluorescence anisotropy with respect to that measured for the free peptide due to the formation of a complex that rotates slower than the free peptide. In the absence of ligands or in the presence of a non-activating ligand, the interaction only occurs at high concentrations of the receptor. In contrast, activating ligands lead to an increase in the anisotropy measured even at low concentrations of PPARγ-LBD indicative of an enhancement of the affinity of the receptor for the peptide coupled to ligand binding.

(a) *Binding of PPARγ-LBD to the labeled peptide in the absence of ligand*

1. Switch on the fluorometer and allow it to equilibrate for 30–45 min before taking the measurements. Set the temperature control at 20°C.
2. Prepare 160 μL of a solution containing PPARγ-LBD at 20 μM and 5 nM of the fluorescein-labeled SRC-1/NR2 peptide in buffer A (solution (c)).
3. Prepare 650–950 μL of a solution containing 5 nM of the fluorescein-labeled SRC-1/NR2 peptide in buffer A (solution (d)).
4. Divide solution (d) into 10–15 eppendorf tubes. Put 60 μL in each tube.
5. Add 90 μL of solution (c) into one of the tubes prepared in step 4 and mix. Withdraw 90 μL from this tube, put them into the second tube containing solution (d), and gently mix. Repeat the procedure with the remaining tubes containing solution (d)
6. Put 60 μL of solution (c) into a 3 × 3 quartz cuvette and measure the anisotropy at λ_{exc} and λ_{em} of 495 and 520 nm, respectively. Measure the anisotropy repeatedly until equilibration and take the average of the last eight points after reaching equilibrium. Subtract the background of the buffer measured under the same conditions.
7. Measure the samples obtained in step 5 as described for solution (c) in step 6.
8. Plot the anisotropy values as a function of the concentration of PPARγ-LBD in the solution to build the binding isotherms.

9. Repeat the entire procedure at least three times.
10. Calculate an average isotherm (±SD) from the three independent titrations.

(b) *Binding of PPARγ-LBD to the labeled peptide in the presence of ligand*

1. Repeat steps 2–10 including a fixed concentration of the ligand in the samples, for example 10–20 μM.
2. Compare the isotherm obtained with that determined in the absence of ligand. If the isotherms overlap, the ligand does not activate PPARγ.

3.4. Cell-Based PPRE Assays

1. RMC cultured as above can be seeded in 6-well culture plates at 200,000 cells/well the day before the assay.
2. Prepare a transfection mixture as specified below:
   - In an eppendorf tube pipet 125 μL of Optimem medium per well and 3 μL of Lipofectamine 2000 and mix thoroughly. Incubate for 5 min at room temperature.
   - To 125 μL of Optimem per well add the appropriate volumes of solutions of each reporter plasmid to be used so that the final amount in the mixture is 1 μg for the PPRE reporter plasmid and 5 ng for the renilla reporter plasmid per well to be transfected. Mix thoroughly.
   - Mix with the Lipofectamine solution by vortexing or tapping the tube. Incubate for 25 min at room temperature.
3. Add 250 μL of this mixture to each well.
4. Culture cells with this mix and after 3 h replace medium with fresh medium without FCS and continue the incubation for 4 h.
5. Add agonists to a final concentration of 5 μM. After 16 h cells should be lysed with the reagents provided in the Dual luciferase assay kit from Promega, according to the instructions of the manufacturer.
6. Cell lysates processed as detailed in the Dual luciferase assay kit are used for determination of renilla and firefly luciferase activities in a luminometer following the instructions of the manufacturer.
7. The assay should be performed in triplicates and the values obtained for firefly luciferase activity should be corrected by those of the renilla luciferase activity for each sample, which provides an internal control of transfection efficiency.

A workflow for the transfection and determination of PPRE procedure is shown in Fig. 4.

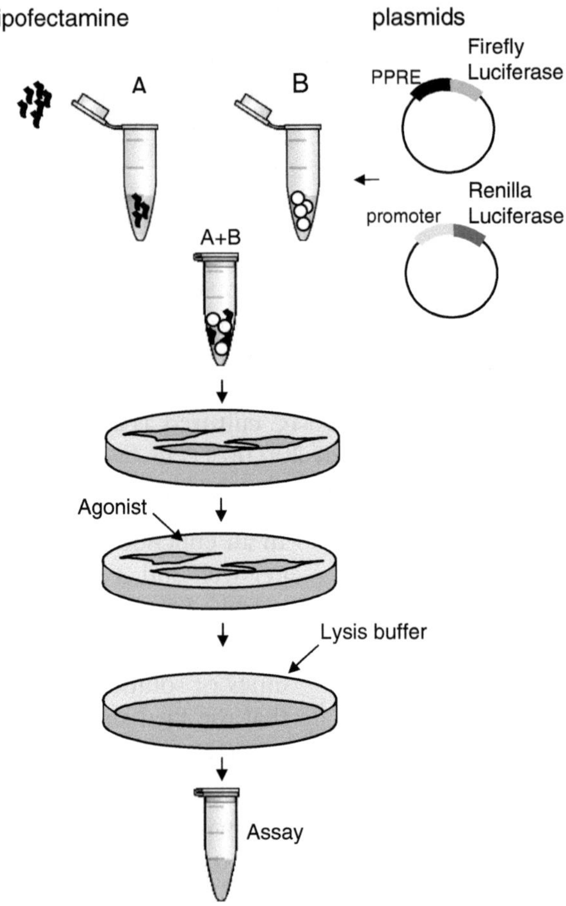

Fig. 4. Scheme of cell-based PPRE reporter assays. Mixtures of transfection medium with Lipofectamine or DNA are prepared separately and then mixed together. The DNA mixture includes a plasmid coding for firefly luciferase, whose expression is driven by a PPRE-containing promoter, plus a plasmid with a PPAR-independent promoter driving the expression of renilla luciferase, which provides an internal control for transfection and non-specific effects of agonists. After mixing Lipofectamine and DNA, mixtures are vortexed and incubated at room temperature as described in Subheading 3, before adding them to cells. After incubation with the transfection mixture, cells are allowed to recover for several hours before addition of agonists. Cell lysis and processing of lysates allows determination of luciferase activities.

4. Notes

ANS Binding and Competition Assays

1. After purification, PPARγ-LBD should be stored in 25–50 μL aliquots at −80°C. The use of freshly thawed PPARγ-LBD in the assays is preferred, however, we note that we have not observed any difference in the ANS binding titrations performed with PPARγ-LBD samples undergoing two freeze-thaw cycles.

2. The buffer for the ANS binding titrations can be modified if necessary but it should be free of substances interacting with ANS such as BSA or HSA, which are commonly employed in the buffers to avoid unspecific sticking of proteins to the tubes.
3. The solutions should be protected from direct light during the experiment.
4. The reference isotherm, that is the ANS binding titration of PPARγ-LBD in the absence of ligand, must be determined at least once during each measurement session. This is because the emission of fluorescence from a sample may vary from day-to-day even if the same spectrofluorometer is used, due to minor changes in the intensity of the excitation beam. The difference may be dramatic when comparing measurements performed very far away in time (weeks or months).
5. The maximum concentration of ANS we have used in our assays is 100 μM. At this concentration and using 3 × 3 mm cuvettes no significant inner filter effects need to be taken into account and the plateau of the isotherm of binding is reasonably well defined. If concentrations of ANS above 100 μM are employed, correction for inner filter effects would need to be included.
6. The concentration of ligand in the assay should be adjusted to observe significant competition with ANS. Hence, lower concentrations would be required for ligands exhibiting high affinity binding to PPARγ-LBD than for those binding less tightly to the receptor.
7. We have used BIOEQS software for the quantitative analysis of the binding isotherms in order to determine the affinity of the ligands for the PPARγ-LBD. This program, together with a detailed manual of its use, can be freely downloaded as described in (24). Alternatively, other software for the thermodynamic analysis of interactions can be used or the equilibrium equations can be written for example in Matlab or Origin packages.

Gel-Based Competition Assays

8. Biotin signals as developed with HRP-Streptavidin can be very intense and saturate the film in a few seconds. Therefore, it is convenient to dilute the samples, if necessary and/or carry out short exposures in order to obtain non-saturated signals.
9. Given the high affinity of the biotin-streptavidin interaction, HRP-streptavidin may remain bound to the membrane even after standard stripping procedures. Therefore, if the membrane is going to be incubated with anti-PPARγ antibody it is convenient to inactivate the remaining HRP. This can be achieved through various procedures, like treatment with an excess hydrogen peroxide (25).
10. It should be taken into account that gel-based assays relying on Western blot detection are semiquantitative. Therefore, only

estimates of the binding parameters can be obtained. In this regard, it is important that exposures are not saturated and the signals obtained fall within the linear range of detection. Ideally, if the competition assay is performed in the presence of several concentrations of the potential ligands, estimates of binding constants can be obtained as described in Shiraki et al. 2005 (26).

11. Certain protocols include a partial denaturation step to ensure that the cysteine residue is accessible to the labeling reagent (26). Under our experimental conditions, PPAR agonists binding covalently to the cysteine residue located in the ligand binding pocket give more consistent results in the competition assays than non-covalently bound agonists. Several hypotheses may explain this result, including the possibility that given the size of the ligand binding cavity, it could accommodate both the ligands and the cysteine reagent.

Fluorescence Anisotropy-Based Ligand Activation Assays

12. In the absence of ligands, PPARγ-LBD interacts with the peptide only at high concentrations and no plateau is reached even at 20 μM protein concentration under our conditions (15). Activating ligands would be those that produce significant enhancement of the binding compared with that obtained in the absence of ligands as it was shown for rosiglitazone and 15d-PGJ_2 (15). Quantitative determination of the affinity of the interaction of PPARγ-LBD with the peptide in the presence of the activating ligands can be achieved through analysis of the binding isotherm.
13. In contrast to the ANS assay, since the method described here is based on anisotropy measurements that are not sensitive to small intensity changes, the isotherm of binding in the absence of ligand does not need to be repeated each time. However, its determination is highly recommendable whenever a new stock of protein or peptide is employed. Moreover, in this case, the performance of positive controls with a well-known activating ligand such as rosiglitazone is a good practice.
14. The buffer for the anisotropy assays can be modified if necessary. However, care should be taken with the pH, since the emission of fluorescence of the dye selected to label the peptide, fluorescein, is sensitive to the pH of the solution. Thus, at pH below 7.0 the predominance of the monoanionic form of fluorescein causes a significant reduction in the observed fluorescence emission. If a pH below this value is required, a peptide labeled with another fluorophore lacking pH sensitivity such as Alexa 488 can be employed.
15. The concentration of labeled peptide can be increased if a spectrofluorometer with lower sensitivity is employed in the measurements.

16. The 3×3 cuvettes requiring 50–60 μL of solution per measurement are particularly useful in the case of expensive ligands, and also for minimal consumption of labeled peptide.
17. Since anisotropy is a parameter very sensitive to temperature changes, the temperature of the cuvette holder of the spectrofluorometer should be very accurately controlled during the whole experiment.
18. Cycles of freeze-thaw of the labeled peptide should be avoided.

Cell-Based Assays

19. Cell lysates obtained for measurement of luciferase activity can also be used for protein determination, in order to ensure that the amount of protein recovered in every experimental condition is similar. A substantial decrease in the amount of protein recovered in any experimental condition may indicate loss of cells due to cytotoxicity or reduction in cell adhesion, and therefore the results of the assay should be taken cautiously.
20. Cell lysates can be also used in Western blot to assess the levels of other proteins of interest, like known PPARγ target genes or the PPAR proteins, as an internal control of the assay.

Acknowledgments

This work was supported by the European Community through the Marie Curie program (MERGT-CT-2006-046474), Grants BFU2006-03905/BMC, SAF2006-03489, SAF2009-11642 and BFU2010-14910 from the Ministerio de Ciencia e Innovación (MICINN, Spain), grant RETICS RIRAAF (RD07/0064/0007) from the Instituto de Salud Carlos III and Grant PIE 200980I186 (CSIC).

References

1. Desvergne B, Michalik L, Wahli W (2006) Transcriptional regulation of metabolism. Physiol Rev 86:465–514
2. Montanari R, Saccoccia F, Scotti E, Crestani M, Godio C, Gilardi F, Loiodice F, Fracchiolla G, Laghezza A, Tortorella P, Lavecchia A, Novellino E, Mazza F, Aschi M, Pochetti G (2008) Crystal structure of the peroxisome proliferator-activated receptor gamma (PPARgamma) ligand binding domain complexed with a novel partial agonist: a new region of the hydrophobic pocket could be exploited for drug design. J Med Chem 51:7768–7776
3. Itoh T, Murota I, Yoshikai K, Yamada S, Yamamoto K (2006) Synthesis of docosahexaenoic acid derivatives designed as novel PPARgamma agonists and antidiabetic agents. Bioorg Med Chem 14:98–108
4. Zoete V, Grosdidier A, Michielin O (2007) Peroxisome proliferator-activated receptor structures: ligand specificity, molecular switch and interactions with regulators. Biochim Biophys Acta 1771:915–925
5. Shiraki T, Kodama TS, Shiki S, Nakagawa T, Jingami H (2006) Spectroscopic analyses of the binding kinetics of 15d-PGJ2 to the PPARγ

ligand-binding domain by multi-wavelength global fitting. Biochem J 393:749–755

6. Itoh T, Fairall L, Amin K, Inaba Y, Szanto A, Balint BL, Nagy L, Yamamoto K, Schwabe JW (2008) Structural basis for the activation of PPARgamma by oxidized fatty acids. Nat Struct Mol Biol 15:924–931
7. Li Y, Zhang J, Schopfer FJ, Martynowski D, Garcia-Barrio MT, Kovach A, Suino-Powell K, Baker PR, Freeman BA, Chen YE, Xu HE (2008) Molecular recognition of nitrated fatty acids by PPAR gamma. Nat Struct Mol Biol 15:865–867
8. Kliewer SA, Xu HE, Lambert MH, Willson TM (2001) Peroxisome proliferator-activated receptors: from genes to physiology. Recent Prog Horm Res 56:239–263
9. Fujimura T, Sakuma H, Ohkubo-Suzuki A, Aramori I, Mutoh S (2006) Unique properties of coactivator recruitment caused by differential binding of FK614, an anti-diabetic agent, to peroxisome proliferator-activated receptor gamma. Biol Pharm Bull 29:423–429
10. Lehmann JM, Moore LB, Smith-Oliver TA, Wilkison WO, Willson TM, Kliewer SA (1995) An Antidiabetic Thiazolidinedione Is a High Affinity Ligand for Peroxisome Proliferator-activated Receptorγ (PPARγ). J Biol Chem 270:12953–12956
11. Forman BM, Tontonoz P, Chen J, Brun RP, Spiegelman BM, Evans RM (1995) 15-Deoxy-delta 12, 14-prostaglandin J2 is a ligand for the adipocyte determination factor PPAR gamma. Cell 83:803–812
12. Royer CA, Scarlata SF (2008) Fluorescence approaches to quantifying biomolecular interactions. Methods Enzymol 450:79–106
13. Kane CD, Bernlohr DA (1996) A simple assay for intracellular lipid-binding proteins using displacement of 1-anilinonaphthalene 8-sulfonic acid. Anal Biochem 233:197–204
14. Collini M, D'Alfonso L, Molinari H, Ragona L, Catalano M, Baldini G (2003) Competitive binding of fatty acids and the fluorescent probe 1-8-anilinonaphthalene sulfonate to bovine beta-lactoglobulin. Protein Sci 12:1596–1603
15. Zorrilla S, Garzon B, Perez-Sala D (2010) Selective binding of the fluorescent dye 1-anilinonaphthalene-8-sulfonic acid to peroxisome proliferator-activated receptor gamma allows ligand identification and characterization. Anal Biochem 399:84–92
16. DeGrazia MJ, Thompson J, Heuvel JP, Peterson BR (2003) Synthesis of a high-affinity fluorescent PPARgamma ligand for high-throughput fluorescence polarization assays. Bioorg Med Chem 11:4325–4332
17. Bouaboula M, Hilairet S, Marchand J, Fajas L, Le Fur G, Casellas P (2005) Anandamide induced PPARgamma transcriptional activation and 3 T3-L1 preadipocyte differentiation. Eur J Pharmacol 517:174–181
18. Jameson DM, Ross JA (2010) Fluorescence polarization/anisotropy in diagnostics and imaging. Chemical reviews 110:2685–2708
19. Pogenberg V, Guichou JF, Vivat-Hannah V, Kammerer S, Perez E, Germain P, de Lera AR, Gronemeyer H, Royer CA, Bourguet W (2005) Characterization of the interaction between retinoic acid receptor/retinoid X receptor (RAR/RXR) heterodimers and transcriptional coactivators through structural and fluorescence anisotropy studies. J Biol Chem 280:1625–1633
20. Bourdoncle A, Labesse G, Margueron R, Castet A, Cavailles V, Royer CA (2005) The nuclear receptor coactivator PGC-1alpha exhibits modes of interaction with the estrogen receptor distinct from those of SRC-1. J Mol Biol 347:921–934
21. Cronet P, Petersen JF, Folmer R, Blomberg N, Sjoblom K, Karlsson U, Lindstedt EL, Bamberg K (2001) Structure of the PPARalpha and -gamma ligand binding domain in complex with AZ 242; ligand selectivity and agonist activation in the PPAR family. Structure 9:699–706
22. Sanchez-Gomez FJ, Cernuda-Morollon E, Stamatakis K, Perez-Sala D (2004) Protein thiol modification by 15-deoxy-Delta12,14-prostaglandin J2 addition in mesangial cells: role in the inhibition of pro-inflammatory genes. Mol Pharmacol 66:1349–1358
23. Royer CA, Smith WR, Beechem JM (1990) Analysis of binding in macromolecular complexes: a generalized numerical approach. Anal Biochem 191:287–294
24. Rosales T, Royer CA (2008) A graphical user interface for BIOEQS: A program for simulating and analyzing complex biomolecular interactions. Anal Biochem 381:270–272
25. Sennepin AD, Charpentier S, Normand T, Sarre C, Legrand A, Mollet LM (2009) Multiple reprobing of Western blots after inactivation of peroxidase activity by its substrate, hydrogen peroxide. Anal Biochem 393:129–131
26. Shiraki T, Kamiya N, Shiki S, Kodama TS, Kakizuka A, Jingami H (2005) Alpha, beta-unsaturated ketone is a core moiety of natural ligands for covalent binding to peroxisome proliferator-activated receptor gamma. J Biol Chem 280:14145–14153

Chapter 19

Exploring PPAR Modulation in Experimental Mice

Anna Rull, Gerard Aragonès, Raúl Beltrán-Debón, Esther Rodríguez-Gallego, Jordi Camps, and Jorge Joven

Abstract

The main concern in exploring modulation of PPARs in experimental animals is probably the choice of the model. Although mechanistic studies may be well designed using knockout and transgenic animals, the interpretation of results with respect to inferred results of PPAR activation into humans should be performed cautiously. This is even more important in this field considering that undesired and unexpected effects have been already described in human epidemiologic studies. Taken together, these observations suggest that a global approach using omic technologies, although expensive, is probably the most suitable to obtain useful data. Also, in this approach, we propose all tissues should be explored, not only those obviously relevant to metabolism, or stored in proper conditions if further assessment is required.

Key words: Animal handling, Atherosclerosis, Genomics, Histology, Inflammation, Metabolism, Metabolomics, Mice, Proteomics, Transcriptomics

1. Introduction

Metabolism and inflammation are closely interrelated, and the combination of overweight or obesity, type 2 diabetes, liver steatosis, and cardiovascular diseases represents a major health concern (1). Peroxisome proliferator-activated receptors (PPARs) are involved in both the regulation of inflammation and energy balance and are currently considered therapeutic targets for treatment of obesity-induced inflammation and associated metabolic disorders (2, 3). PPARα is expressed in metabolically active tissues involved in fatty acid oxidation (4) and immunologic responses (although this not completely understood) that may suppress the expression of certain important genes in the inflammatory response (5). Additionally, PPARγ and PPARβ/δ are potential regulators of adipogenesis in obesity and are associated with the development

Mostafa Z. Badr and Jihan Youssef (eds.), *Peroxisome Proliferator-Activated Receptors (PPARs): Methods and Protocols*, Methods in Molecular Biology, vol. 952, DOI 10.1007/978-1-62703-155-4_19, © Springer Science+Business Media New York 2013

of anti-inflammatory responses in macrophages and other cells (6, 7). However, conflicting results in humans and animals preclude firm conclusions on the relevance of PPARs in the overall immune response and consequently on clinical implications. To further assess the role of PPARs, animal models are paramount. Despite multiple differences between men and other animals, mice are currently considered valuable models due to our ability to alter and maintain genomic changes, the relatively low cost and shorter lifespan, as well as the possibility to easily alter environmental factors (8). However, obtaining reliable results in exploring PPAR modulation is difficult. This research not only requires extensive experience in handling animals but also requires suitable facilities as well as well-trained personnel and probably extramural collaboration. Ethically, the number of animals involved in experiments should be kept into a minimum, and the results should be obtained with low variability and high reproducibility. A substantial factor responsible for repeated experiments is basically the absence of adequate strategic planning that should necessarily include possible future hypotheses and proper storage of all available material irrespective of the needs for the current hypothesis. We also believe the global approach may spare the lives of many animals.

For instance, recent and expected advances in omic technologies (genomic, transcriptomic, proteomic, and metabolomic) hold great promise for the increased understanding of complex metabolic disorders, especially those in which PPAR modulation play a significant role, particularly liver steatosis and atherosclerosis. This research requires capturing information from all targets: DNA, RNA, metabolites, and proteins (see Fig. 1). Few laboratories have experience with all of these technologies, but commercial services are readily available. Genomic and transcriptomic approaches identify genes and pathways (9), but proteomic and metabolomic studies may yield specific biomarkers and real-time information on pathogenesis (10). Therefore, potential benefits of such approaches include better tools for diagnosis and identification of therapeutic targets (11, 12). We have focused on methods that may add significant value to experimental studies directed to explore the metabolic and anti-inflammatory role of PPARs and the expected effects of continuous or regulated activation.

2. Materials

It is important to use analytical grade reagents as well as prepare and properly store all materials to be used before performing the desired experiment. This preparation is usually done the same day. In this regard, we strongly recommend following accredited and certified quality procedures such as ISO 9001:2008, even for small

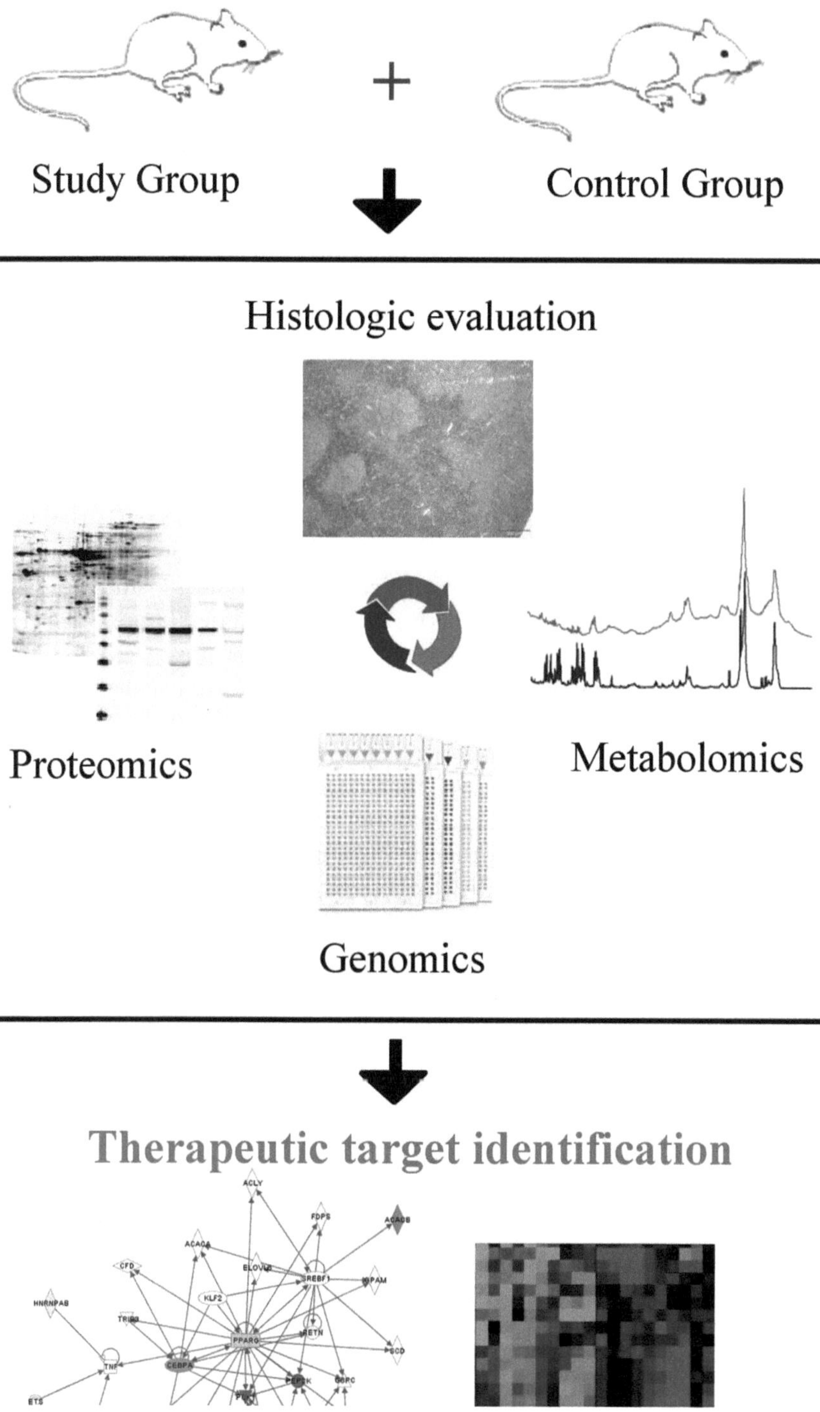

Fig. 1. Proposed approach to aid in designing experiments functionally exploring PPARs in metabolism and inflammation.

and well-managed laboratories. Limitations should be always considered before designing the study. Choice of the experimental model requires previous familiarization. Unfortunately, there are no available models for human diseases, and inferring results is a risky procedure. However, knockout (KO) or transgenic animals may provide insights into mechanisms of action. For instance, there is no accepted model for metabolic syndrome. Obesity is easily obtained via the KO of leptin or its receptor, and the resulting mice develop substantial insulin resistance and consequently liver steatosis; however, this condition is extremely unusual in humans. Dietary-induced obesity normally requires feeding animals with unnatural or even toxic elements. Attempts to include hyperlipidemia into the equation result in extremely high plasma cholesterol values. Atherosclerosis is evident in KO mice for LDL receptor or ApoE and easily assessed, but dietary requirements to produce a significant lesion are different in both models. Particularly when exploring metabolism and/or inflammation, all precautions should be followed to minimize or avoid stress, and animal handling should only be performed by trained personnel.

2.1. Chemicals Commonly Used for PPAR Modulation

1. Several providers are reliable, but we usually rely on Sigma, for which reference numbers are provided. Fenofibrate (F6020) is a specific PPARα agonist that is commonly used in clinical practice, but similar results may be obtained with WY14643 (C7081). Rosiglitazone may be used to activate PPARγ and can be directly obtained from GlaxoSmithKline. However, recently implemented restrictions in humans may represent difficulties in its supply and dissemination of results. Similar effects may be obtained with pioglitazone (E6910) or GW1929 (G5668). For experiments in which a PPAR pan-agonist is preferred, bezafibrate (B7273) is an available and acceptable choice.
2. Dosage and route of administration should be previously predetermined, preferably with the aid of pharmacokinetic data or alternatively with thorough assessment of response in simple laboratory measurements. We strongly recommend the administration via selected chow, a method which requires industrial equipment and reliable providers. All required chow for the experiment should be prepared before the experiment. Discard all surplus materials. Other routes of injection may represent an unacceptable degree of stress in experimental animals.

2.2. Routine Metabolic Assessment

1. Intraperitoneal glucose tolerance test (IGTT), oral fat test (OFT), and simple glucose measurements are preferred. Materials required include isofluorane inhalation anesthesia, scalpel blades, timer, 1 mL syringes, and 25-Gx 5/8-in. needles for IGTT and 22G ball-tip needle for OFT.

2. For IGTT, additional materials include 1 g D-glucose/2 mL sterile physiologic solution and a blood glucose monitor and associated test strips for glucose measurement (e.g., Accu-Check Sensor Comfort, Roche Diagnostics).
3. For OFT, oil olive and a blood triglycerides monitor and associated test strips for triglycerides measurement (e.g., AccuTrend GCT, Roche Diagnostics) are needed.
4. Although these are commonly perceived as simple measurements, intra-day and inter-day variations may occur if they not performed with extreme caution.
5. Also, falsely high or low results are possible in the presence of previously untested chow. Also, sample contamination results in unreliable results.

2.3. Blood and Tissue Collection

1. For the anesthetic mixture, required materials include 12.5 μL 2% Xylazine (Xylagesic), 50 μL 50 mg/mL Ketamine (Ketolar), 2-mL eppendorf tubes, 1-mL syringes, and 25-Gx 5/8-in. needles. Contamination should be avoided during blood collection.
2. Sodium-heparinized 1 mL syringes with 25-Gx 5/8-in. needles should be used with EDTA-containing tubes for plasma and untreated tubes for serum (see Note 1).
3. Other materials include a refrigerated centrifuge, a cork board and pins, and 70% ethanol. Always proceed in an extremely clean environment.
4. Once exsanguinated, mouse necropsy should be performed under sterile conditions when possible, and all materials should be previously prepared, including dissection instruments (e.g., scissors, surgical blade), perfusion pump and catheter, plastic tubes, Petri dishes, liquid nitrogen, appropriate containers, isopentane (2-methylbutane), dry ice, filter paper, tissue freezing medium (Tissue-Tek (O.C.T.), Sakura Finetechnical) and plastic and metallic embedded moulds.
5. PBS solution is commercially available. However, it may be easily prepared in a 10× solution before experiments (80 g NaCl, 2 g KCl, 14.4 g Na_2HPO_4 and 2.4 g K_2HPO_4/L at pH 7.4). To prepare a PBS working solution, mix 100 mL with 900 mL deionized water.
6. 4% (w/v) paraformaldehyde solution and a 20% sucrose solution in phosphate-buffered saline (PBS) (100 mM) should be prepared and stored at 4°C, before use.
7. Prepare labeled tubes or containers necessary for the collection of all tissues. To proceed rapidly, which is important, collecting tissues should be performed by at least two individuals.

2.4. Tissue Histopathological Evaluation

1. Common materials include slides and cover slides, 100% ethanol, deionized water, Harris hematoxylin, Oil Red O, acetic acid, xylol, eosin, and mounting medium.
2. For the Oil Red O stock solution, gently mix 2.5 g Oil Red with 500 mL of isopropanol. For Oil Red O working solution, gently mix 24 mL Oil Red stock solution with 16 mL deionized water and wait 10 min before use.
3. Harris hematoxylin solution (Sigma Aldrich) should be filtered before use. For the Eosin Y working solution (Sigma Aldrich), slowly add up to 0.5 mL acetic acid to 100 mL of the stain solution (see Note 2).

2.5. Atherosclerotic Lesion Assessment

1. Note that this procedure should be performed using frozen tissue to preserve lipids in lesions. For convenience, some investigators assume that unoccupied spaces in paraffin-embedded sections may represent accurate assessment of lipids. When possible, we recommend lipid staining, which will at least require a cryostat, slides and cover slides, isopropanol, Sudan IV, Harris hematoxylin, lithium carbonate, and aqueous mounting medium.
2. For Sudan IV working solution, use 10 g Sudan IV, 500 mL 70% ethanol, and 500 mL acetone. Dissolve Sudan IV in the solvent, filter and store in a dark bottle at room temperature. Discard surplus material to avoid reusing.
3. 1.35% lithium carbonate solution: mix 9.4 g lithium carbonate with 700 mL deionized water and store at room temperature.

2.6. Immunohistochemical Analyses

1. Although simple, procedures to expose antigens in tissue sections should be very carefully performed. It is important to ensure complete tissue fixation. Conditions may vary from tissue to tissue, but limiting factors are those related to the choice of the antibody. For these procedures, use 100% ethanol, Mayer's hematoxylin, histological poly-lysine treated glass slides and cover slides, permanent mounting medium and PBS (100 mM)-Glycine (20 mM), which is made by dissolving 1.5 g glycine in 1 L PBS.
2. 3% hydrogen peroxide.
3. Acid alcohol: 0.37% HCl in 70% ethanol.
4. 2% bovine serum albumin (BSA): dissolve 0.2 g BSA in 10 mL PBS and store as 1 mL aliquots at −20°C. To prepare the 1% BSA solution, mix 500 μL of 2% BSA with 500 μL of 100 mM PBS.
5. There are many combinations of antibodies and revealing methods. We recommend goat polyclonal antibodies (Santa Cruz BioTech): PPARα (N–19) (sc-1985), PPARβ (K29) (sc-1987) and anti-PPARγ (N-20) (sc-1984) and biotinylated

anti-goat IgG (H+L) (Atom) in combination with Vectastain ABC complex kit (Vector PK-4000) and chromogen 3, 3′ diaminobenzidine (DAB) substrate (Dako, K3466).

2.7. Homogenization and Lipid or Protein Extraction from Tissues Using Precellys 24 System

1. Most protocols and methods of extraction work perfectly, but a limiting factor is how and when to mince the collected tissues. In our hands, the Precellys 24 system (Bertin Technologies, France) offers considerable advantages, mainly the potential to process multiple samples in a few minutes.
2. 2-mL Precellys microtubes, glass tubes, and the preferred extraction solvent.
3. The preferred extraction solvent in this protocol is a hexane/isopropanol mixture (3:2) at 4°C and 0.47 M sodium anhydrous sulfate solution prepared by mixing 6.7 g sodium anhydrous sulfate in 100 mL distilled water. Additional materials needed include 0.5% sodium cholate and orbital shaker.

2.8. Assessment of mRNA Expression in Tissue Extracts

1. A refrigerated centrifuge with adaptors for 96- or 384-well plates (MicroAmp, Applied Biosystem). Both plate sizes are useful and should be considered before proceeding with the analyses.
2. 2-mL Precellys microtubes, 2-mL eppendorf tubes and the RNeasy Plus mini kit (QIAgen); note that the procedure requires the addition of 10 μL β-mercaptoethanol to the provided RLT buffer.
3. High Capacity cDNA Reverse Transcription kit (Applied Biosystems, Foster City, USA) is very convenient for this protocol.
4. Other recommended instruments include a NanoDrop 1000 spectrophotometer (Thermo Fisher Scientific Inc., Washington, DE, USA), a thermal cycling apparatus (GeneAmp PCR System 9700), and the ABI Prism 7900HT Sequence Detection System (Applied Biosystem). Taqman primers and probes are obtained from validated Assays-on-Demand products (Applied Biosystem): Mm00440939-ml for PPARα, Mm00803186-gl for PPARδ, and Mm00440945-m1 for PPARγ. It is important to note the PPAR expression in many tissues is low, and consequently, considerable amounts of mRNA should be prepared.

2.9. Other Procedures to Assess Protein Expression (Western Blot or Protein Immunoblotting)

1. Materials include all those related to nuclear extraction, electrophoresis, transfer to special blotting material, blocking with a generic protein, and antibodies designed to convert a substrate to be seen or photographed.
2. Among others, 2-mL Precellys microtubes, 2-mL eppendorf tubes, Lysis Buffer (20 mM Tris–HCl, 150 mM NaCl, 1 mM EDTA, 1 mM EGTA, 1% CHAPS, 1 mM Pefabloc® SC, Nref

76307 (Sigma) and 1% phosphatase inhibitor cocktail 2, Nref P5726 (Sigma) in 100 mL deionized water with Pefabloc and phosphatase inhibitor added immediately before use), and the 2D Quant kit (GE Healthcare) are required.

3. For electrophoresis, required materials include the following: NuPage Novex Bis-Tris 4–12% Gel 1.0 mm 12 well, Nref NP0322BOX (Invitrogen), SeeBlue Plus2 Pre-Stained Standard, Nref LC5925 (Invitrogen), MagicMark XP Western Protein Standard, Nref LC5602 (Invitrogen), iBlot Gel Transfer Device, Nref 1001EU and iBlot Transfer Stack Regular, Nref IB3010-01 (Invitrogen).
4. Several combinations of antibodies may be useful, but we suggest goat polyclonal antibodies (Santa Cruz BioTech): anti-PPARα (N–19) (sc-1985), anti-PPARβ (K29) (sc-1987), and anti-PPARγ (N-20) (sc-1984) and polyclonal Rabbit anti-Goat Immunoglobulins/HRP (DAKO).
5. Also prepare a TBS solution (10×) by dissolving 30 g Tris, 80 g NaCl, 2 g KCl in 1 L deionized water (pH 7.4) and a working solution of 100 mL 10× TBS and 900 mL deionized water.
6. The blocking agent (4% TBS) is prepared by dissolving 0.8 g commercial blocking agent (Amersham, GE Healthcare) in 20 mL TBS and then adding 20 μL Tween 20 (Sigma, Ref: P9416). The blocking agent (2% TBS) is prepared by dissolving 0.8 g commercial blocking agent in 40 mL TBS and then adding 40 μL Tween 20.
7. For the wash buffer (TBS-SDS-NP40), mix 200 mL TBS, 1 mL 10% SDS and 2 mL Nonidet P40 (Fluka, Ref: RB 10355).
8. We recommend the ECL Advance Western Blotting Detection Kit (Amersham, GE Healthcare, Ref: RPN2135) and the VersaDoc system (Bio-Rad) for final assessment.

3. Methods

In vivo mouse experiments should be designed and performed in accordance with relevant laws and after approval by the local animal care and use committee. Relaxation in these guidelines frequently leads to poor results. Experimental procedures must be performed only by trained personnel to minimize stress-induced effects on the animals. Previous strategies for animal model development should be taken into account before final design of a project (see Note 3). If mice are ordered from commercial providers, we recommend generating the experimental population in your own facilities.

3.1. PPAR Modulation

PPAR modulation depends entirely on the design of the study. Agonists are most commonly used, but antagonists are also available (http://www.scbt.com/chemicals-table-ppar_agonists_and_antagonists.html). It is important to note certain xenobiotics, including antibiotics, natural fatty acids, and food contaminants may have substantial effects in an experiment. Consequently, feeding animals is a major issue, and quality controls should be routinely implemented. As stated before, the route of administration, when possible, should be limited to mixing drugs with chow to minimize stress; in this case, differences in food intake should be taken into account. Most researchers choose to administer experimental reagents in excess, but this is only possible if pharmacokinetic data are previously obtained and considered. Dietary-induced obesity is a major problem when PPARs are explored because chow composition usually alters results. The use of sham or control groups does not eliminate the possibility of false data. Finally, PPARs are involved in both metabolism and the generation of inflammatory responses; moreover, metabolism and inflammation are frequently interrelated processes. Therefore, it is even more important to avoid factors that may represent additional and undesired low-grade inflammation and /or infection, and this should be checked in all experiments. The choice of adequate plasma biomarkers depends on the analytical experience of each laboratory.

3.2. Routine Methods for Whole Body Metabolic Assessment in Experimental Animals

This step is ideally performed in metabolic chambers that are not usually available to most laboratories. However, beside energy expenditure and respiratory quotient, useful data may be acquired with simple methods in nonspecialized facilities. Changes in body weight may be assessed without substantial stress, but we recommend limiting such monitoring to once a week. Measurement of food intake should be performed twice weekly to detect false effects. Use grids in cages, and measure liquid intake frequently (see Note 4):

1. Start recording body weight at the beginning of the experiment
2. Place a container on a balance and calibrate to zero. Place the mouse in the container and record the weight.
3. Construct the body weight curve by plotting the weekly body weight on the y-axis against the week on the *x*-axis.
4. Insert a metal floor grid in clean cages and wash it frequently.
5. Weigh the appropriate feed for animals in the cage for the feeding interval proposed (2–3 days).
6. Remove the grid and weigh the remaining food.
7. Calculate cumulative food intake. Multiple measurements ensure that the assumption that mice consume similar amounts is correct.
8. To measure liquid intake, accurately weigh the liquid bottle.
9. Fill the bottle with the chosen liquid and record total weight.

10. Change the liquid every day if possible, although it is acceptable to repeat measurements every 2–3 days.
11. To measure IGTT, be sure that mice are fasted at a previously decided time, usually between 12 and 16 h. Weigh the animals and prepare the required glucose working solution in individual prefilled syringes.
12. Prepare all required materials before manipulation.
13. Anesthetize mice using isofluorane. Measure the basal glucose concentration in tail blood (using minimal incisions) by employing the recommended strips and measurement devices. Set aside samples for quality control if required. Once the basal glucose of all mice has been recorded, intraperitoneally inject 2 mg glucose/g body weight while avoiding stress. Proceed rapidly.
14. Repeat measurements at 15, 30, 60, and 120 min using the same incision.
15. For OFT, prepare the required materials and ensure that all animals are in the same fasting state before proceeding to gavages of 150 μL oil olive. Other sources of fat can be used, but it is important to avoid toxicity.
16. Anesthetize mice using isofluorane and proceed to basal triglyceride concentration measurements in the same manner as described for glucose measurements.
17. Once the basal TG of all mice has been recorded, administer the oral gavages to each animal at intervals of 175 s, the time that it takes for the device to deliver reliable data. Obviously, the investigative team may proceed more rapidly if more devices are available.
18. Repeat measurements at 1.30, 3, 5, and 7 h using the same incision.

3.3. Blood and Tissue Collection: Mouse Necropsy

It is important to design a strategic plan to rapidly and efficiently process samples. Tissues may undergo autolysis that can affect the subsequent molecular analyses and the robustness of data (Fig. 2).

1. Prepare the lethal dose of anesthesia required for all the animals by mixing ketamine and xylazine, usually in separated syringes.
2. For a mouse weighing 25–30 g, the lethal dose is 100 μL, and intraperitoneal injection is preferred. Before proceeding further, make sure the animal is not reactive.
3. Prepare 1-mL sodium-heparinized syringes with 25-Gx 5/8-in. needles for blood collection by cardiac puncture. Prepare two 1-mL tubes to collect the blood: one with EDTA for plasma and the other untreated for serum (see Note 5).
4. Proceed to cardiac puncture; allow for blood to flow slowly through the syringe. An assistant should process the blood samples while the researcher continues with the sample collection.

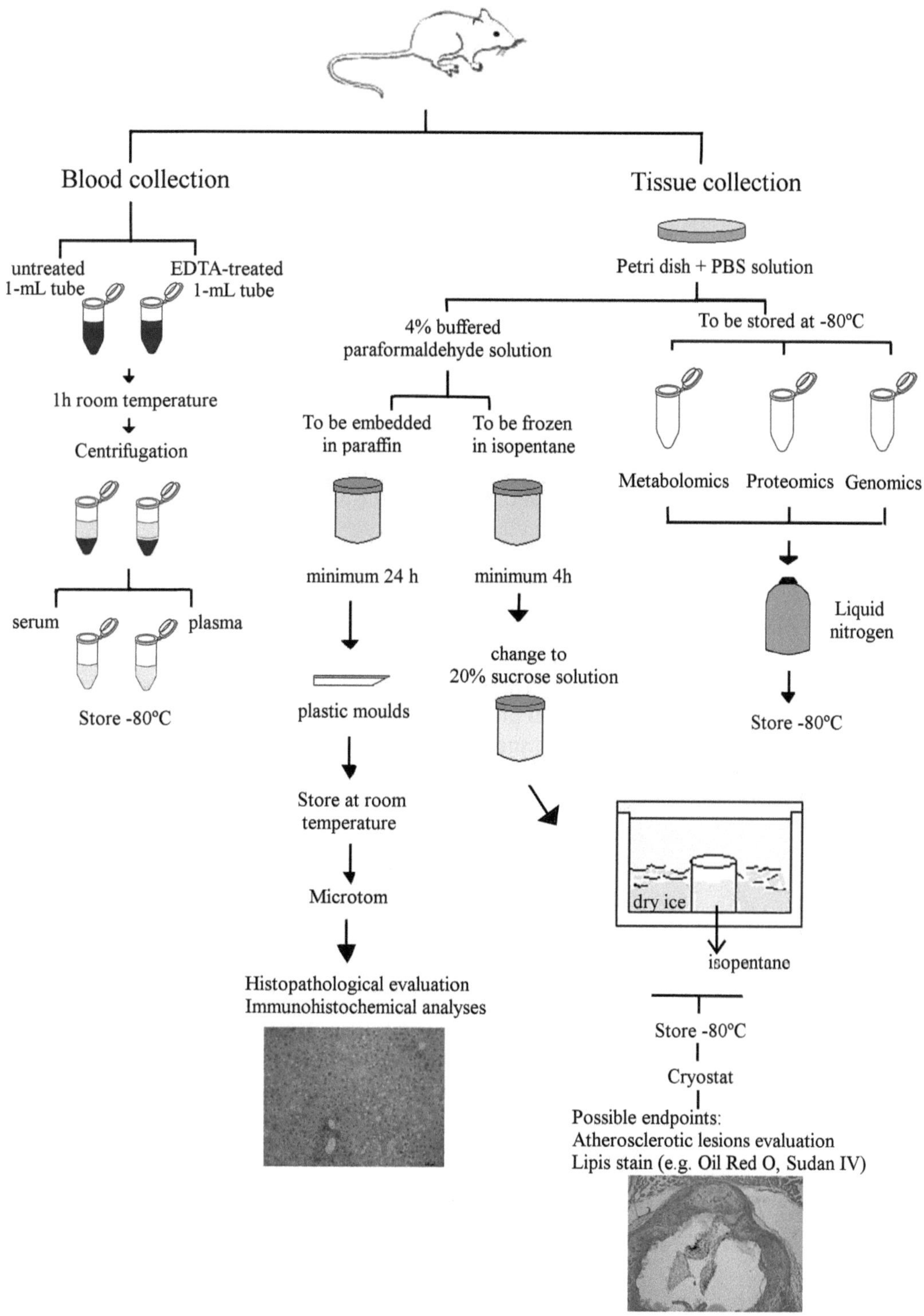

Fig. 2. Proposed flow chart to assist in the correct acquisition of tissue and biological samples.

5. Collect blood in previously prepared tubes. To obtain serum, keep blood at room temperature for 1 h before centrifugation (15 min at 12,000 × *g* and 4°C for 1-mL tubes). To obtain plasma, proceed immediately as previously described to obtain serum samples.
6. Samples are ready to use, but if the analyses are not being performed immediately, aliquots should be stored at –80°C.
7. Fixation by perfusion is recommended to allow the optimal preservation of the tissues and organs for subsequent molecular analyses. Applying this technique is recommended on deeply anesthetized mice rather than in euthanized mice because the pumping heart permits a better distribution of the PBS solution.
8. Be sure that all required materials and the PBS working solution are available and functioning correctly.
9. Open the abdomen of the animal from the pubic symphysis to the sternum taking care not to damage internal organs. Pin the skin to the cork board. Insert the perfusion needle into the left ventricle, start the perfusion pump and open the right auricle. Stop the perfusion after 5 min (see Note 6).
10. Dissect desired tissues and rinse in a Petri dish containing PBS to ensure there is no remaining blood. Before cutting the organs into pieces and placing them into the appropriate tubs, the liquid surrounding the organ must be blotted on filter paper and then the organ should be weighed. Samples to be used for genomic, metabolomic, or proteomic analyses must be immediately placed in tubes and flash-frozen in liquid nitrogen without any pretreatment.
11. For samples to be embedded in paraffin, fix the tissue by immersion in a 4% buffered paraformaldehyde solution for a minimum of 24 h. Put the organ in the plastic mould to start the embedding process into paraffin. Proceed to cutting into the desired sections when necessary.
12. For samples to be frozen in isopentane (see Note 7), fix the tissue by immersion in a 4% buffered paraformaldehyde solution for a minimum of 4 h. Replace the 4% buffered paraformaldehyde with a 20% sucrose solution to cryoprotect the tissues. Prepare a Styrofoam box with dry ice, place a cold-resistant glass container with isopentane and wait until the isopentane reaches –50°C (liquid). Orient the organ in the metallic mould and fill with tissue freezing medium (OCT), avoiding air bubbles. Immerse the mould into the isopentane, and close the lid of the box until the freezing medium turns completely white. Recover the organ with aluminum foil and place it on dry ice. Immediately transfer the samples to –80°C.

13. To cut sections with a cryostat requires practice. Be sure to practice with samples not included in your experiment. This step is especially important when assessing atherosclerotic lesions (see below).
14. Frequently, a complete assessment of metabolic conditions requires relatively simple laboratory measurements that may include components in plasma or serum. This may be a problem because the available amount of sample is limited. Be sure to use analyzers with low sample requirements. Same arguments may be applied for the measurements of more sophisticated analytes, such as cytokines. When possible, use commercially available arrays. To assess the effect of PPAR agonists, it is convenient to proceed to FPLC lipoprotein fractionation.

3.4. Histopathology Evaluation, Atherosclerosis Lesion Assessment, and Basic Immunochemistry

The methods change slightly from tissue to tissue. Herein, we describe handling of liver and aorta, but the important issue is the availability of observers and the ability of double-blind assessments and final interpretation. This step is only important when effects are not dramatic and differences among groups require statistical significance. Minor changes may be highlighted if they correspond to what is expected and are difficult to detect in the final evaluation. A basic method that is absolutely required is hematoxylin and eosin staining:

1. Prepare sections and solutions.
2. Dry the histological sections by placing the slides into an oven for approximately 5 min at 58°C. The paraffin will melt.
3. Remove the paraffin thoroughly in two steps with xylol (10 and 5 min each, respectively) and follow with successive immersions of 5 min in 100% ethanol, 95% ethanol, 90% ethanol, 70% ethanol, and a final wash for 3 min in deionized water.
4. Add Harris hematoxylin for 3 min, immediately wash in tap running water for 3 min, add acid alcohol for 2 s, and immediately wash for another 3 min.
5. Add Eosin Y working solution for 30 s, then wash in running water for another 30 s.
6. Dehydrate the sections by immersing them in 95% ethanol in two different containers for 2 min each and two immersions in 100% ethanol for 3 min each. Place sections consecutively in clean xylol solution twice (2 min each).
7. Mount the sections using permanent mounting medium and place a coverslip on the slide. Dry slides at room temperature.
8. Assess liver steatosis as the percentage of hepatocytes containing macrovesicular fat droplets using a scale graded as 0 (<5%), 1 (5–33%), 2 (34–66%) or 3 (>66%). Also assess the degree of

fibrosis, usually none, inflammation and the lipid droplet content as percentages of lipid droplet area/cell. Ideally, the values should be obtained automatically using commercially available algorithms, such as the Soft Imaging System, Münster, Germany.

9. The amount of neutral fat lipid content in the liver can be detected with Oil Red O staining, which can determine whether there is an increased accumulation of glycogen. To do this, leave frozen sections at room temperature for 10 min and rehydrate with PBS working solution for 2 min.
10. Stain in Oil Red O working solution for 10 min and wash the sections in deionized water for 30 s.
11. Stain sections in Harris hematoxylin for 1 min and immediately wash in running water for 2 min.
12. Mount the sections using aqueous mounting medium, such as 78% glycerol, and place on a coverslip.
13. For proper atherosclerosis assessment (see Note 8), place frozen sections in a 60% isopropanol solution and remove excess solution with soft tissue paper.
14. Stain sections in Sudan IV for 3 min. Wash the sections with two changes in deionized water (30 s each).
15. Stain sections in Harris hematoxylin for 40 s and immediately wash with two changes in deionized water (30 s each).
16. Place the slides in 1.35% lithium carbonate solution for 30 s and wash the sections with two changes in deionized water (30 s each).
17. Before assessment, dry slides properly (at least 48 h at room temperature), mount the sections using aqueous mounting medium, such as 78% glycerol, and place on a coverslip.
18. For immunochemistry, proceed as in Subheading 3.4, steps 1–3, but rehydrate sections with a PBS-glycine solution for 10 min at room temperature.
19. Place the slides in a 3% H_2O_2 solution for 25 min at room temperature. Avoid exposure to light.
20. Remove the excess 3% H_2O_2 solution with soft tissue paper and wash with PBS-glycine solution. Add 50 μL of 2% BSA for 20 min at room temperature.
21. Remove the excess 2% BSA with soft tissue paper and incubate with 50 μL primary antibody overnight. Concentration varies among antibody suppliers, different tissues, and the PPARs to be examined. Starting conditions for the antibody described should be: 1/1,000 for anti-PPARα and anti-PPARβ and 1/500 for anti-PPARγ.

22. Wash sections with a PBS-glycine solution and add 50 μL secondary antibody working solution (1/200) for 90 min at room temperature.
23. Wash sections with PBS-glycine solution and incubate with 50 μL ABC complex for 45 min at room temperature.
24. Detect the desired protein under the microscope with DAB substrate and then wash with distilled water.
25. Stain sections in Mayer's hematoxylin for 1 min and immediately wash in running tap water for 1 min.
26. Proceed as in step 6 with 95% ethanol twice for 2 min each, 100% ethanol twice for 3 min each and xylol twice for 2 min each. Mount the sections using permanent mounting medium and dry before assessment.

3.5. Measurement of Lipid Content in Tissues

Handling of tissue and extraction procedures are fundamental steps to obtain reliable data using reagents commonly used in clinic laboratories. We describe here an extraction protocol for lipids, but hydrosoluble analytes will also require careful assessment. Note that tissues should be used once available. Alternatively, tissues should be immediately flash-frozen in liquid nitrogen, then stored in sealed containers at −80°C.

1. Fill 2-mL Precellys microtubes with 1 mL hexane/isopropanol mixture (3:2).
2. Add frozen samples (usually 50 mg).
3. Homogenize tissues using the Precellys 24 system.
4. Transfer the homogenized solution to a 10–15-mL glass tube and add 4 mL hexane/isopropanol mixture (3:2).
5. Place tubes into an orbital shaker overnight. Add 1.5 mL of 0.47 M sodium anhydrous sulfate solution. Place tubes into the orbital shaker again for 15 min.
6. Centrifuge (200 g for 5 min at room temperature), transfer the supernatant to a clean tube and wait until solvent evaporation.
7. Rehydrate samples with 500 μL of 0.5% sodium cholate solution and shake for an additional 24 h. Samples are ready to use directly in automatic analyzers.

3.6. Rapid and Efficient mRNA Expression in Tissue Extracts

Several methods may be used to efficiently isolate RNA, but we strongly recommend the use of methods proposed for the QIAcube (Qiagen) using a column-based method from Qiagen (IZASA). It represents a hands-free protocol and allows for isolation of multiple samples in a limited time, avoiding contamination.

1. Fill 2-mL Precellys microtubes with 350 μL Buffer RTL Plus.

2. Remove pre-weighed samples (20–30 mg) from freezer and immediately transfer to the previously prepared Precellys microtubes.
3. Homogenize tissues using the Precellys 24 system for only a few seconds.
4. Transfer the homogenized solution to a clean 2-mL microtube. Centrifuge the samples at 12,000 × *g* for 3 min at room temperature.
5. Switch on QIAcube and choose the RNeasy Plus Mini kit from the menu. Put the samples in the correct place following the manufacturer's instructions and start the protocol.
6. RNA quality and concentration must be assessed using the $OD_{260/280}$ ratio. Note that the amount required for PPAR evaluation is usually higher than for other common genes.
7. Samples may be stored at −80°C until further analyses, starting with conversion of RNA into cDNA.
8. Prepare the master mix solution from the High Capacity cDNA Reverse Transcription kit needed for all the experiments into a 2-mL eppendorf tube. Note that for PPARs, at least 1,000 ng will be required, and the maximal capacity is 2,000 ng.
9. Transfer samples to the chosen plates: ideally 10 μL RNA sample and 10 μL master mix, which requires calculation and in some cases concentration of samples with available kits. It is advisable to use the same RNA concentration in all samples.
10. Gently mix and spin down the plates. Proceed to PCR with the following parameters: 10 min at 25°C, 60 min at 37°C, and 5 min at 85 °C. Store samples properly (preferably −40 °C) until further analyses.
11. Transfer samples into a new plate to proceed with thermal cycling and fluorescence detection in any of the ABI Prism Sequence Detection Systems (preferably 7900HT). Add a considerable amount of RNA, between 50 and 100 ng, to ensure optimal results in several tissues. Always duplicate or triplicate the results.
12. Run 40 cycles using the following parameters: 2 min at 50°C, 10 min at 95°C, and for each cycle, 15 s at 95°C for denaturation and 1 min at 60°C for transcription.
13. A cycle threshold (C_t) value is generated by setting the threshold during the geometric phase of the cDNA sample amplification. Relative transcript levels are quantified as a comparison of measured threshold cycles numbers (C_t) for each reaction to a designated endogenous control via the $2^{-\Delta\Delta C_t}$ method. Use the appropriate and supplied software to calculate significant values using at least two selected house-keeping genes (β-actin and GAPDH).

3.7. Protein Immunoblot

Protein blotting is an analytical method that involves the immobilization of proteins on membranes before detection using monoclonal or polyclonal antibodies. In immunoblotting, prior to protein immobilization, sample proteins are separated using SDS-PAGE to provide information about molecular weight and the existence of isoforms. This is a common procedure, and multiple protocols can be found in the literature. As SDS-PAGE is a widely used technique, the only limiting factor is the nuclear extraction of PPARs. Although several suppliers provide adequate reagents, when possible, use only fresh tissue samples because nuclear fractionation with frozen tissue is inefficient and commonly requires whole-cell extracts procedures, which may present a number of drawbacks. Mainly, it is important to avoid excess heat and foam formation during the homogenization because nuclear proteins may be lost in the cytoplasmic fraction during the rupture of nuclei. In this case, the force of vortex and centrifugation should be checked.

1. Transfer 20–30 mg tissue to prechilled Precellys microtube.
2. Homogenize tissues using lysis buffer and Precellys 24 system. Transfer the samples to a clean tube and place it in an orbital shaker for 30 min at 4°C.
3. Centrifuge the samples at 10,000 × *g* for 10 min at 4°C. Transfer the supernatant to a clean tube.
4. Measure the amount of protein using the 2D Quant kit and store samples at −80°C until Western blot analyses.
5. Proceed to electrophoreses using NuPAGE 4–12% Bi-Tris Gradient polyacrylamide gels with MES buffer.
6. Transfer proteins to nitrocellulose membranes using the iBlot transfer system.
7. Place the nitrocellulose membrane in 4% TBS blocking agent (10–20 mL blocking agent/membrane) for 1 h at room temperature.
8. Place the nitrocellulose membranes in the detection antibody solution overnight at 4°C: anti-PPARα (1/1,000), anti-PPARβ (1/1,000), and anti-PPARγ (1/1,000).
9. Wash the nitrocellulose membrane six times thoroughly with wash buffer. Add secondary antibody solution (1/20,000) for 30 min at room temperature.
10. Wash the nitrocellulose membrane six times thoroughly with wash buffer and place into the TBS solution.
11. Proceed to chemiluminescent detection using the ECL Advanced Western Blotting Detection kit in a VersaDoc system.

3.8. Analyses in Transcriptomic, Proteomic, and Metabolomic Platforms: Tips for Handling and Interpretation

These analyses are usually performed in extramural facilities, and there are two limiting factors: the quality of samples and the final interpretation. These two conditions should be further discussed with providers before the experiment begins. Best results are obtained when data are compared across the experimental and control groups. We describe here considerations that have been repeatedly tested in our metabolomic platform (11, 12).

For nuclear magnetic resonance (^{1}H NMR)-based analyses, one of the most commonly used methods in metabolomic analyses, the extraction procedure is simple and can be performed, as follows, in any laboratory prior to sending samples to the specialized platform:

1. Fill 2-mL Precellys cryotubes with 1 mL H_2O/CH_3CN (1/1) buffer.
2. Add frozen tissue, usually 50 mg (should be accurately weighed), to be homogenized.
3. Centrifuge samples at 5,000 g for 15 min at 4°C. Transfer supernatant to a clean tube as a hydrophilic extract. The pellets represent the lipophilic extracts.
4. Lyophilize to remove water and store at –80°C.

Using other equipment and automated extraction using different combinations of organic and aqueous solvents are usually required. Matrix effects are usually considered by pooling small amounts of each sample. Process variability should be continuously checked, and instrumental variation should be less than 5% when using internal standards added to each sample. Samples for liquid chromatography/mass spectrometry (LC/MS, LC/MS/MS) are analyzed in duplicates: one using acidic positive ion optimized conditions and the other using basic negative ion optimized conditions in separate columns. Samples for gas chromatography/mass spectrometry (GC/MS) are dried under vacuum desiccation and derivatized under dried nitrogen using bistrimethyl-silyl-triflouroacetamide. It is important to guarantee correct tracking and identification of samples as well as a control of all processes via a Laboratory Information Management System that will include all the statistical and graphic software or may easily transfer data to acceptable software. Bioinformatics is the most valuable tool in the final interpretation. The resulting peaks are identified by comparison to library entries of purified standards; also, the combination of chromatographic properties and mass spectra gave an indication of a match to the specific compound or an isobaric entity. The choice of the statistical tests depends on the experimental design; in omic technologies, in which there are a large number of tests, it is important to recognize the number of false positives.

For instance, random noise only provides five results out of 100 compounds with *p*-values less than 0.05 and is consequently considered significant. Among methods to correct for multiple testing, discard those too conservative and consider the false discovery rates (FDR) with adjustments to ensure, as much as possible, *zero* false discoveries. The FDR for a given set of compounds can be estimated using the *q*-value (13). For functional association analyses, the Ingenuity Pathways Analysis software can combine metabolomic and genomic analyses.

3.9. Notes

1. This is particularly important because of the concentration of circulating cytokines, mainly chemokines, differs considerably between plasma and serum, probably indicating physiologically relevant effects in the procedure.
2. Adding 0.1–0.0.5 mL acetic acid to a 100 mL of the stain usually gives the stained tissue a deeper red color. The appearance of a precipitate in the stain solution upon the addition of acetic acid indicates that you have exceeded the amount of acetic acid needed.
3. The time needed to acquire animal protocol approval and to generate the experimental mouse cohorts should be taken into consideration when planning the study; considerable delays are possible and mice are expensive. When generating genetically modified mice, the choice of genetic background to control the variability of phenotypic results is important (14). The most widely used inbred strain is the C57BL/6, a model susceptible to diet-induced obesity, type 2 diabetes, and atherosclerosis. Mouse models offer the advantage of relative ease of the generation of genetic alterations and are a powerful tool for the study of specific metabolic dysfunctions. For instance, the hyperlipidemic LDL-receptor deficient mice have provided valuable data on how genetic alterations may interact with the environment in the pathogenesis of complex diseases (8, 15). Although both male and female animals can be used, the groups must be sex-matched to minimize inter-animal variability. We recommend the use of male animals to reduce metabolic variations caused by the hormone-derived changes present in female animals. All mice should be derived from age-matched littermates according to the experimental design. In general, for metabolic exploration, mice between 6 and 8 weeks are already suitable to begin experiments. It is important to have a comprehensive knowledge of the phenotypic peculiarities present in the inbred strain used to generate the genetically modified mouse line. Diets must be prepared and labeled by an independent investigator according to the randomization

schedule to ensure allocation concealment. To minimize animal stress, standard conditions (constant room temperature at 22°C, air humidity at 55%, and light/dark cycle of 12 h) should be maintained throughout the experiment. The allocation of mice into cages is limited by the mouse cohort, but we recommend not allowing more than five mice in the same cage. Do not mix gender in study groups and take note that males from different littermates should not be housed together. The duration of the actual experiment is determined by preliminary experiments designed to familiarize personnel with methods and to assess possible stress in experimental animals. The number of mice may represent a limiting factor, and when possible, should be restricted to low numbers. To test the viability of an experimental procedure and to establish the optimal number of animals per group is an important issue that may result either in false-negative or false-positive results; this requires previous and careful consideration. Particularly for PPAR assessment, 6–8 animals per group should be sufficient to observe statistically significant differences between groups because expected effects are usually consistent and strong. Avoid excluding animals from analyses.

4. Researchers may decide to obtain body weight twice a week or even more frequently. Perform the weighing at the same hour of day. Normally, mice consume 3–5 g/food a day; outlier values should be further checked. For both IGGT and OFT, one test at the beginning and another at the end of the experiment is sufficient to obtain reliable data.
5. We recommend the assistance of at least two individuals when sacrificing animals. Routinely test that deep anesthesia is obtained, i.e., no response to stimuli. Cardiac puncture is the best procedure to obtain blood more efficiently (approximately 900 μL), and this should be performed with the body in full extension by pinning all four extremities to a cork board. Use heparin in the syringe. Both serum and plasma should be obtained, and whether results from serum and plasma are interchangeable for a given analyte should be checked.
6. Perfusion is correct if the liver color rapidly turns from red to pale pink.
7. Fixation procedures should be preselected. Usually a mistaken choice results in irreparable loss of data. Please refer to Mark et al. (16).
8. For the assessment of atherosclerosis, hearts and regions 2 mm below the ascending aorta should be used. Use aortic valve leaflets serves as an anatomic reference. Store the slides properly.

Acknowledgments

AR, RBD, and GA are recipients of fellowships from the Comissionat per a Universitats i Recerca del Departament d'Innovació, Universitats i Empresa de la Generalitat de Catalunya i del Fons Social Europeu. This work was financially supported by the Fondo de Investigación Sanitaria (FIS PI08/1381 and FIS PI08/1175).

References

1. Medzhitov R (2008) Origin and physiological roles of inflammation. Nature 454:428–435
2. Gregoire FM, Kresten S, Harrington W (2007) PPARs and obesity. PPAR Res 2007:78475
3. Stienstra R, Duval C, Muller M et al (2007) PPARs, obesity, and inflammation. PPAR Res 2007:95974
4. Mandard S, Muller M, Kersten S (2004) Peroxisome proliferator-activated receptor α target genes. Cell Mol Life Sci 61:393–416
5. Vanden Berghe W, Vermeulen L, Delerive P et al (2003) A paradigm for gene regulation: inflammation. NF-kappaB and PPAR. Adv Exp Med Biol 544:181–196
6. Lazar MA (2005) PPPAR γ, 10 years later. Biochimie 87:9–13
7. Barish GD, Narkar V, Evans R (2006) PPAR delta: a dagger in the heart of the metabolic syndrome. J Clin Invest 116:590–597
8. Rodríguez-Sanabria F, Rull A, Aragones G et al (2010) Differential response of two models of genetically modified mice fed with high fat and cholesterol diets: relationship to the study of non-alcoholic steatohepatitis. Mol Cell Biochem 343:59–66
9. Rull A, Beltran-Debon R, Aragones G et al (2010) Expression of cytokine genes in the aorta is altered by the deficiency in MCP-1: effect of a high-fat, high-cholesterol diet. Cytokine 50:121–128
10. Rull A, Rodriguez F, Aragones G et al (2009) Hepatic monocyte chemoattractant-1 is upregulated by dietary cholesterol and contributes to liver steatosis. Cytokine 48:273–279
11. Vinaixa M, Rodgriguez M, Rull A et al (2010) Metabolomic assessment of the effect of dietary cholesterol in the progressive development of fatty liver disease. J Proteome Res 9:2527–2538
12. Rull A, Vinaixa M, Angel Rodriguez M et al (2010) Metabolic phenotyping of genetically modified mice: An NMR metabonomic approach. Biochimie 91:1053–1057
13. Storey J, Tibshirani R (2003) Statistical significance for genomewide studies. Proc Natl Acad Sci USA 100:9440–9445
14. Antal C, Teletin M, Wendling O et al (2007) Tissue collection for systematic phenotyping in the mouse. Curr Protoc Mol Biol Chapter 29:Unit 29A.4
15. Joven J, Rull A, Ferre N et al (2007) The results in rodent models of atherosclerosis are not interchangeable: the influence of diet and strain. Atherosclerosis 195:e85–92
16. Mark M, Teletin M, Antal C et al (2007) Histopathology in mouse metabolic investigations. Curr Protoc Mol Biol Chapter 29:Unit 29B.4

Chapter 20

Induction of Adipogenic Differentiation in Three-Dimensional Culture Model on a Novel Microfabricated Scaffold

Yoshitaka Miyagawa, Hajime Okita, and Nobutaka Kiyokawa

Abstract

The development of a three-dimensional (3D) culture system is very important for regenerative medicine and drug discovery applications of stem cell technology because the 3D culture condition could mimic the stem cell environment in vivo and support accurate differentiation. This chapter describes a novel 3D culture technique for efficient induction of adipogenic differentiation. This 3D culture system provides an easy way to allow cells to form a 3D spheroid structure without any matrix derived from animal and chemical substances. We firstly describe the details of the 3D culture technique using human mesenchymal stem/progenitor cells (MPCs) and its optimization. Then we elaborate on the protocol of efficient induction for adipogenic differentiation and adipocyte-specific gene expression, including peroxisome proliferator-activated receptor-γ, with our culture system in human MPCs.

Key words: Human mesenchymal stem/progenitor cells, Microfabricated scaffold, Three-dimensional culture, Adipogenesis, Peroxisome proliferator-activated receptor-γ

1. Introduction

Recent advances of stem cell research show that a variety of cell sources including embryonic stem (ES) cells and mesenchymal stem/progenitor cells (MPCs) derived from bone marrow, adipose, and umbilical cord blood can be available for cell therapy and regenerative medicine. In addition, Yamanaka group has generated induced pluripotent stem (iPS) cells that can differentiate into various lineages as ES cells (1, 2). These provide us with a real possibility of undertaking regenerative medicine. On the other hand, it is hoped that stem cells and their derivatives will be an ideal cell source for drug discovery. This is because, unlike limited tissue-derived cell sources, some kinds of stem cells, such as ES cells, iPS

Mostafa Z. Badr and Jihan Youssef (eds.), *Peroxisome Proliferator-Activated Receptors (PPARs): Methods and Protocols*, Methods in Molecular Biology, vol. 952, DOI 10.1007/978-1-62703-155-4_20, © Springer Science+Business Media New York 2013

cells, and tumor-derived stem cells are immortal and have the capacity to differentiate into specific lineages. However to realize these approaches with stem cells, the gap between cell culture and live tissue is a serious obstacle. Traditional culture environment forces cells to behave differently from cells in vivo. One of the contributing factors has been the difference of the environment between traditional in vitro culture and in vivo. In most cases, stem cells isolated from human tissues can lose their real characteristics as stem cells including differentiation capacity through in vitro two-dimensional (2D) culture because of the lack of a 3D microenvironment in vivo (3). In vivo, cells are supported by the 3D microenvironment, such as extracellular matrix, and communicate with each other via the microenvironment and through the secretion of cytokines. Several researchers have demonstrated that the signals from the stem cell microenvironment play a crucial role in maintaining the "stemness" and determining the cell fate of stem cells (4–6). Therefore, an in vitro cell culture environment mimicking in vivo more closely is required for maintaining stem cells and ensuring their accurate differentiation.

In vitro 3D culture system has been developed to resolve these problems. Currently various materials and devices are used for 3D culture environment. Some 3D culture environments can support cellular attachment, proliferation, and lineage-specific differentiation of stem cells in vitro. In addition, they can also induce change of the gene expression pattern by promoting cell–cell communication and cell-microenvironment interaction. Current 3D culture technology, at least in part, can mimic the microenvironment in vivo; however, it is important to study the pros and cons and to procure materials and develop devices for this purpose. In this chapter, we describe a reproducible and simple protocol for 3D culture that allows highly efficient adipogenic differentiation of human MPCs without any animal-derived protein sources. In our protocol, 3D spheroid formation is induced using a Nanoculture plate® (NCP) (Fig. 1a). Instead of traditional basement membrane matrix that has been used for in vitro 3D culture, NCP has a specific nanostructure on the culture surface. The nanostructure consists of uneven microfabrications with a diameter of approximately 2–3 μm arranged in a honeycomb or square pattern. The characteristic features of NCP-based 3D culture are as follows; firstly, the handling is simple. When human MPCs are simply seeded on NCP, they easily form an adhesive spheroid. Since NCP does not include any matrix but can support adherent spheroid formation efficiently, harvesting the cells is extremely easy. Second, NCP contains no animal compounds. To date, several approaches have been investigated for stem cell culture and differentiation under xeno-free conditions (7). Combining the technique with NCP system, in vitro 3D stem cell culture and the lineage-specific differentiation can be carried out under "complete" xeno-free conditions.

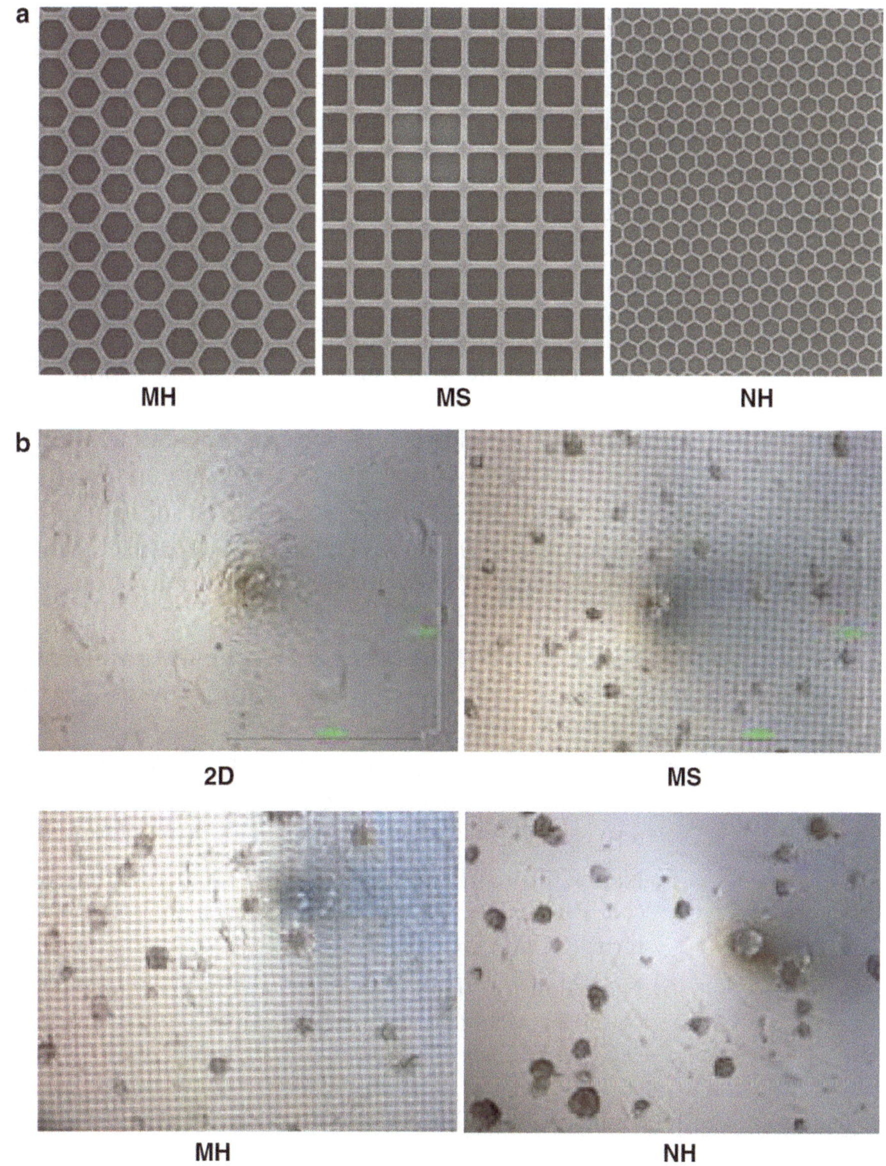

Fig. 1. (**a**) Photographs of microfabricated scaffold on NCP-L-MH, NCP-L-MS, and NCP-L-NH. (**b**) Spheroid formation of UET-13 cells on NCP-L-MH, NCP-L-MS, and NCP-L-NH. UET-13 cells form round adhesive spheroids on all of NCP but the spheroids on NCP-L-NH are larger than the others (reproduced from (8) with permisson from Mary Ann Libert, Inc, publishers).

We previously demonstrated that NCP system is useful for the spheroid formation and differentiation of human MPCs into specific lineages, in particular, the adipogenic lineage (8). During the differentiation, the transcription of key factors for adipogenesis, such as peroxisome proliferator-activated receptor-γ (PPAR-γ) and CCAAT-enhancer-binding protein α (C/EBP-α), rapidly increases compared with that under 2D conditions. Here we describe the protocol in detail. This protocol includes the preparation of NCP, optimization of NCP-based culture condition, induction of adipogenic differentiation on NCP, and confirmation of the differentiation by specific staining.

2. Materials

2.1. Cells

1. UET-13 cells (Human MPCs derived from bone marrow with an extended life span using retroviral transduction).

2.2. 3D-Culture Materials

1. NanoCulture® Plate NH pattern, Low Attachment, 96-well (NCP-L-NH) (96-well NCPs with a nano-honeycomb pattern) (SCIVAX Corp., Kanagawa).
2. NanoCulture® Plate MH pattern, Low Attachment, 96-well (NCP-L-MH) (96-well NCPs with a microhoneycomb pattern) (SCIVAX Corp.).
3. NanoCulture® Plate MS pattern, Low Attachment, 96-well (NCP-L-MS) (96-well NCPs with a microsquare pattern) (SCIVAX Corp.).
4. NanoCulture® Plate NH pattern, High Attachment, 96-well (NCP-H-NH) (96-well NCPs with a nano-honeycomb pattern) (SCIVAX Corp.).
5. NanoCulture® Plate MH pattern, High Attachment, 96-well (NCP-H-MH) (96-well NCPs with a microhoneycomb pattern) (SCIVAX Corp.).
6. NanoCulture® Plate MS pattern, High Attachment, 96-well (NCP-H-MS) (96-well NCPs with a microsquare pattern) (SCIVAX Corp.).

2.3. Tissue Culture

1. Dulbecco's modified Eagle's medium (DMEM) (Sigma).
2. Fetal bovine serum.
3. Phosphate buffered saline (PBS).
4. 0.25% (w/v) Trypsin-EDTA (IBL, Tokyo).
5. 10 cm Culture dish (Corning).
6. 96-Well tissue culture plate (Corning).
7. 15 mL Tissue culture tube (Corning).
8. Sealing tapes (Nunc).
9. Hemocytometer.

2.4. Adipogenic Differentiation of Human MPCs

1. hMSC adipogenic differentiation medium (Lonza).
2. Adipogenic maintenance medium (Lonza).
3. PBS.

4. 0.25% (w/v) Trypsin-EDTA.
5. 10 cm Culture dish.
6. 96-Well tissue culture plate.
7. 15 mL Tissue culture tube.
8. Sealing tapes.
9. Hemocytometer.

2.5. Isolation of Total RNA from Differentiated UET-13 Cells

1. PBS.
2. ISOGEN (Nippon Gene Corp., Osaka).
3. 1.5 mL Eppendorf tube.
4. Isopropanol.
5. 70% Ethanol.
6. RNase-free water.
7. 21-G needle.
8. RNase-free syringe.
9. Etachinmate (Nippon Gene Corp.).

2.6. Triglyceride Staining

1. PBS.
2. 10% formalin.
3. Oil-red O solution (3 mg/mL Oil-red O in 60% isopropanol).
4. 60% Isopropanol.
5. AdipoRed™ Assay Reagent (Lonza).

3. Methods

3.1. Propagation and Maintenance of UET-13 Cells

In our experiment, human bone marrow-derived MPCs, UET-13 cells are used as a human MPC model (see Note 1). UET-13 cells can be obtained from RIKEN Cell Bank in Japan (RCB2161). We usually propagate UET-13 cells in DMEM supplemented 10% heat-inactivated fetal bovine serum at 37°C under a humidified 5% CO_2 atm with conventional tissue culture dish. For propagation of UET-13 cells, the split ratios should be 1:5–1:10. Over-confluent conditions should be avoided. We recommend 90% confluence on passage. Different types of MPCs (i.e., mouse MPCs and other tissue-derived MPCs) might be used for differentiation experiments into specific lineages on NCP. Before experiments, the cells should be examined for differentiation potential.

3.2. Optimization of NCP-Based 3D Culture

The cell density and type of NCP are critical for efficient cell attachment, spheroid formation, and differentiation. To obtain the highest efficiency, we strongly recommend first determining the optimal cell number and NCP type with the cell type for NCP-based 3D culture (see Note 2). The following is an example to optimize the NCP-based 3D culture conditions with UET-13 cells.

3.3. Seeding UET-13 Cells on NCP

1. Thaw UET-13 cells in UET-13 culture media for at least 1 week before starting 3D culture to ensure that cells are growing properly. Maintain the UET-13 cells in 10 cm tissue culture dish with UET-13 culture media, and passage every 2–3 days.
2. Prepare different types of NCP (Low attachment type: NCP-L-NH, NCP-L-MH, and NCP-L-MS, High attachment type: NCP-H-NH, NCP-H-MH, and NCP-H-MS) (see Note 3).
3. Add 100 μL of UET-13 culture medium to cover the plate surface for pre-incubation (see Note 4).
4. Incubate the plate for at least 15 min at room temperature in the tissue culture hood.
5. During the pre-incubation of NCP, remove a UET-13 cell culture dish from the incubator and aspirate UET-13 culture medium from the dish and then wash UET-13 cells once with PBS.
6. Add 2 mL of 0.025% Trypsin/EDTA to dissociate the cells from the dish. Incubate in a CO_2 incubator at 37°C for 5 min.
7. Add 8 mL of UET-13 culture medium and then pipette the cell suspension to break up the cell aggregate.
8. Transfer UET-13 cells into a 15 mL tissue culture tube and pellet by centrifugation at $200\times g$ for 5 min at room temperature.
9. Aspirate the supernatant and add 10 mL of UET-13 culture medium. Resuspend the cell pellets.
10. Count the cells using a hemocytometer. To evaluate optimal cell density, make several dilutions of cell suspension (i.e., 2.5×10^3, 5×10^3, 1×10^4, 2.5×10^4, 5×10^4, 1×10^5 cells/mL).

3.4. 3D Culture on NCP

1. Aspirate the media for pre-incubation from NCPs and then seed 100 μL of the cell suspension into different types of NCP immediately (see Note 5). When cells are seeded into NCP, place the tip along one side of the well wall so as not to make air bubbles in the well (see Note 6). As a control, also seed the cells into the conventional 96-well tissue culture plate in the same manner.
2. Cover the top of NCP with sealing tape and return the NCP into the CO_2 incubator at 37°C (see Note 7).

3. Next day after the seeding, check the spheroid formation under a microscope (Fig. 1b) (see Note 8). Once cells form a spheroid, observe the cells to see whether they maintain the spheroid structure.

3.5. Induction of Adipogenic Differentiation of UET-13 Cells on NCP

After optimization of the 3D culture condition on NCP, try to induce the differentiation into adipogenic lineage with the optimal cell density and NCP.

1. Add 100 μL of pre-warmed hMSC adipogenic differentiation medium into NCP-L-NH and pre-incubate the NCP for at least 15 min at room temperature in the tissue culture hood.
2. Prepare the cell suspension as step 5–10 in Subheading 3.3 with hMSC adipogenic differentiation medium instead of UET-13 culture medium. Adjust the cell density to 5×10^4 cells/mL. Aspirate the media for pre-incubation from the NCP and then seed 100 μL of the cell suspension into NCP immediately (optionally, cells might be harvested for expression analysis on adipogenic-specific markers at this point see Subheading 3.6). Return the NCP gently to the CO_2 incubator at 37°C. As a control, also seed the cells into a conventional 96-well tissue culture plate in the same manner.
3. Next day after the seeding, check the spheroid formation under a microscope.
4. Three days after the induction of adipogenic differentiation, remove the media from NCP carefully using a micropipette and add 100 μL of fresh hMSC adipogenic differentiation medium (see Note 9) (optionally, cells might be harvested for expression analysis on adipogenic-specific markers at this point).
5. Six days after the induction, repeat step 4 in Subheading 3.5 (optionally, cells might be harvested for analysis on adipogenic-specific markers at this point).
6. Eight days after the induction, aspirate the media from NCP carefully and add 100 μL of adipogenic maintenance medium instead of hMSC adipogenic differentiation medium. Change the media every 3 days until the lipid droplet accumulation is observed in the cells (see Note 10).

3.6. RNA Extraction from Differentiated UET-13 Cells

To confirm the adipogenic differentiation of UET-13 cells, the gene expression is investigated by quantitative real-time RT-PCR (qPCR). We usually examine PPAR-γ C/EBP-α, adiponectin, and aP2 expression for adipocyte-specific markers (Fig. 2) (see Table 1).

1. Wash the cells with PBS once carefully (see Note 11).
2. Remove the PBS and add 100 μL of ISOGEN solution into the well directly and pipette the solution up and down to lyse the cells.

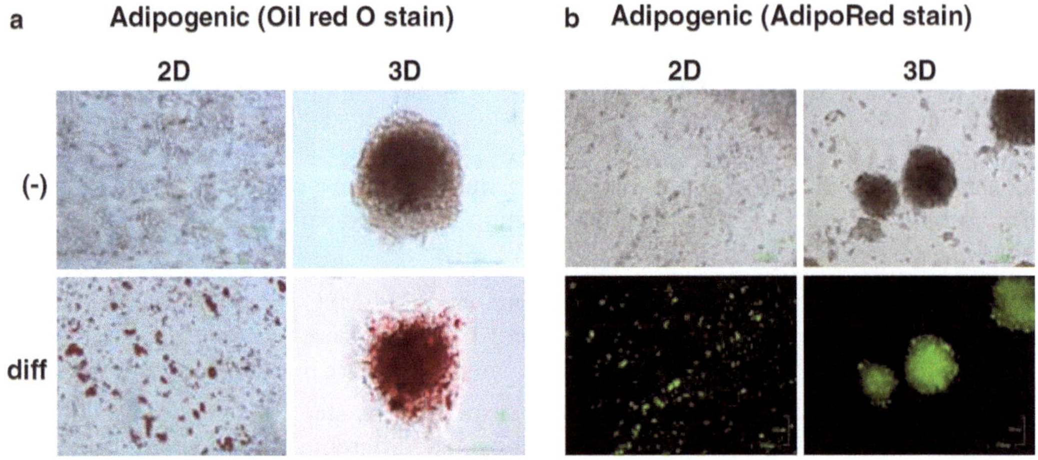

Fig. 2. Adipogenic-specific marker expression of differentiated UET-13 cells by quantitative real-time RT-PCR (qPCR). UET-13 cells are cultured in hMSC adipogenic differentiation medium on conventional 2D tissue culture dish or NCP. After differentiation, RNA is extracted and the expression levels of adipogenic-specific markers, (**a**) PPAR-γ, (**b**) adiponectin, (**c**) C/EBP-α, and (**d**) aP2 are examined by qPCR. The signal intensities were normalized using that of a control housekeeping gene (human GAPDH gene). Data are the relative values with the SD of triplicate wells and normalized to the mRNA level at 0 h, which was arbitrarily set to one in the graphical presentation. Differentiated UET-13 cells on NCP exhibit more rapid and higher expression of adipogenic-specific genes than those on 2D culture. This suggests that NCP-mediated 3D culture promotes not only adipogenic differentiation but also induction of adipogenic-specific gene expression in UET-13 cells. (reproduced from (8) with permisson from Mary Ann Libert, Inc, publishers).

Table 1
Primers used for PCR

Gene		Sequence
PPARγ	Sense	5′-GACCTGAAAC TTCAAGAGTA CCAAA-3′
	Antisense	5′-TGAGGCTTAT TGTAGAGCTG AGTC-3′
C/EBPα	Sense	5′-GCAAATCGTG CCTTGTCAT-3′
	Antisense	5′-CTCATGGGGG TCTGCTGTAG-3′
Adiponectin	Sense	5′-AGAGATGGCA CCCCTGGT-3′
	Antisense	5′-CACCGATGTC TCCCTTAGGA-3′
aP2	Sense	5′-CCTTTAAAAA TACTGAGATT TCCTTCA-3′
	Antisense	5′-GGACACCCCC ATCTAAGGTT-3′

3. Transfer the lysate into a 1.5 mL Eppendorf tube and scrape the cells completely by passing the lysate through a 21-G needle fitted to an RNase-free syringe. Then store the tube at room temperature for 5 min.
4. Add 20 μL of chloroform, vortex the tube for 15 s and then store the tube at room temperature for 3 min.
5. Centrifuge the tube at 12,000 × *g* for 15 min at 4°C.

6. Transfer the upper clear aqueous layer to a fresh 1.5 mL Eppendorf tube. Be careful not to take any of the white interphase and blue organic phase including DNA and protein.
7. Add 0.5 μL of Etachinmate and 50 μL of isopropanol to the tube, mix the sample and then incubate the sample for 10 min at room temperature.
8. Centrifuge the tube at 12,000 × *g* for 10 min at 4°C.
9. Remove the supernatant and add 100 μL of 70% ethanol to the tube.
10. Centrifuge the tube at 7,500 × *g* for 5 min at 4°C.
11. Remove the supernatant and dry the pellet.
12. Resuspend the pellet with an appropriate amount of RNA-free water. Use the RNA sample for examination of mRNA level for adipogenic-specific markers by real-time RT-PCR. Figure 2 shows the sequential changes in expression levels of adipogenic-specific markers after adipogenic differentiation in UET-13 cells. Primer sequences for examination of expression for adipogenic-specific markers are listed in Table 1.

3.7. Triglyceride Staining of Differentiated UET-13 Cells

To further confirm the adipogenic differentiation, specific staining is required (see Note 12).

1. Wash the cells with PBS once carefully and then fix the cells with 10% formalin for 1 h.
2. Wash the cells with PBS once carefully and add 100 μL of Oil-red O solution.
3. Store the plate at room temperature for 20 min.
4. Wash the cells with 60% isopropanol once very carefully.
5. Observe lipid droplet accumulation by microscopy. Lipid droplet is stained in red color (Fig. 3a) (see Note 13).

4. Notes

1. UET-13 cells have an extended the life span but maintain human MPC characteristics in terms of surface marker expression (9), growth factor reactivity (10), and differentiation potential into osteoblasts (8), adipocytes (8), hepatocytes (11), and neurogenic cells (12).
2. In our experience, UET-13 cells on NCP-L-NH can form bigger adhesive spheroids than those on other types of NCP. They can maintain the 3D structure (Fig. 1b). The optimal cell density is 5×10^3 cells/well.

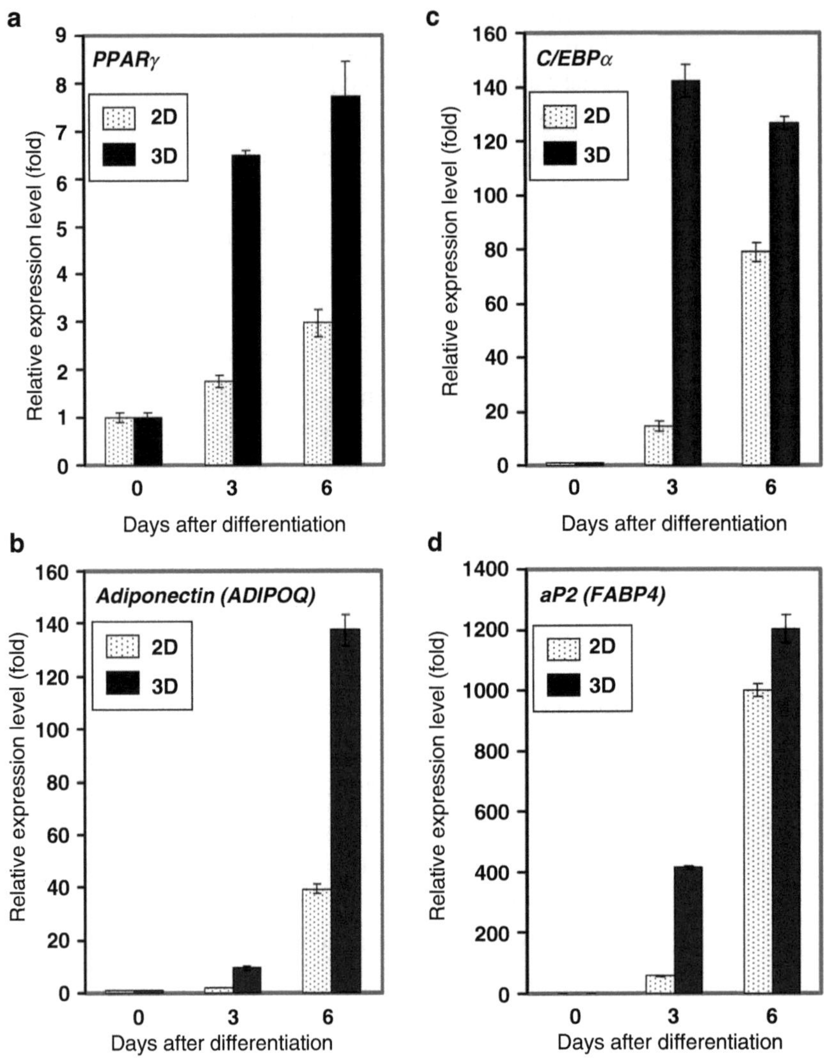

Fig. 3. Differentiation of adipocytes from human MPCs (**a**) Oil-red O, (**b**) AdipoRed staining of differentiated cells on conventional 2D tissue culture plate or NCP. The lipid droplet accumulations of UET-13 cells on NCPs are greater than those on 2D culture. (reproduced from (8) with permisson from Mary Ann Libert, Inc, publishers).

3. Currently, several different types of NCP are commercially available. Check the information at http://www.scivax.com/cell/english/index.html.

4. The medium pre-warmed at 37°C should be used for pre-incubation of NCP. Minimize bubbles by scraping and pipetting gently. Formation of air bubbles is a serious obstacle to successful spheroid formation. The pre-incubation of NCP will help to avoid air bubbles during culture. If any air bubbles remain after the pre-incubation, they should be removed from the well by pipetting or changing the medium. Alternatively,

add 50 μL of medium into the wells, and centrifuge the plate at 300–500×*g* for 1–3 min to remove bubbles. Adjust the cell density to twice that in step 10 in Subheading 3.3. At step 1 in Subheading 3.4, seed 50 μL of the cell suspension without aspirating pre-incubation medium.

5. To examine the effect of culture on NCP correctly, cells should be seeded at least in duplicate wells. The seeding procedure must be performed as quickly as possible to ensure optimal cell attachment. Drying the well surface up might decrease the cell attachment to NCP.
6. Be careful not to scratch the microfabricated surface in the well with a tip.
7. Covering the top of NCP with sealing tape is required to avoid medium evaporation. Be careful not to shake the plate for uniform spheroid formation.
8. Generally cells should form spheroid within day 1–3 after seeding. This depends on the cell type.
9. During adipogenic differentiation, the attachment of the spheroid will be weak. Be careful not to disturb the spheroid from the NCP by pipetting.
10. After exchanging hMSC adipogenic differentiation medium for adipogenic maintenance medium, the lipid droplet accumulation can be observed (Fig. 3).
11. Since the amount of RNA isolated from one well is small, we usually isolate the RNA from 6 to 12 wells.
12. In this step, be careful not to disturb the spheroid by pipetting. Alternatively, the staining can be carried out after transferring the spheroids into the tissue culture tube.
13. We also recommend staining with AdipoRed™ Assay Reagent to observe intracellular lipid accumulation (Fig. 3b). Optionally, we also measure the amount of triglyceride using TG E-test Wako (Wako).

Acknowledgments

This work was supported in part by Health and Labour Sciences Research Grants, the third Term Comprehensive 10-Year Strategy For Cancer Control and H22-011, and a Grant-in Aid for Scientific Research (JSPS. KAKENHI 22790366). We respectfully thank Y. Toda for her secretarial work and M. Itagaki for many helpful discussions and support.

References

1. Takahashi K, Yamanaka S (2006) Induction of pluripotent stem cells from mouse embryonic and adult fibroblast cultures by defined factors. Cell 126(4):663–676
2. Takahashi K, Tanabe K, Ohnuki M, Narita M, Ichisaka T, Tomoda K, Yamanaka S (2007) Induction of pluripotent stem cells from adult human fibroblasts by defined factors. Cell 131(5):861–872
3. Fehrer C, Lepperdinger G (2005) Mesenchymal stem cell aging. Exp Gerontol 40:926–930
4. Moore KA (2004) Recent advances in defining the hematopoietic stem cell niche. Curr Opin Hematol 11(2):107–111
5. Calvi LM, Adams GB, Weibrecht KW, Weber JM, Olson DP, Knight MC, Martin RP, Schipani E, Divieti P, Bringhurst FR, Milner LA, Kronenberg HM, Scadden DT (2003) Osteoblastic cells regulate the haematopoietic stem cell niche. Nature 425(6960):841–846
6. Doetsch F (2003) A niche for adult neural stem cells. Curr Opin Genet Dev 13(5):543–550
7. Ross PJ, Suhr ST, Rodriguez RM, Chang EA, Wang K, Siripattarapravat K, Ko T, Cibelli JB (2010) Human-induced pluripotent stem cells produced under xeno-free conditions. Stem Cells Dev 19(8):1221–1229
8. Miyagawa Y, Okita H, Hiroyama M, Sakamoto R, Kobayashi M, Nakajima H, Katagiri YU, Fujimoto J, Hata JI, Umezawa A, Kiyokawa N (2010) A microfabricated scaffold induces the spheroid formation of human bone marrow-derived mesenchymal progenitor cells and promotes efficient adipogenic differentiation. Tissue Eng Part A 17:513–521
9. Miyagawa Y, Okita H, Nakajima H, Horiuchi Y, Sato B, Taguchi T, Toyoda M, Katagiri YU, Fujimoto J, Hata J, Umezawa A, Kiyokawa N (2008) Inducible expression of chimeric EWS/ETS proteins confers Ewing's family tumor-like phenotypes to human mesenchymal progenitor cells. Mol Cell Biol 28(7):2125–2137
10. Toyoda M, Takahashi H, Umezawa A (2007) Ways for a mesenchymal stem cell to live on its own: maintaining an undifferentiated state ex vivo. Int J Hematol 86(1):1–4
11. Shimomura T, Yoshida Y, Sakabe T, Ishii K, Gonda K, Murai R, Takubo K, Tsuchiya H, Hoshikawa Y, Kurimasa A, Hisatome I, Uyama T, Umezawa A, Shiota G (2007) Hepatic differentiation of human bone marrow-derived UE7T-13 cells: effects of cytokines and CCN family gene expression. Hepatol Res 37(12):1068–1079
12. Mori T, Kiyono T, Imabayashi H, Takeda Y, Tsuchiya K, Miyoshi S, Makino H, Matsumoto K, Saito H, Ogawa S, Sakamoto M, Hata J, Umezawa A (2005) Combination of hTERT and bmi-1, E6, or E7 induces prolongation of the life span of bone marrow stromal cells from an elderly donor without affecting their neurogenic potential. Mol Cell Biol 25(12):5183–5195

Chapter 21

Analyzing PPARα/Ligand Interactions by Chemical Cross-Linking and High-Resolution Mass Spectrometry

Mathias Q. Müller, Christian H. Ihling, and Andrea Sinz

Abstract

The combination of chemical cross-linking and high-resolution mass spectrometry is an emerging technique for monitoring conformational changes in proteins induced by drug binding. In this chapter, we describe this approach for gaining insights into the conformational changes of the peroxisome proliferator-activated receptor alpha after binding of low-molecular weight ligands. Our strategy provides a basis to efficiently characterize target protein-drug interactions.

Key words: PPARα/ligand interaction, Chemical cross-linking, Isotope labeling, SDS-PAGE, *In-gel* digestion, Nano-HPLC, High-resolution mass spectrometry

1. Introduction

The development and improvement of rapid and reliable analytical tools for screening protein/drug interactions is indispensable for developing novel drugs. One important drug target for antidiabetic and antidyslipidemic drugs is the peroxisome proliferator-activated receptor alpha (PPARα) (1–4). Here, we describe an alternative approach for rapidly analyzing conformational changes that are induced in PPARα upon binding of potential drugs. Our strategy is based on building up a set of structurally defined interactions by covalently connecting pairs of functional groups within a protein using chemical cross-linking. After the chemical cross-linking reaction, the protein is proteolytically digested and the resulting peptide mixtures are subsequently analyzed by high-resolution mass spectrometry (5–9).

Mostafa Z. Badr and Jihan Youssef (eds.), *Peroxisome Proliferator-Activated Receptors (PPARs): Methods and Protocols*, Methods in Molecular Biology, vol. 952, DOI 10.1007/978-1-62703-155-4_21, © Springer Science+Business Media New York 2013

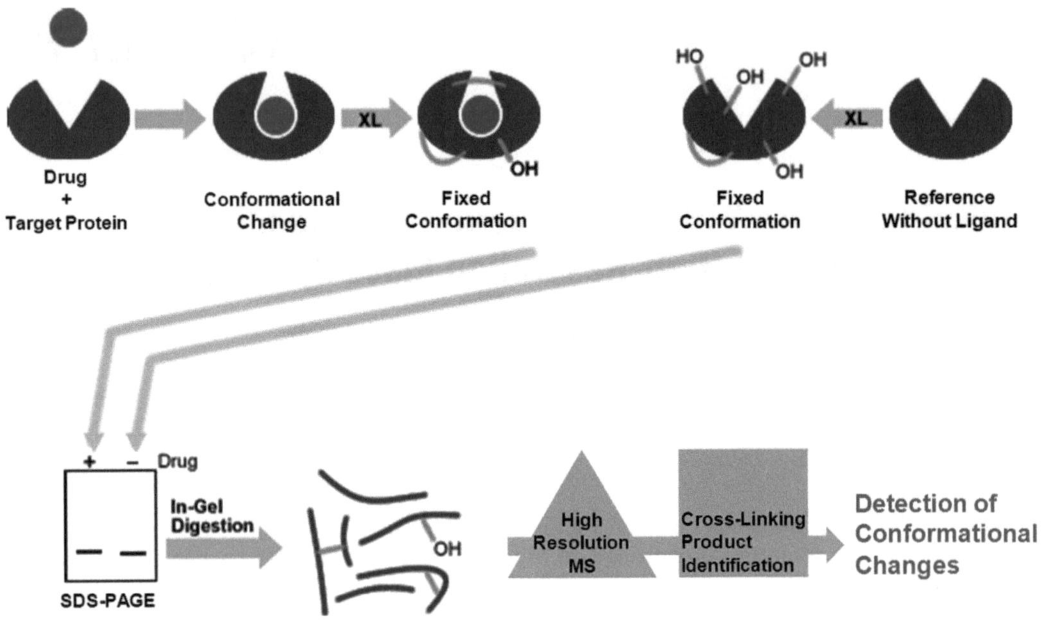

Fig. 1. Analytical strategy for analyzing conformational changes in proteins upon ligand binding by chemical cross-linking and high-resolution mass spectrometry.

For PPARα, conformational changes are monitored in the presence of an antagonist, an agonist, and compared to the cross-links obtained for free PPARα (10). The cross-linking/MS strategy is shown in Fig. 1. The cross-linking reactions are conducted using homobifunctional isotope-labeled amine-reactive *N*-hydroxysuccinimide (NHS) esters bis(sulfosuccinimidyl) glutarate (BS^2G) and bis(sulfosuccinimidyl) suberate (BS^3) with distances of ca. 7.7 and 11.4 Å, which are employed as 1:1 mixture of non-deuterated (D_0) and four times deuterated (D_4) derivatives in order to facilitate a subsequent MS identification of cross-linking products (11, 12). The reaction mechanism of the cross-linkers is shown in Fig. 2. The cross-linkers will covalently connect ε-amino groups of lysine residues as well as α-amino groups of a free N-terminus, in case they are in the correct distance from each other to be cross-linked.

Cross-linked PPARα is separated by one-dimensional gel electrophoresis (SDS-polyacrylamide gel electrophoresis (SDS-PAGE), Fig. 3). Gel bands of interest are excised from the gel and *in-gel* digested with trypsin (cleaves C-terminal of arginines and lysines) and GluC (cleaves C-terminal to glutamic and aspartic acid). Extracted peptide mixtures are analyzed by LC/MS.

The cross-links that are identified within PPARα by high-resolution mass spectrometry allow drawing conclusions on conformational changes that are induced by binding of low-molecular weight ligands.

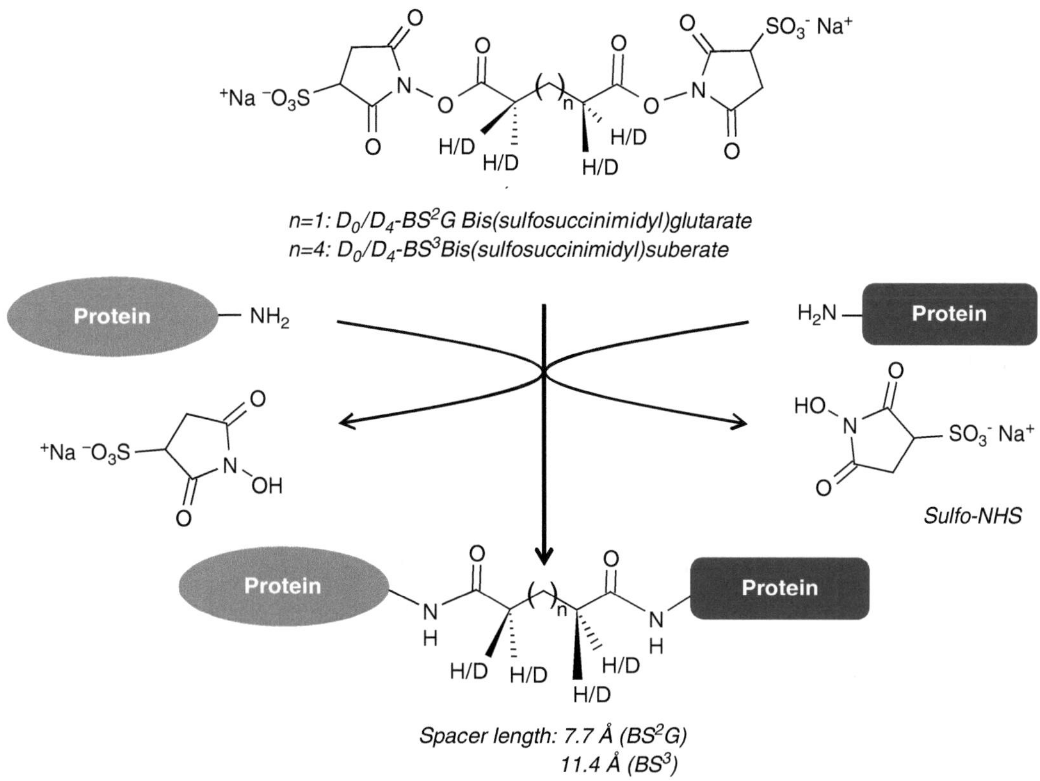

Fig. 2. Reaction mechanism of homobifunctional cross-linking with *N*-hydroxysuccinimide (NHS) esters.

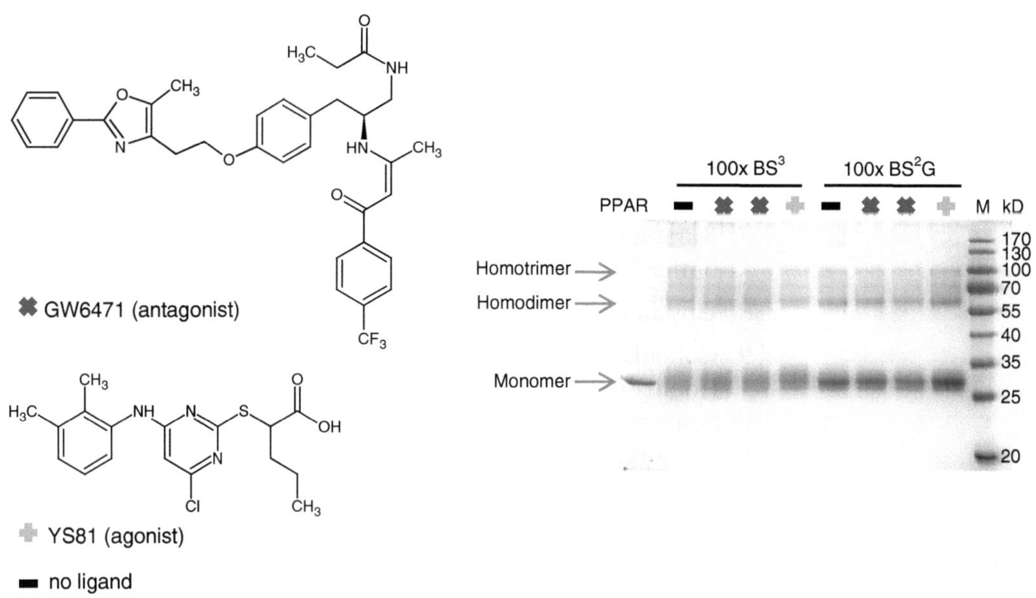

Fig. 3. Separation of cross-linking reaction mixtures (60 min reaction time, 200 pmol protein per lane) by SDS-polyacrylamide gel electrophoresis (SDS-PAGE), exemplified for peroxisome proliferator-activated receptor alpha (PPARα)/ligand complexes (10). M: molecular weight marker; PPAR: PPARα before cross-linking experiments (17); *minus sign*: cross-linking in the absence of ligands; *cross sign*: cross-linking in the presence of an antagonist GW6471 (18); *plus sign*: cross-linking in the presence of the putative agonist YS81 (19).

2. Materials

Prepare all solutions using ultrapure water, i.e., MilliQ water, and use reagents at the highest available purity. Chemical cross-linkers should be stored dry under inert gas (argon or nitrogen) and stock solutions have to be prepared in dimethylsulfoxide (DMSO) directly before use.

2.1. Chemical Cross-Linking

1. Cross-linking reagents: BS^3-D_0, BS^3-D_4, BS^2G-D_0, and BS^2G-D_4 (Thermo Fisher Scientific, see Note 1).
2. Neat DMSO (see Note 2).
3. Cross-linking buffer: 20 mM HEPES buffer, pH 7.4 (see Note 3).
4. Quenching solution: 1 M ammonium bicarbonate.
5. Recombinant PPARα stored in cross-linking buffer, concentration should be between 1 and 10 μM.
6. 5 mM ligand stock solution (dissolved in a suitable solvent).

2.2. SDS-Polyacrylamide Gel Electrophoresis

1. MilliQ water (see Note 4).
2. Resolving gel: 1.5 M Tris–HCl, pH 8.8.
3. Stacking gel: 0.5 M Tris–HCl, pH 6.8.
4. 40% (w/v) acrylamide/bisacrylamide solution (37.5:1 (w/w)) and *N,N,N′,N′*-Tetramethyl-ethylenediamine (TEMED, both VWR, stored at 4°C).
5. Laemmli buffer and 10% (w/v) SDS solution (store at room temperature, both Bio-Rad).
6. Ammonium persulfate (APS, Sigma): prepare 10% (w/v) solution in MilliQ water and immediately freeze in single use (100 μL) aliquots at −20°C.
7. Running buffer: 10× concentrated TGS (Tris–Glycine–SDS) buffer (Roth, store at room temperature).
8. Unstained molecular weight markers (SM0661, Bio-Rad).
9. Colloidal Coomassie staining solution: Solution A: 5% (w/v) Coomassie Brilliant Blue G250 suspended in MilliQ H_2O. Solution B: 2% (w/w) *ortho*-phosphoric acid, 10% (w/w) ammonium sulfate, dissolved in MilliQ H_2O (see Note 5) (13).
10. Fixing solution: 40% (v/v) methanol, 10% (v/v) acetic acid.

2.3. Proteolytic In-Gel Digestion

1. Trypsin sequencing grade (Promega). Storage: dissolve 25 μg lyophilized trypsin in 50 μL 1 mM HCl, prepare aliquots containing 1 μg/2 μL trypsin, quick-freeze in liquid nitrogen, and store at −80°C.

2. GluC sequencing grade (Promega, see Note 6).
3. Destaining solutions: acetonitrile (ACN) gradient grade (VWR), 50% (v/v) ACN, 100 mM ammonium bicarbonate (see Note 7).
4. Reducing buffer: freshly prepared 10 mM dithiothreitol (DTT) in 100 mM ammonium bicarbonate.
5. Alkylating buffer: freshly prepared 55 mM iodoacetamide (Sigma) in 100 mM ammonium bicarbonate.
6. ACN gradient grade (VWR).
7. Trifluoroacetic acid (TFA, Sigma).
8. Digestion buffer, freshly prepared: dissolve one trypsin or GluC aliquot in 80 μL 50 mM ammonium bicarbonate to obtain a solution of 13 ng/μL and store on ice.
9. Extraction solution: ACN/5% (v/v) TFA 2:1.

2.4. High-Resolution Mass Spectrometry

1. In our lab we are using the instrumental setup described below, but other nano-HPLC systems or mass spectrometers, such as FTICR or Q-TOF instruments equipped with a nano-ESI source can be used as well. *Nano-HPLC*: Ultimate nano-HPLC system (Dionex Corporation, Idstein, Germany) equipped with reversed phase C18 columns (precolumn: Acclaim PepMap, 100 μm × 2 mm, 5 μm, 100 Å, separation column: Acclaim PepMap, 75 μm × 250 mm, 3 μm, 100 Å, Dionex Corp., Idstein, Germany). *Mass spectrometer*: LTQ-Orbitrap XL hybrid mass spectrometer (Thermo Fisher Scientific, Bremen, Germany) equipped with nano-ESI source (Proxeon, Odense, Denmark).
2. Solvent A: 5% (v/v) ACN, 0.1% (v/v) formic acid (FA) prepared in MilliQ H_2O (see Note 4).
3. Solvent B: 80% (v/v) ACN, 0.08% (v/v) FA.

2.5. Software for Data Acquisition and Analysis

1. Data acquisition: LTQ-Tune 2.4 and XCalibur 2.0.7 (Thermo Fisher Scientific) in combination with DCMS link 2.0 (Dionex).
2. Preparing mass lists: Proteome Discoverer 1.2 (Thermo Fisher Scientific).
3. Identification of cross-linking products: GPMAW (General Protein Mass Analysis for Windows, version 8.2, Lighthouse Data, Odense, Denmark, http://www.gpmaw.com) (14) or *in-house* StavroX software, version 2.0.3 (written by Michael Goetze, Martin-Luther University Halle-Wittenberg; for further information regarding this software please contact the authors).

3. Methods

3.1. Chemical Cross-Linking

1. Prepare an equimolar mixture of target protein and drug (ranging from 1 to 10 μM each, final concentration) in the cross-linking buffer (e.g., 20 mM HEPES buffer) in case a 1:1 protein/ligand complex is studied. For equilibration of the protein/ligand complex, the mixture is incubated for 30 min at room temperature using an incubation shaker. For each reaction condition at varying protein/ligand concentrations, cross-linking reagents or cross-linker excesses, a new Eppendorf tube should be used. The minimal final volume should be 100 μL per reaction condition.
2. Freshly prepare cross-linker stock solutions (200 mM) in neat DMSO.
3. Add 1:1 mixtures of non-deuterated (D_0) and four times deuterated (D_4) cross-linkers BS^3 and BS^2G (see Note 8), respectively, in 50 and 100 M excess of cross-linker over the target protein/ligand complex. Allow the reaction to proceed for 30 and 60 min (see Note 9).
4. For reaction quenching, take an aliquot from the cross-linking reaction mixture after 30 min, transfer it into a new Eppendorf tube, and add 1 M ammonium bicarbonate solution (20 mM final concentration). The remaining cross-linking reaction mixture is quenched in the same fashion after 60 min.
5. For storage, quick-freeze the samples in liquid nitrogen and store at −20°C before MS analyses are performed.

3.2. SDS-PAGE

1. The instructions for SDS-PAGE are based on the use of a Mini-PROTEAN 3-cell or Mini-PROTEAN Tetra cell system (Bio-Rad). Other systems can be used as well.
2. Prepare two 0.75-mm resolving gels (12%) by mixing 3 mL acrylamide/bisacrylamide solution, 2.5 mL 1.5 M Tris–HCl, pH 8.8, 100 μL 10% (w/v) SDS, and adding MilliQ H_2O to a total volume of 10 mL. Start the polymerization reaction by adding 10 μL TEMED and 50 μL 10% (w/v) APS. Pour the gel into the tray, leaving space for a stacking gel, and overlay with MilliQ H_2O. The gel should be polymerized after 30 min.
3. Remove the water and rinse the top of the gel twice with MilliQ H_2O.
4. Prepare the stacking gel (4%) by mixing 625 μL acrylamide/bisacrylamide solution, 1,250 μL 0.5 M Tris–HCl, pH 6.8, and 50 μL 10% (w/v) SDS, and by adding MilliQ H_2O to a total volume of 5 mL. Polymerization is initiated by adding 10 μL TEMED and 25 μL 10% (w/v) APS.

5. Pour the stacking gel onto the polymerized resolving gel and immediately insert a 10-well comb. The gel should be polymerized after 45 min (see Note 10).
6. Prepare the Colloidal Coomassie Blue staining solution at least 2 h before use (13). Per gel, suspend 500 μL of solution A into 25 mL of solution B and shake gently.
7. Take 20 μL of each cross-linking sample (absolute protein amount should be at least 5 μg per sample), reduce the sample volume in a vacuum concentrator to 10 μL, add 10 μL Laemmli buffer containing 5% (v/v) BME, and incubate at 95°C for 5 min.
8. Prepare the running buffer by diluting 100 mL of the ten times concentrated running buffer with 900 mL MilliQ H_2O and mix gently.
9. Add the running buffer to the upper and lower chambers of the gel unit and load 20 μL of each sample in a well. In each of the two outer wells, 5 μL of unstained molecular weight marker is prepared.
10. Connect the gel chamber to a power supply. Run the gel at 100 V for 15 min, then at 200 V for 35–40 min until the tracking dye has reached the rim of the gel.
11. Remove the gels from the glass plates, transfer each into a single staining box, and rinse twice with MilliQ H_2O. Add 25 mL of fixing solution and incubate for 1 h at room temperature on a shaker.
12. Discard the fixing solution and replace it by 25 mL of Colloidal Coomassie Blue staining solution. Incubate overnight at room temperature on a shaker.
13. Discard the staining solution and destain with MilliQ H_2O.

3.3. In-Gel Digestion

Working under a laminar flow box wearing gloves and sleeve protectors is recommended to avoid keratin and dust contamination of the samples. The laminar flow box should include an incubation shaker suitable for Eppendorf tubes.

A. Excise protein bands from the SDS gel.
   1. Wash the Coomassie-stained gel twice for 10 min with MilliQ H_2O on a shaker.
   2. Excise gel bands of interest using a sterile scalpel and transfer gel bands to a clean glass plate. Cut the gel band into cubes of ca. 1 mm^3 and transfer them into an Eppendorf tube.

B. Reduce and alkylate proteins.
   3. Add 100 μL MilliQ H_2O to the gel pieces, shake for 10 min.

4. Add 500 μL ACN. Shake for 5 min and discard the liquid.
5. Add 50 μL reduction buffer. Incubate at 56°C for 30 min, and then let it cool to room temperature.
6. Add 500 μL ACN. Shake for 5 min and discard the liquid.
7. Add 50 μL alkylation buffer. Incubate at room temperature in the dark for 30 min.
8. Add 500 μL ACN. Shake for 5 min and discard the liquid.

C. Wash and destain gel pieces.

9. Add 100 μL ACN/100 mM ammonium bicarbonate 1:1, shake for 10–30 min to destain gel pieces.
10. Add 500 μL ACN. Shake for 5 min and discard the liquid.
11. Repeat steps 9 and 10 if necessary.

D. *In-gel* digestion.

12. Bring the gel pieces to complete dryness before adding the enzyme solution (trypsin or GluC in ammonium bicarbonate).
13. Per gel band, add 25 μL trypsin buffers and incubate gel pieces for 30 min on ice. In case a double digestion is performed, add the second protease (e.g., GluC) and incubate gel pieces for another 60 min on ice.
14. Add 50 mM ammonium bicarbonate. The gel pieces have to be covered completely.
15. Incubate overnight at 37°C.

E. Extract peptides.

16. Stop the digestion by adding the double volume extraction solution (see Note 11).
17. Shake for 15 min at 37°C, remove the supernatant, and transfer it to a new Eppendorf tube.
18. Concentrate supernatants in a vacuum concentrator to ca. one fourth of the original volume. Do not concentrate to dryness in order to avoid sample loss.
19. Peptide mixtures are ready to be analyzed by nano-HPLC/nano-ESI-LTQ-Orbitrap mass spectrometry. Before MS analysis, samples might be stored at –20°C.

3.4. Nano-HPLC/Nano-ESI-LTQ-Orbitrap-MS

In our studies, fractionation of tryptic peptide mixtures of PPARα is carried out using an Ultimate nano-HPLC system (Dionex Corporation, Idstein, Germany) directly coupled to the nano-ESI source (Proxeon, Odense, Denmark) of an LTQ-Orbitrap XL hybrid mass spectrometer (ThermoFisher Scientific, Bremen, Germany).

After washing the peptides on the precolumn for 15 min with water containing 0.1% (v/v) TFA, they are eluted and separated using gradients from 0 to 50% B (90 min), 50–100% B (1 min), and 100% B (5 min), where solvent A is 5% (v/v) ACN containing 0.1% (v/v) FA and solvent B is 80% (v/v) ACN containing 0.08% (v/v) FA. MS data are acquired over 122 min in data-dependent MS^2 mode: Each high-resolution full scan (*m/z* 300–2,000, resolution 60,000) in the orbitrap is followed by five product ion scans in the linear ion trap (LTQ) for the three most intense signals in the full-scan mass spectrum (isolation window 2 u). Dynamic exclusion (exclusion duration 180 s, exclusion mass width ±5 ppm) is enabled to allow detection of less abundant ions.

3.5. Data Analysis

Data analysis of cross-linked peptides is performed using the GPMAW software (14) or our in-house software StavroX allowing lysines and protein N-termini as reaction sites of NHS-ester based cross-linkers. As NHS-esters have also been found to react with serines, threonines, and tyrosines to a large extent, these reaction sites should also be taken into account (15, 16). Please note that trypsin will cleave at modified lysines with low frequency, therefore the number of missed cleavage sites has to be set to a higher value.

Application of isotope-labeled D_0/D_4 cross-linking reagents allows filtering of putative cross-linked candidates based on their characteristic isotope patterns (Fig. 4a). All monoisotopic deconvoluted *m/z* values (obtained by Proteome Discoverer software) showing a mass difference of 4 amu remain in the mass list. The remaining *m/z* values, which do not fulfill this criterion, are removed. The resulting theoretical mass lists are subsequently compared with experimental mass lists obtained by high-resolution mass spectrometry, allowing a maximum mass deviation of 3 ppm. As a result, a list of potential cross-linked peptides is automatically created, for which tandem MS data have to be evaluated. The StavroX software supports the calculation of theoretical fragment ions of cross-linking products (Fig. 4b). Nevertheless, full mass spectra as well as tandem MS data have to be checked manually.

Depending on the different conformational states of the target protein induced by ligands, i.e., agonists or antagonists, different cross-linking products are identified within the protein as is shown for PPARα/ligand complexes in Table 1 (10). For example, Lys-222 is connected with Lys-449 (*m/z* 1,379.7401)—this cross-link is exclusively found in presence of the antagonist GW6471 and demonstrates a large conformational change upon drug binding. Lys-449 is positioned close to the C-terminal activation function helix AF2 (amino acids 458–467) flips over in direction towards Lys-222 and closes the ligand binding pocket of PPARα (10).

Additionally, the different patterns of lysines that are modified by partially hydrolyzed cross-linkers yield valuable information on the surface accessibilities of specific residues in the target protein.

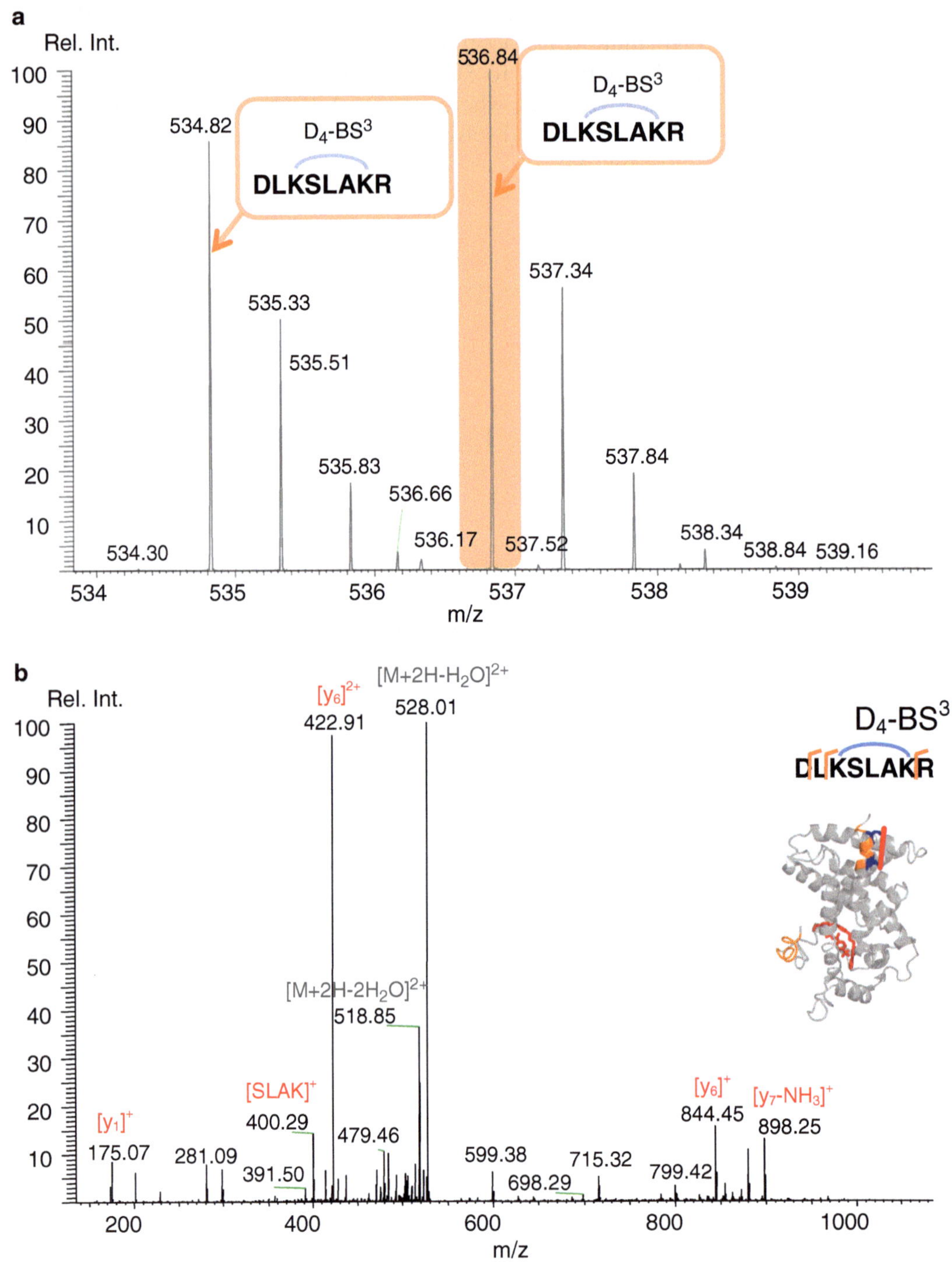

Fig. 4. (**a**) Full mass spectrum (analyzed in the orbitrap at high-resolution $R = 60,000$) of a putative cross-linking product showing a characteristic D_0/D_4 isotopic pattern. (**b**) Tandem mass spectrum (fragment ions analyzed in the LTQ) of the doubly charged intrapeptide cross-linked product (amino acids 202–209; *m/z* 536.84), in which lysine-204 is connected with lysine-208 of PPARα in the presence of the antagonist GW6471. Fragment ions are indicated as b-type (charge on the N-terminal fragment) and y-type (charge on the C-terminal fragment) ions. The cross-linked lysines are shown in *blue*; the antagonist GW8471 is shown as *red sticks*.

Table 1
Interpeptide cross-linked products in PPARα/antagonist GW6471 and PPARα/agonist YS81 complexes with the amine-reactive cross-linkers BS^3 and BS^2G

$[M+H]_{exp}^+$	$[M+H]_{theo}^+$	Ligand	XL[a]	Sequence[b]
1,152.5601	1,152.5605	–	BS^3	K_{349}PFC + K_{449}TES
1,152.5600	1,152.5605	GW6471	BS^3	K_{349}PFC + K_{449}TES
1,152.5600	1,152.5605	YS81	BS^3	K_{349}PFC + K_{449}TES
1,379.7401	1,379.7352	GW6471	BS^3	NFNMNK_{222}VK + K_{449}T
1,389.7843	1,389.7835	GW6471	BS^3	EK_{252}TLVAK + K_{449}TES
1,110.5127	1,110.5136	–	BS^2G	K_{349}PFC + K_{449}TES
1,110.5123	1,110.5136	GW6471	BS^2G	K_{349}PFC + K_{449}TES
1,110.5129	1,110.5136	YS81	BS^2G	K_{349}PFC + K_{449}TES
1,347.7360	1,347.7366	GW6471	BS^2G	EK_{252}TLVAK + K_{449}TES

Masses are given for D_0 cross-linkers. All peptide sequences were confirmed by MS/MS data
[a]XL: cross-linker
[b]All cysteines (C) are carbamidomethylated

4. Notes

1. Cross-linking reagents are sensitive to hydrolysis and therefore should be stored dry under inert gas (argon or nitrogen).
2. Neat DMSO is required to prepare the NHS cross-linker stock solutions immediately before use in order to prevent hydrolysis of the reagents.
3. In most cases, the buffer conditions for storing the protein are suitable. The pH value should range between 7.4 and 8.5. Tris buffers interfere with amine-reactive cross-linking reagents and should be avoided. Please note that amine-reactive NHS esters also react with hydroxyl groups of serines, threonines, and tyrosines, albeit with a lower frequency compared to amine groups (15, 16).
4. MilliQ H_2O and HPLC grade solvents are recommended in order to reduce contamination of cross-linking samples.
5. Complete destaining of the gels is important to obtain high-quality LC/MS data. Colloidal Coomassie Blue stain is more sensitive compared to conventional staining protocols with Coomassie Brilliant Blue R250.
6. Trypsin cleaves with lower frequency at lysines that are modified by cross-linkers. Therefore, a second protease is required to

achieve good proteolytic digestion yields, e.g., endoproteinase AspN (cleaves N-terminally to aspartic and also glutamic acid) or GluC (cleaves C-terminally to glutamic and also aspartic acid). AspN and GluC solutions in MilliQ H_2O are stable for 1–2 days at 2–8°C. Freezing is not recommended.

7. Organic solvents must not be kept in plastics due to a potential contamination by polymers.
8. Using 1:1 mixtures of deuterated (D_0) and non-deuterated (D_4) cross-linking reagents facilitates identification of cross-linking products in high-resolution mass spectra by characteristic mass shifts of 4 amu. Nano-HPLC retention time shifts are observed between non-deuterated and deuterated species.
9. The following parameters are recommended to be optimized: molar excess of cross-linking reagents ranging from 20-fold to 200-fold for maximal yields of cross-linking products. Highly cross-linked proteins could hamper subsequent proteolysis and increase the possibility of yielding homooligomers of the protein. Allow the cross-linking reaction to proceed between 5 and 120 min. A reaction time of 60 min gives optimal results in most cases.
10. Remaining SDS gels wrapped in wet precision wipes and packed in plastic bags keep fresh in the refrigerator at 4°C for 2 weeks. SDS gels that have been allowed to polymerize overnight at 4°C give optimal separation results of cross-linking products.
11. In case *offline*-ESI-MS experiments are planned, the use of formic acid (FA) should be preferred over TFA.

References

1. Michalik L, Wahli W (1999) Peroxisome proliferator-activated receptors: three isotypes for a multitude of functions. Curr Opin Biotechnol 10:564–570
2. Plutzky J (2003) The potential role of peroxisome proliferator-activated receptors on inflammation in type 2 diabetes mellitus and atherosclerosis. Am J Cardiol 92:34J–41J
3. Evans RM, Barish GD, Wang Y (2004) PPARs and the complex journey to obesity. Nat Med 10:355–361
4. Issemann I, Prince RA, Tugwood JD, Green S (1993) The peroxisome proliferator-activated receptor: retinoid X receptor heterodimer is activated by fatty acids and fibrate hypolipidaemic drugs. J Mol Endocrinol 11:37–47
5. Young MM, Tang N, Hempel JC, Oshiro CM, Taylor EW, Kuntz ID et al (2000) High throughput protein fold identification by using experimental constraints derived from intramolecular cross-links and mass spectrometry. Proc Natl Acad Sci U S A 97:5802–5806
6. Sinz A (2003) Chemical cross-linking and mass spectrometry for mapping three-dimensional structures of proteins and protein complexes. J Mass Spectrom 38:1225–1237
7. Sinz A (2006) Chemical cross-linking and mass spectrometry to map three-dimensional protein structures and protein-protein interactions. Mass Spectrom Rev 25:663–682
8. Lamos S, Krusemark C, Mcgee C, Scalf M, Smith L, Belshaw P (2006) Mixed isotope photoaffinity reagents for identification of small-molecule targets by mass spectrometry. Angew Chem Int Ed 45:4329–4333
9. Sinz A (2006) Isotope-labeled photoaffinity reagents and mass spectrometry to identify protein-ligand interactions. Angew Chem Int Ed 46:660–662
10. Müller MQ, de Koning LJ, Schmidt A, Ihling C, Syha Y, Rau O et al (2009) An innovative

method to study target protein-drug interactions by mass spectrometry. J Med Chem 52:2875–2879

11. Ihling C, Schmidt A, Kalkhof S, Schulz DM, Stingl C, Mechtler K et al (2006) Isotope-labeled cross-linkers and Fourier transform ion cyclotron resonance mass spectrometry for structural analysis of a protein/peptide complex. J Am Soc Mass Spectrom 17:1100–1113
12. Schmidt A, Kalkhof S, Ihling C, Cooper DMF, Sinz A (2005) Mapping protein interfaces by chemical cross-linking and Fourier transform ion cyclotron resonance mass spectrometry: application to a calmodulin/adenylyl cyclase 8 peptide complex. Eur J Mass Spectrom 11:525–534
13. Neuhoff V, Arold N, Taube D, Ehrhardt W (1988) Improved staining of proteins in polyacrylamide gels including isoelectric focusing gels with clear background at nanogram sensitivity using Coomassie Brilliant Blue G-250 and R-250. Electrophoresis 9:255–262
14. Peri S, Steen H, Pandey A (2001) GPMAW—a software tool for analyzing proteins and peptides. Trends Biochem Sci 26:687–689
15. Kalkhof S, Sinz A (2008) Chances and pitfalls of chemical cross-linking with amine-reactive *N*-hydroxysuccinimide esters. Anal Bioanal Chem 392:305–312
16. Mädler S, Bich C, Touboul D, Zenobi R (2009) Chemical cross-linking with NHS esters: a systematic study on amino acid reactivities. J Mass Spectrom 44:694–706
17. Müller MQ, Roth C, Sträter N, Sinz A (2008) Expression and purification of the ligand-binding domain of peroxisome proliferator-activated receptor alpha (PPARalpha). Protein Expr Purif 62:185–189
18. Xu HE, Stanley TB, Montana VG, Lambert MH, Shearer BG, Cobb JE et al (2002) Structural basis for antagonist-mediated recruitment of nuclear co-repressors by PPARalpha. Nature 415:813–817
19. Rau O, Syha Y, Zettl H, Kock M, Bock A, Schubert-Zsilavecz M (2008) Alpha-alkyl substituted pirinixic acid derivatives as potent dual agonists of the peroxisome proliferator activated receptor alpha and gamma. Arch Pharm 341:191–195

Chapter 22

Synthesis, Mass Spectrometric Characterization, and Analysis of the PPARδ Agonist GW1516 and Its Major Human Metabolites: Targets in Sports Drug Testing

Mario Thevis, Ines Möller, Simon Beuck, and Wilhelm Schänzer

Abstract

The elucidation of metabolic pathways and the detection of emerging therapeutics potentially enhancing athletic performance are of paramount importance to doping control authorities to protect the integrity of elite sports. A new drug candidate belonging to the family of the peroxisome proliferator-activated receptor-delta agonists termed GW1516 (also referred to as GW501516) has been prohibited by the World Anti-Doping Agency in 2009 due to its potential to artificially increase endurance. Consequently, sports drug testing laboratories need to establish detection methods enabling the identification of the intact substance and/or its metabolite(s) that unambiguously prove the presence or absence of the target substances in doping control specimens. Simulating human metabolic reactions using liver microsomal preparations, minute amounts of possible urinary metabolites were obtained that were characterized by mass spectrometry-based methods. Subsequently, the most abundant metabolic products were chemically synthesized and as well characterized by mass spectrometry and nuclear magnetic resonance spectroscopy. Finally, GW1516 and two oxidized metabolites were implemented in a routine doping control analytical assay based on liquid chromatography-(tandem) mass spectrometry (LC-MS/MS), which was tested for its fitness-for-purpose using spiked urine samples.

Key words: Gene doping, Sports, Mass spectrometry, PPARδ

1. Introduction

Although designed and developed for the treatment of severe diseases and health status amelioration, the misuse of approved therapeutics as well as emerging drug candidates has been an undesirable facet of sports for many decades (1). A great variety of compounds has been prohibited by the World Anti-Doping Agency (WADA) due to their proven or assumed performance enhancing properties

Mostafa Z. Badr and Jihan Youssef (eds.), *Peroxisome Proliferator-Activated Receptors (PPARs): Methods and Protocols*, Methods in Molecular Biology, vol. 952, DOI 10.1007/978-1-62703-155-4_22, © Springer Science+Business Media New York 2013

as well as harmful nature if administered without medical indication (2, 3). With constantly improving detection capabilities of doping control laboratories, the cheating athletes' search for "alternatives" (that presumably remain below the drug testing radar) has comprised the arena of non-approved tailor-made designer drugs (4) as well as newly developed therapeutics, which are commonly still in early or advanced clinical trials (5). One of these drug candidates is the peroxisome proliferator-activated receptor (PPAR)δ agonist GW1516 (Fig. 1) that has been categorized as gene doping agent due to its mechanism of action and assumed beneficial effects on athletic performance (6). Being developed and tested for the treatment of dyslipidemia and metabolic syndrome in phase II clinical trials (7–11), the administration of GW1516 to laboratory rodents significantly reduced the susceptibility to weight gain and increased their physical performance at a daily dosage of 2–5 mg/kg (6, 12). Combined with exercise stimuli, the application of the PPARδ agonist resulted in an improved endurance performance that exceeded control cohorts by ~70%. The effect was particularly attributed to the induction of oxidative genes and the increased preference for lipid consumption with concurrently decreased carbohydrate utilization (6, 11).

In order to provide doping control laboratories with information that enables the detection of GW1516 or its metabolites in sports drug testing samples, the compounds of interest (i.e., abundant metabolites) are prepared using in vitro assays with human liver microsomal preparations and identified and characterized by mass spectrometry-based approaches (13). Upon assignment of structural properties, the target compounds (i.e., the sulfoxide and sulfone of GW1516, Fig. 1) are chemically synthesized, characterized by high resolution/high accuracy mass spectrometry and finally used in a doping control analytical method.

2. Materials

All aqueous solutions were prepared using deionized water (MilliQ grade).

2.1. In Vitro Metabolism Assay

1. Reaction buffer: Two separate solutions are prepared of (A) 50 mM Na_2HPO_4 (710 mg in 100 mL of deionized water) and (B) 50 mM KH_2PO_4 (680 mg in 100 mL of deionized water). B is added to A (under constant stirring) until the pH of 7.4 is reached. Finally, 48 mg of $MgCl_2$ are added and the resulting reaction buffer is stored at RT.
2. 50 mM NADPH solution: To 100 μL of reaction buffer (see Subheading 2.1, item 1) 4.2 mg of NADPH tetrasodium salt are added (store at −20°C).

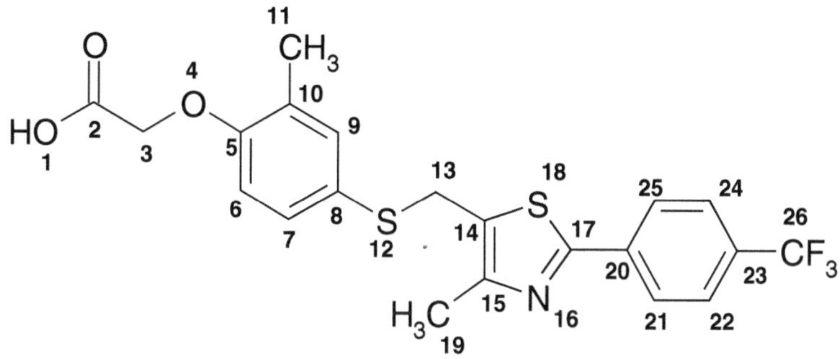

1

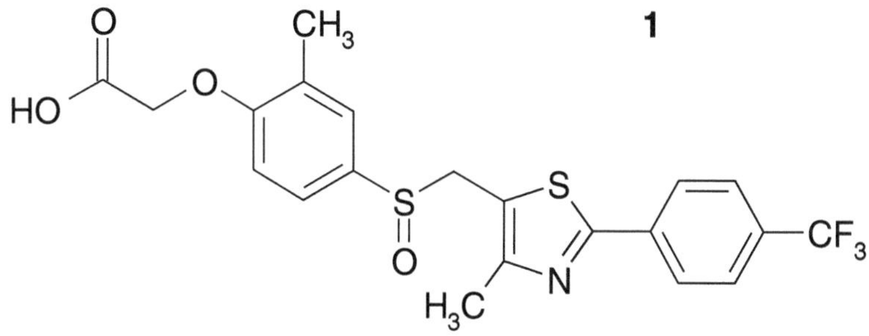

2

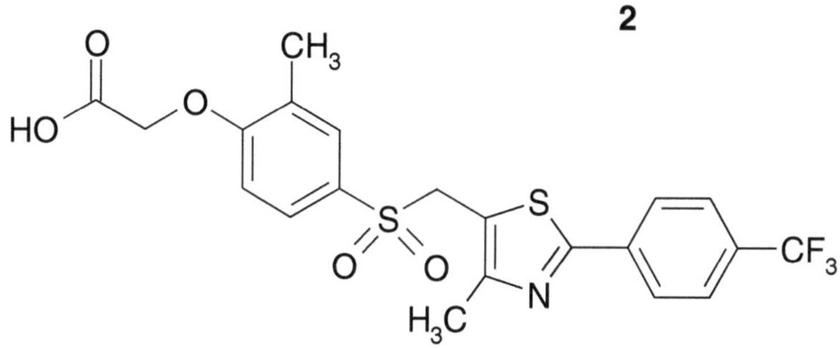

3

Fig. 1. Structures of GW1516 (**1**) and its corresponding oxidized metabolites **2** (sulfoxide) and **3** (sulfone).

3. GW1516 working solution: 9 mg of GW1516 are dissolved in 2 mL of DMSO yielding a 10 mM stock solution of the substrate. Ten microliter of the stock solution are diluted with 1 mL of the reaction buffer (see Subheading 2.1, item 1) to yield a 100 μM working solution of GW1516 (store at +4°C).
4. Glucuronidase inhibitor solution: 10 mg of D-saccharic acid 1,4-lactone monohydrate are dissolved in 1 mL of reaction buffer (see Subheading 2.1, item 1) to prepare a 50 mM solution of D-saccharic acid 1,4-lactone (store at +4°C).
5. Human liver microsomal fraction (Roche Diagnostics, Mannheim, Germany).

2.2. Chemical Synthesis of Sulfoxide and Sulfone Metabolites of GW1516

1. GW1516 (Axxora, Lörrach, Germany).
2. 35% Hydrogen peroxide (Aldrich, Steinheim, Germany).
3. 2% Acetic acid is prepared by adding 2 mL of glacial acetic acid to 98 mL of deionized water.
4. 0.2 M Sodium hydroxide is prepared by dissolving 800 mg of NaOH pellets in 100 mL of deionized water.
5. Solvent A for LC fractionation: 5 mM ammonium acetate, pH 3.5. A total of 385 mg of ammonium acetate is dissolved in 1 L of deionized water followed by the addition of 1 mL of glacial acetic acid.
6. Solvent B: 100% methanol.

2.3. Mass Spectrometric Characterization

1. Solvent A: 0.1% formic acid is prepared by adding 1 mL of formic acid to 1 L of deionized water.
2. Solvent B: acetonitrile containing 0.1% formic acid is prepared by adding 1 mL of formic acid to 1 L of acetonitrile.
3. D_2O is used as solvent A in case of H/D exchange experiments (see Note 1).
4. Pure acetonitrile is used as solvent B in case of H/D exchange experiments (see Note 1).

2.4. Doping Control Analytical Assay

1. 0.8 M Phosphate buffer, pH 7.0 for enzymatic hydrolysis. To 230 mL of deionized water are added 17.3 g of Na_2HPO_4 and 8.8 g of NaH_2PO_4 monohydrate (see Note 2).
2. β-Glucuronidase (*Escherichia coli*) from Roche Diagnostics (Mannheim, Germany).
3. Internal standard solution: 10 mg of methyltestosterone are dissolved in 10 mL of acetonitrile. This stock solution is further diluted by mixing an aliquot of 100 μL with 10 mL of acetonitrile to yield the working solution with 10 ng/μL.
4. 2% Carbonate buffer, pH = 9.6. To 240 mL of deionized water are added 30 g of K_2CO_3 and 30 g of $KHCO_3$ (see Note 2).
5. Solvent for sample reconstitution. Deionized water (40 mL) and acetonitrile (10 mL) are combined to yield the sample reconstitution solution.

3. Methods

3.1. In Vitro Metabolism Assay

1. Reaction mixtures of 100 μL are composed in 1.5 mL Eppendorf tubes: 10 μL of the working solution of GW1516 are combined with 10 μL of the 50 mM NADPH solution, 10 μL of the 50 mM glucuronidase inhibitor solution, and 60 μL of the

reaction buffer. Finally, 10 μL of the human liver microsomal preparation is added (see Notes 3 and 4).

2. The incubation is conducted at 37°C for 2 h with continuous shaking.
3. Incubations are terminated by adding 100 μL of ice-cold acetonitrile, and the samples are centrifuged at 10,000 × *g* for 10 min to pelletize precipitated protein.
4. The supernatant is transferred via pipette to a fresh Eppendorf tube and evaporated under reduced pressure in a vacuum centrifuge (see Note 5).
5. The dry residue is reconstituted in 20 μL of 2% acetic acid and target analytes are further purified by liquid–liquid extraction. A volume of 500 μL of ethyl acetate is added, the tube shaken for 10 min using a rotating mixer, and subsequently centrifuged at 10,000 × *g* for 3 min (see Note 6).
6. The organic layer is transferred to a new Eppendorf tube, evaporated to dryness using a vacuum centrifuge (see Note 5), and the dry residue is reconstituted in 50 μL of acetonitrile and deionized water (1:1, v/v) for subsequent LC-MS/MS analysis (see Note 7).

3.2. Chemical Synthesis of Sulfoxide and Sulfone Metabolites of GW1516

1. The preparation of the sulfoxide of GW1516 is accomplished by adding 75 μL of 35% hydrogen peroxide to 200 μL of methanol containing 2 mg of GW1516 in a glass tube (see Note 8).
2. The reaction mixture incubates for 5 h under gentle stirring at room temperature, before the pH is adjusted to 7 using an aqueous solution of 1 M sodium hydroxide (see Note 9).
3. Purification of the sulfoxide of GW1516 is done by HPLC fractionation employing, e.g., an Agilent 1100 Series liquid chromatograph (Waldbronn, Germany) equipped with a Zorbax Eclipse XDB-C8 column (9.4 × 250 mm, particle size 5.0 μm) and an Applied Biosystems API2000 QTrap mass spectrometer (Darmstadt, Germany) in positive electrospray ionization mode. The flow rate is set to 3 mL/min (for solvents A and B see items 5 and 6 of Subheading 2.2, respectively). The injection volume is 20 μL. The sulfoxide elutes under isocratic conditions (45/55, v/v) between 24.5 and 28.5 min and is collected from an open split capillary at the 3-way junction connecting the LC column exit capillary with the electrospray interface (see Note 10).
4. The combined fractions are diluted with deionized water (1:6, v/v) and concentrated by solid-phase extraction on an Oasis HLB cartridge (Waters, Milford, MA, USA) (see Note 11).
5. The sulfone derivative of GW1516 is obtained analogously but modified by adding 5 mL of 35% hydrogen peroxide

instead of 75 μL (see Note 8) and extending the incubation duration to 20 h.

6. After neutralization of the reaction mixture to pH 7 with aqueous sodium hydroxide (1 M), the target analyte is extracted twice into 4 mL of ethyl acetate (see Note 12).
7. The combined organic layer is evaporated to dryness under reduced pressure using a rotary evaporator at 40°C and the dry residue is reconstituted in 200 μL of MeOH (1:1, v/v).
8. The GW1516 sulfone is subsequently purified under the above mentioned LC conditions by collecting the substance-containing fraction between 31.7 and 34.5 min. The injection volume is 20 μL.
9. The combined fractions are evaporated to dryness using a rotary evaporator at 40°C, reconstituted in 200 μL of solvents A and B (1:1, v/v, see items 5 and 6 of Subheading 2.2, respectively) and further purified by HPLC fractionation using the identical equipment but gradient elution. Starting at 30% A, the percentage of A is decreased to 20% within 10 min, maintained at 20% A for 2 min, followed by re-equilibration for 5 min at 30% a prior to the next injection. The GW1516 sulfone is collected between 6.6 and 7.1 min (see Note 10).
10. The fractions are combined in a glass tube and evaporated to dryness under reduced pressure at 40°C to yield the desired product as a colorless solid.

3.3. Mass Spectrometric Characterization

1. In vitro metabolism study samples are analyzed and metabolic products characterized using liquid chromatography—high resolution/high accuracy (tandem) mass spectrometry. Different instruments can be used; one option is the Thermo Accela liquid chromatograph (LC) interfaced to the Thermo Exactive mass spectrometer (Thermo Fisher Scientific, Bremen, Germany). The LC is equipped with a Thermo Fisher Scientific Hypersil Gold column (2.1 × 50 mm, particle size 1.9 μm), and solvents are 0.1% formic acid (A) and acetonitrile containing 0.1% formic acid (B). The LC flow rate is set to 200 μL/min, and gradient elution is conducted starting at 90% A, decreasing to 20% A in 12 min, then to 0% A in further 3 min, maintaining 0% A for 2 min followed by re-equilibration at 90% A. The mass spectrometer (MS) is operated in positive ionization mode and calibrated using the manufacturer's calibration mixture (yielding a total of seven reference masses). Mass accuracies <5 ppm are specified and accomplished for the period of analysis. The ionization voltage is +3.5 kV, the capillary temperature is set to 290°C, and three MS settings are used throughout the analytical runs: (a) full scan MS from *m/z* 50 to 2,000 at a resolution of 25,000 (FWHM), (b) full scan MS (*m/z* 50–2,000, resolution

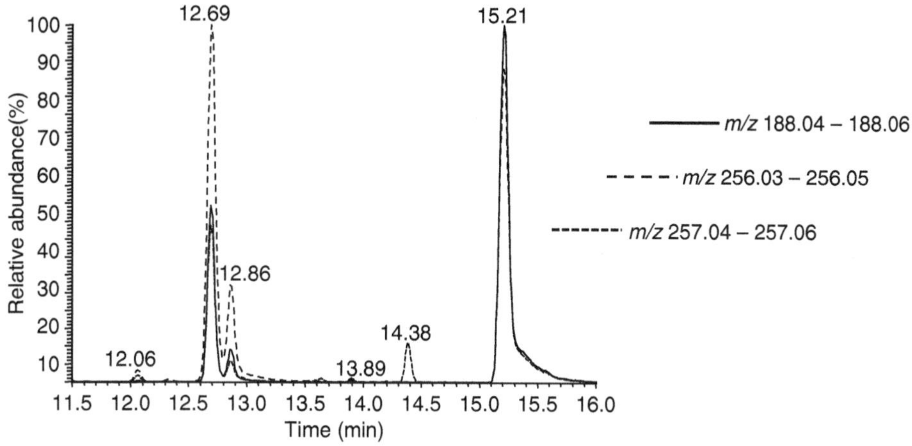

Fig. 2. LC-MS analysis of an in vitro metabolism experiment showing three diagnostic ion traces of GW1516 at *m/z* 188, 256, and 257. The signals at 12.69, 12.86, 13.89, and 15.21 correspond to the sulfoxide, the sulfone, a (currently not further characterized) hydroxylated metabolite, and GW1516, respectively.

set to 10,000) with higher energy collision-induced dissociation (HCD) set to 20 V, and (c) full scan MS (*m/z* 50–2,000, resolution set to 10,000) with HCD set to 50 V (see Note 13). The extraction of diagnostic ion traces of GW1516 at *m/z* 188, 256, and 257 yields signals at 12.69, 12.86, 13.89, and 15.21 corresponding to the sulfoxide, the sulfone, a hydroxylated metabolite, and GW1516, respectively (Fig. 2).

2. Further characterization of in vitro-derived and chemically synthesized metabolites by tandem mass spectrometry (MS/MS) is conducted using identical LC conditions but, e.g., an LTQ-Orbitrap MS (Thermo Fisher Scientific, Bremen, Germany) to allow MS*n* experiments from selected precursor ions (isolation width 1 Da) of the generated metabolites. Resulting product ion spectra are depicted in Fig. 3 including the intact compound GW1516, the sulfoxide, the sulfone, and a hydroxyl metabolite with a not entirely characterized structure.

3. Experiments using D_2O and acetonitrile as eluents are conducted to differentiate the formation of hydroxylated and oxygenated products of GW1516. Also here, identical LC-MS conditions are applied except for the exchange of solvent A by deuterium oxide (see Note 1). In cases of the sulfoxide and sulfone metabolites, two deuterium atoms are present in the measured analytes (one by the exchange of the mobile proton at the carboxyl residue and one introduced during the ionization process); in contrast, the hydroxylated metabolite contains three deuterium atoms using the H/D exchange conditions.

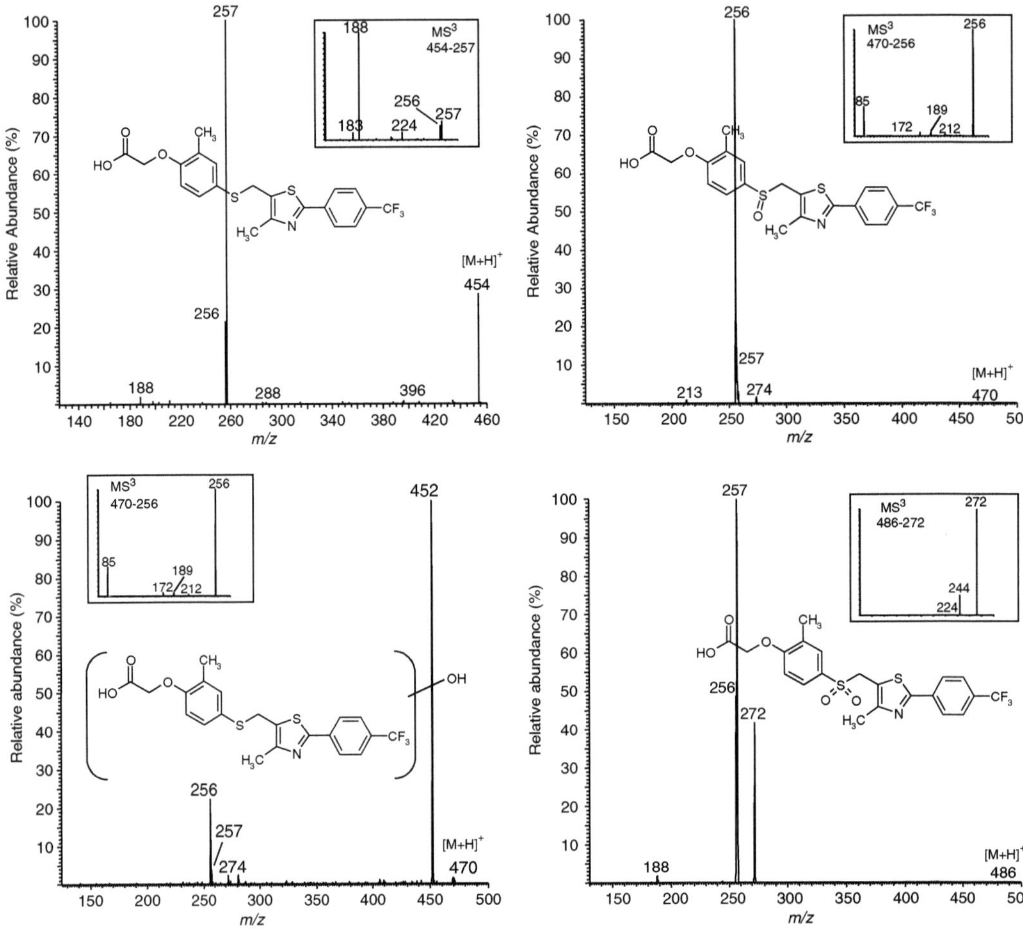

Fig. 3. ESI product ion mass spectra generated from protonated molecules $[M+H]^+$ of (**a**) *m/z* 454 (GW1516); (**b**) *m/z* 470 (sulfoxide); (**c**) *m/z* 470 (hydroxylated GW1516), and (**d**) *m/z* 486 (sulfone).

4. Based on the product ions generated by collisional activation of the precursor ion, their determined elemental composition (as deduced from respective accurate masses), diagnostic dissociation pathways are elucidated to provide characteristic patterns that uniquely identify the location of introduced molecular changes. The fragmentation pathway of GW1516 and the corresponding sulfone yielding the characteristic ions at *m/z* 256, 257, and 272, respectively, are shown in Fig. 4.

3.4. Doping Control Analytical Assay

The sample preparation is performed in accordance to established multi-analyte doping control screening procedures to allow for comprehensive detection protocols.

1. A volume of 3 mL of urine is placed in a glass tube and buffered to pH 7.0 by adding 1 mL of 0.8 M phosphate buffer for enzymatic hydrolysis.

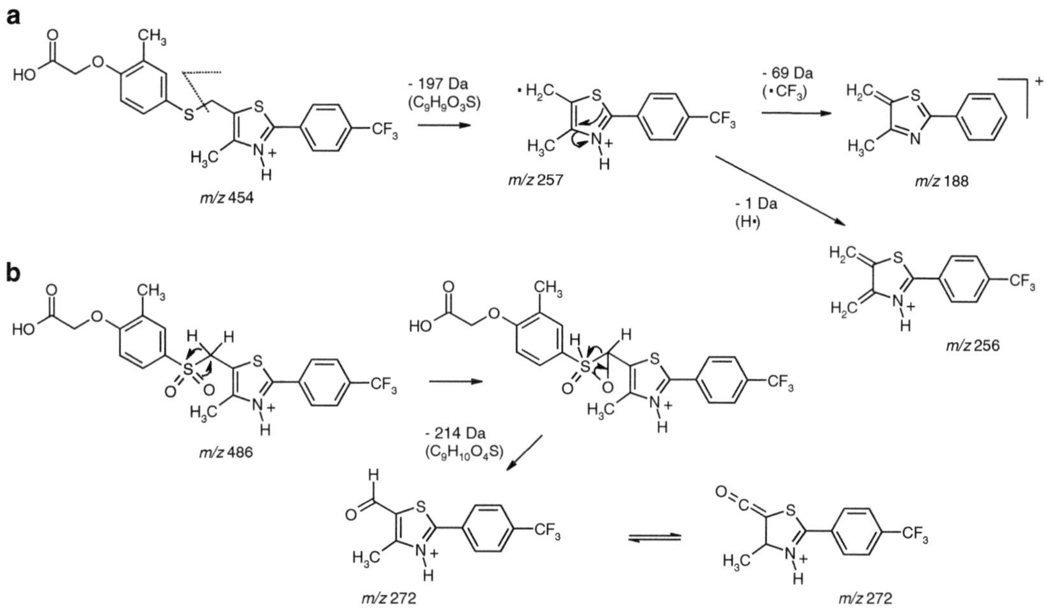

Fig. 4. Proposed dissociation routes of (**a**) GW1516 and (**b**) its major metabolite.

2. The sample is fortified with 200 ng of the internal standard methyltestosterone (20 μL of the 10 ng/μL working solution) and 20 μL of β-glucuronidase.
3. The mixture is incubated at 50°C for 1 h.
4. After cooling to ambient temperature, 0.75 mL of the carbonate buffer, pH 9.6, and 4 mL of *t*-butyl methyl ether are added. The sample is shaken for 10 min and centrifuged at 600×*g* for 5 min.
5. The organic layer is subsequently transferred into a fresh test tube and evaporated to dryness by means of a rotary evaporator under reduced pressure.
6. The dry residue is finally reconstituted in 200 μL of water/acetonitrile (4:1, v:v), the solution is placed in HPLC vials and 10 μL are injected into the LC-MS(/MS) system.
7. Routine doping control analysis can be performed using an Agilent 1100 Series LC coupled to an Applied Biosystems API4000 Qtrap mass spectrometer (Darmstadt, Germany) with electrospray ionization. As an example, the LC is equipped with a Macherey-Nagel Pyramid column (4×70 mm), and the eluents used are 5 mM ammonium acetate containing 0.1% acetic acid (mobile phase A) and acetonitrile (mobile phase B). A gradient is employed starting at 90% A decreasing to 0% A within 10 min followed by re-equilibration at 90% A for 2 min. The flow rate is set to 800 μL/min and the ion source operated in positive mode at 550°C using a spray voltage of 5,500 V.

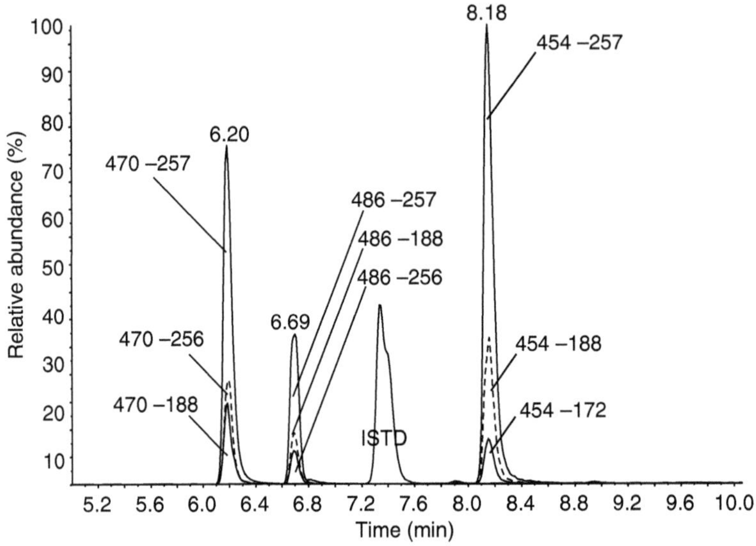

Fig. 5. Extracted ion chromatograms of a urine sample containing GW1516 (8.18 min) and the two metabolites (sulfoxide at 6.20 min and sulfone at 6.69 min).

8. GW1516 as well as its metabolites and the internal standard (methyltestosterone) are detected by means of characteristic product ions formed from protonated molecules by collision-induced dissociation (CID) utilizing the multiple reaction monitoring mode (Fig. 5) (see Note 14).

4. Notes

1. The addition of deuterated 0.1% formic acid is an option to improve the ionization efficiency.
2. The mixture is vigorously stirred for ~15 min to accelerate the solution process.
3. Samples are mixed gently prior to the addition of the human liver microsomal preparation.
4. Two blank samples are prepared and analyzed with each batch of incubation reactions, i.e., one blank specimen containing all reaction components without the substrate GW1516 to probe for potential contaminants mimicking metabolites, and a second blank sample that contains all reaction elements but not the human liver microsomal fraction to monitor non-enzymatic reactions during the incubation.

5. In order to avoid artifact generation during the evaporation process, the temperature should not exceed 40°C.
6. The extraction step can be repeated and organic layers combined to increase the recovery of the metabolites.
7. In case of H/D experiments, the residue is dissolved in 50 μL of acetonitrile and deuterium oxide (1:1, v:v).
8. Precipitate potentially forming under the addition of the peroxide solution is dissolved by the drop-wise addition of 2% acetic acid followed by pH adjustment to 4 by the drop-wise addition of 0.2 M sodium hydroxide.
9. Only few microliter are required.
10. The split ratio should be roughly 99:1 and is regulated by the length and inner diameter of the employed capillaries, i.e., the collector capillary and the ESI interface capillary. Further, initial tests concerning a possible time delay/dead volume between the MS detection and the collection effluent should be done by analyzing the first fractions for accurateness of the collection window.
11. The extraction cartridge is preconditioned with 2 mL of methanol and 2 mL of deionized water. After gravity flow loading of the diluted fractions, the target substance is eluted with 2 mL of methanol into a glass tube and concentrated under reduced pressure without heating to avoid further oxidation.
12. The mixture is shaken for 10 min, centrifuged at 660 × *g* for 5 min, and the organic layer transferred to a fresh glass tube.
13. The use of different collision energies allows the detection of the intact molecules (at an offset voltage of 0 V) as well as diagnostic product ions of different abundance depending on the intensity of collisional activation.
14. Typical ion transitions for the target analytes are: (a) GW1516: *m/z* 454–257, 454–188, and 454–172; (b) GW1516 su: *m/z* 470–257, 470–256, and 470–188; (c) GW1516 sulfone: *m/z* 486–257, 486–256, and 486–188. Collision energies for optimal abundances of each ion transition should be determined for the individually employed mass spectrometer.

Acknowledgments

The presented work was supported by Antidoping Switzerland (Berne, Switzerland) and the Manfred-Donike-Institute for Doping Analysis (Cologne, Germany).

References

1. Thevis M (2010) Mass spectrometry in sports drug testing—characterization of prohibited substances and doping control analytical assays. Wiley, Hoboken
2. WADA (2011) World Anti-Doping Agency, vol 2010
3. Thevis M, Kuuranne T, Geyer H, Schänzer W (2011) Annual banned-substance review: analytical approaches in human sports drug testing. Drug Test Anal 3:1–14
4. Kazlauskas R (2010) Designer steroids. Handb Exp Pharmacol 195:155–185
5. Thevis M, Thomas A, Kohler M, Beuck S, Schänzer W (2009) Emerging drugs: mechanism of action, mass spectrometry and doping control analysis. J Mass Spectrom 44:442–460
6. Narkar VA, Downes M, Yu RT, Embler E, Wang YX, Banayo E, Mihaylova MM, Nelson MC, Zou Y, Juguilon H, Kang H, Shaw RJ, Evans RM (2008) AMPK and PPARdelta agonists are exercise mimetics. Cell 134:405–415
7. US National Institutes of Health (2009) Clinical trials including GW501516. http://www.clinicaltrials.gov/ct2/results?term=gw501516. Accessed 20 Jun 2009
8. Wang YX, Lee CH, Tiep S, Yu RT, Ham J, Kang H, Evans RM (2003) Peroxisome-proliferator-activated receptor delta activates fat metabolism to prevent obesity. Cell 113:159–170
9. Krämer DK, Al-Khalili L, Guigas B, Leng Y, Garcia-Roves PM, Krook A (2007) Role of AMP kinase and PPARdelta in the regulation of lipid and glucose metabolism in human skeletal muscle. J Biol Chem 282:19313–19320
10. Luquet S, Lopez-Soriano J, Holst D, Gaudel C, Jehl-Pietri C, Fredenrich A, Grimaldi PA (2004) Roles of peroxisome proliferator-activated receptor delta (PPARdelta) in the control of fatty acid catabolism. A new target for the treatment of metabolic syndrome. Biochimie 86:833–837
11. Brunmair B, Staniek K, Dorig J, Szocs Z, Stadlbauer K, Marian V, Gras F, Anderwald C, Nohl H, Waldhausl W, Furnsinn C (2006) Activation of PPAR-delta in isolated rat skeletal muscle switches fuel preference from glucose to fatty acids. Diabetologia 49:2713–2722
12. Wang YX, Zhang CL, Yu RT, Cho HK, Nelson MC, Bayuga-Ocampo CR, Ham J, Kang H, Evans RM (2004) Regulation of muscle fiber type and running endurance by PPARdelta. PLoS Biol 2:e294
13. Thevis M, Möller I, Thomas A, Beuck S, Rodchenkov G, Bornatsch W, Geyer H, Schänzer W (2010) Characterization of two major urinary metabolites of the PPARdelta-agonist GW1516 and implementation of the drug in routine doping controls. Anal Bioanal Chem 396:2479–2491

Chapter 23

LC-MS-Based Method for the Qualitative and Quantitative Analysis of the Novel PPARγ Agonist KR-62980

Min-Sun Kim and Myung Ae Bae

Abstract

A sensitive liquid chromatography/tandem mass spectrometry (LC/MS/MS) method was developed for a novel peroxisome proliferator-activated receptor γ (PPARγ) agonist, KR-62980, in rat plasma. It involves liquid–liquid extraction (LLE) followed by HPLC separation and electrospray ionization tandem mass spectrometry. The linear ranges of the assay were 0.01–10 μg/mL with a correlation coefficient (R^2) greater than 0.99 and the lower limit of quantification was 0.01 μg/mL. The average recovery was 90.1 and 98.4% from rat plasma for KR-62980 and imipramine, respectively. The coefficients of variation of intra- and inter-assay were 1.2–10.6% and the relative error was 0.8–13.2%. The method was validated and successfully applied to the pharmacokinetic study of KR-62980 in rat.

Key words: KR-62980, Method validation, Pharmacokinetics, LC/MS/MS, Rat plasma, PPARγ

1. Introduction

"Peroxisome proliferator-activated receptor-γ" (PPARγ) is a member of the nuclear hormone receptor (NHR) superfamily of transcrip tion factors. Activation of PPARγ is critical in several metabolic pathways such as adipocyte differentiation, insulin resistance, atherosclerosis, and hypertension (1–7). Thus, some PPARγ agonists are used clinically as anti-diabetic agents. KR-62980, 1-(*trans*-methylimino-*N*-oxy)-6-(2-morpholineoethoxy-3-phenyl-1H-indene-2-carboxylic acid ethyl ester, is functionally active as a selective PPARγ agonist, with an EC50 of 15 nM by a transactivation assay (8). KR-62980 displays anti-hyperglycemic activity, including in vivo glucose lowering activity, with little weight gain (8) and anti-adipogenic function, through the suppression of PPARγ-mediated adipocyte differentiation by activating nuclear

Mostafa Z. Badr and Jihan Youssef (eds.), *Peroxisome Proliferator-Activated Receptors (PPARs): Methods and Protocols*, Methods in Molecular Biology, vol. 952, DOI 10.1007/978-1-62703-155-4_23, © Springer Science+Business Media New York 2013

TAZ (transcriptional coactivator with PDZ-binding motif) (9). Thus, these findings suggest that KR-62980 may be a good drug candidate for the treatment of type 2 diabetes.

Thus, our study was to validate an LC/MS/MS method using simple liquid–liquid extraction for the quantitative analysis of KR-62980 in rat plasma. The method was evaluated with regard to its accuracy, precision, selectivity, sensitivity, reproducibility, and stability. It was used successfully in a preliminary pharmacokinetic study of intravenously administered KR-62980 (10).

2. Materials

KR-62980 was synthesized by the Medicinal Science Division at the Korea Research Institute of Chemical Technology (Daejeon, Korea). Imipramine (Sigma-Aldrich). All chemicals were of the highest analytical grade available. Organic solvents were of HPLC grade and distilled water was obtained from a Milli-Q system.

2.1. Standard Solution

1. Standard solution of KR-62980: Prepare the stock solution at 1 mg/mL. Weigh 1 mg KR-62980 and transfer to 1.5 mL microfuge tube. Add methanol to a volume of 1 mL. Dilute the stock solution in methanol:water (1:1 v/v). Make the working standard solution equivalent to levels of 0.1, 0.3, 10, 50, 80, and 100 μg/mL. Store at −20°C.
2. Standard solution of internal standard (IS): Dissolve 1 mg imipramine (IS) in 1 mL methanol. Prepare the internal standard working solution at 3 μg/mL with methanol:water (1:1 v/v). Store at −20°C.

2.2. Calibration and Quality Control Samples

1. Pipet 45 μL blank plasma in 1.5 mL microfuge tube and add 5 μL working standard solution (see Note 1). Make up each calibration sample as in previous step. Prepare KR-62980 concentrations of 0.03, 0.5, and 8 μg/mL in rat plasma for quality control (QC) samples.

2.3. Instruments

1. Liquid chromatography (LC).
   - Optimize LC conditions: Column, Hypersil GOLD C18 column (100 × 2.1 mm i.d., 3 μm; Thermo, Waltham, MA, USA); the mobile phase, aetonitrile-ammonium formate (10 mM; 80:20 v/v); flow rate, 0.2 mL/min; total run time is 4 min; the temperatures of the autosampler, 4°C; column oven, 20°C.
2. Mass spectrometry (MS/MS): API 4000 Q trap mass spectrometer (Applied Biosystems, Foster City, CA, USA).

– Optimize MS conditions: source temperature, 350°C; curtain gas, 20 psi; nebulizing (GS1), 50 psi; heating (GS2), 50 psi; collision energy (CE), 59 V for KR-62980 and 25 V for imipramine. A turbo electrospray interface; positive ionization mode; the multiple reaction monitoring (MRM) mode.

3. Methods

3.1. Sample Preparation

1. Place rat plasma sample (50 μL) in a 1.5 mL microfuge tube and add 20 μL of IS solution (3 μg/mL imipramine).
2. Vortex for 1 min and add 1 mL cold ethyl acetate. Vortex for 5 min and centrifuge 15,000 × *g* for 5 min at 4°C (see Note 2).
3. Transfer 1 mL of the supernatant to another tube and dry by centrifugal evaporator.
4. Add 200 μL mobile phase to the dried residue and vortex for 1 min. Inject 2 μL of the supernatant onto the analytical column.

3.2. Mass Spectra and Chromatography

1. Fragmentation ion of KR-62980: *m/z* 114.2 (C-O bond cleavage) and *m/z* 391.0 (ester bond cleavage). Choose the ion at *m/z* 114.2 as the quantitative ion and at *m/z* 391.0 as the confirmative ion (see Note 3) (Fig. 1).
2. Fragmentation ion of Imipramine (IS): produce intense product-ion signals at *m/z* 281.3 → 86.1 (Fig. 1).
3. Chromatogram: Elute KR-62980 and IS at 2.3 and 2.7 min (see Note 4) (Fig. 2).

3.3. Validation

1. Calibration curves: construct by linear regression of the peak area-ratios (y) of KR-62980 to internal standard, versus the concentration (x) in μg/mL (see Note 5).
2. Intra- and inter-day precisions: estimate by analyzing the spiked samples at four different concentrations (0.01, 0.03, 0.5, and 8 μg/mL) in a single day and for 5 days (see Note 6) (Table 1).

3.4. Matrix Effect and Recovery

1. Recovery: Compare the peak areas of analyte spiked before extraction (set 1) with those of analyte spiked post-extraction matrix (set 2).
2. Matrix effect: Calculate peak areas of analyte spiked post-extraction matrix (set 2) with reference standards prepared by spiking with the same concentration of KR-62980 in mobile phase (set 3).

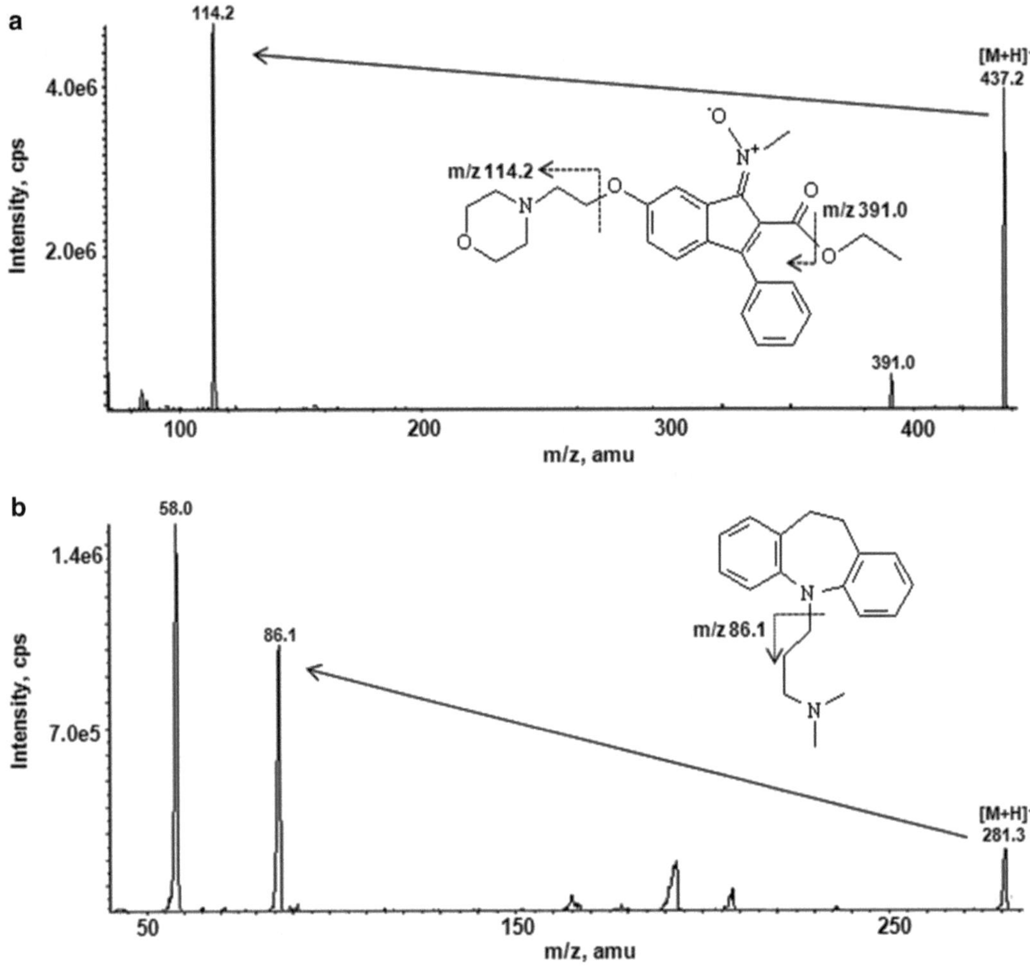

Fig. 1. The structures and product-ion scan spectra of: (**a**) KR-62980 and (**b**) imipramine (IS).

3. Process efficiency: Compare reference standards prepared by spiking with the same concentration of KR-62980 in mobile phase (set 3) with the peak areas of analyte spiked before extraction (set 1) (see Note 7) (Table 2).

3.5. Stability

1. Stock solution stability: Store at –20°C for 3 weeks (see Note 8).
2. KR-62980 stability in rat plasma.

Short-term stability includes: (a) freeze-thaw cycle stability; (b) exposure of samples to room temperature for 1, 3, 6 h, and 1 day; (c) exposure to 4°C for 1 day; (d) exposure to –20°C for 1 day; (e) exposure to –80°C for 1 day; (f) exposure in mobile phase at 4°C for 1 day and 1 week after preparation, and (g) exposure in mobile phase at room temperature for 1 day after preparation. Long-term stability includes: (a) exposure of samples to –80°C for

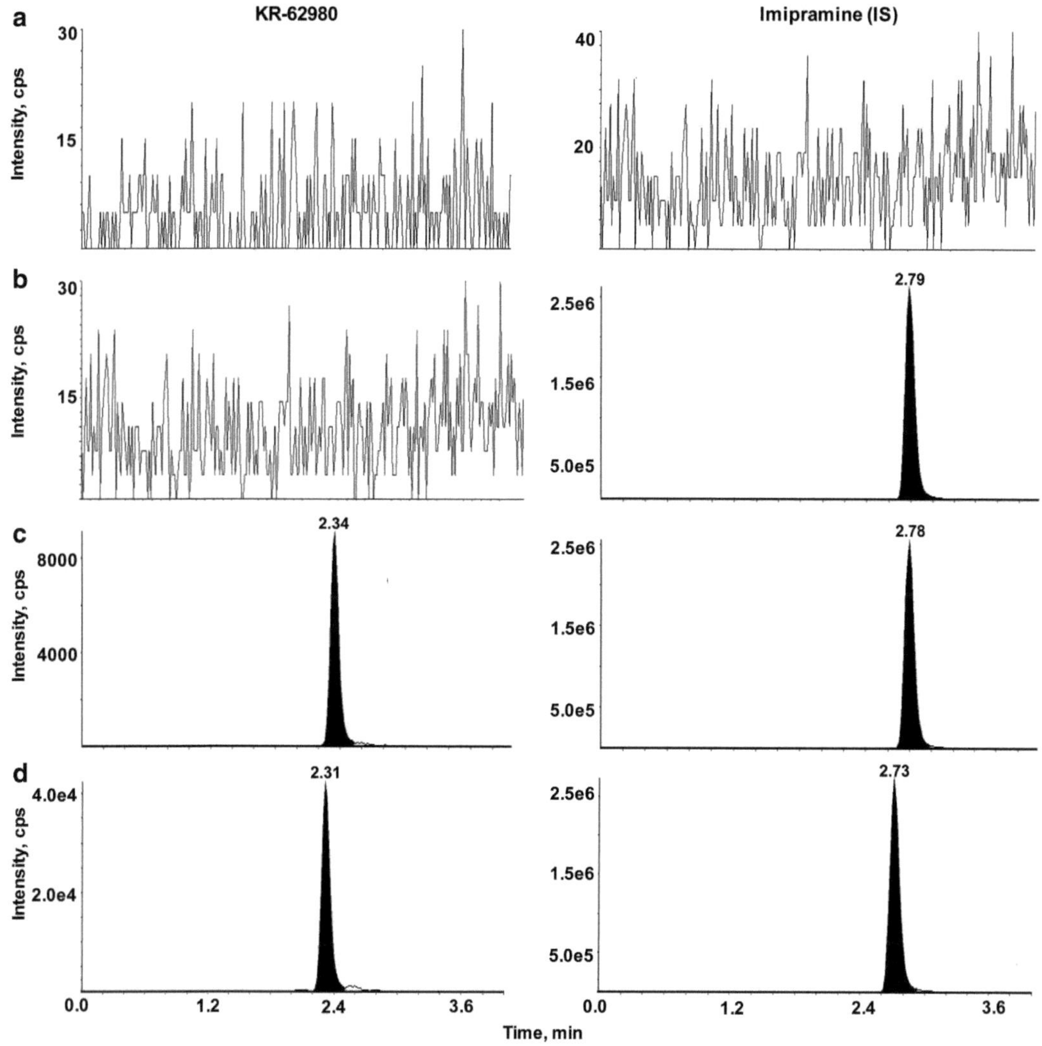

Fig. 2. Representative MRM chromatograms of (**a**) double blank rat plasma, (**b**) blank rat plasma spiked with IS, (**c**) blank rat plasma spiked 0.01 μg/mL (LLOQ) of KR-62980 and IS, (**d**) a plasma sample obtained at 1 h after intravenous administration of KR-62980 at 10 mg/kg to rats.

Table 1
Reproducibility and accuracy for KR-62980 in rat plasma ($n = 5$)

Theoretical concentration (μg/mL)	Intra-day Concentration found (μg/mL)	RSD[a] (%)	RE[b] (%)	Inter-day Concentration found (μg/mL)	RSD (%)	RE (%)
0.01	0.01	3.4	−6.6	0.01	10.6	1.1
0.03	0.03	1.6	10.8	0.03	9.0	−5.3
0.5	0.57	1.2	13.2	0.50	9.2	−0.8
8	7.43	2.2	−7.1	7.57	8.5	−5.4

[a]RSD (%) = standard deviation of the concentration/mean concentration × 100

[b]RE (%) = (calculated concentration − theoretical concentration)/theoretical concentration × 100

Table 2
Recovery, matrix effect, and process efficiency ($n=3$)

	Concentration (μg/mL)	Matrix effect[a] (%)	Recovery[b] (%)	Process efficiency[c] (%)
KR-62980	0.03	90.0	96.4	86.8
	0.5	94.1	88.6	83.3
	8	96.5	85.4	82.4
	Mean	93.5	90.1	84.2
IS (Imipramine)	3	89.2	98.4	87.8

[a]Matrix effect expressed as the ratio of the mean peak area of an analyte added post-extraction (set 2) to the mean peak area of the same analyte standards (set 3) multiplied by 100
[b]Recovery calculated as the ratio of the mean peak area of an analyte added before extraction (set 1) to the mean peak area of an analyte spiked post-extraction (set 2) multiplied by 100
[c]Process efficiency calculated as the ratio of the mean peak area of an analyte added before extraction (set 1) to the mean peak area of the same analyte standards (set 3) multiplied by 100

30 days and (b) exposure to −20°C for 30 days. The effect of esterase on KR-62980 stability is conducted by adding NaF (10 mg/mL) and PMSF (0.18 mg/mL) to the rat plasma, respectively. These samples are to be exposed to room temperature for 1, 3, 6, 24 h (see Notes 9, 10, 11, 12, and 13) (Table 3).

3.6. Pharmacokinetic Study

1. Three male Sprague-Dawley rats, 8 weeks old. Keep rats in plastic cages with free access to standard rat diet and water. Maintain the room temperature at 23 ± 3°C, relative humidity of 50 ± 10%, and an approximately 12/12-h light/dark cycle.
2. Formulate intravenous dose solutions (10 mg/kg) in PEG 400:distilled water:dimethyl sulfoxide (DMSO) at a ratio of 40:55:5.
3. Collect blood (about 0.2 mL) at predose, 0.033, 0.167, 0.5, 1, 2, 4, 8, 12, and 24 h after intravenous administration. Centrifuge blood samples immediately, collect plasma and store at −70°C until analysis.
4. Use a non-compartmental method using the nonlinear least squares regression program WinNonlin (Pharsight, Mountain View, CA) to calculate the pharmacokinetic parameters. Calculate the area under the plasma concentration-time curve from time zero to the last measured concentration ($AUC_{0\rightarrow last}$) and to infinite time ($AUC_{0\rightarrow\infty}$) by adding extrapolated area. Obtain the terminal elimination half-life ($t_{1/2}$), total body clearance (CL), volume of distribution at steady state (Vss), and mean residence time (MRT) for KR-62980 using individual plasma concentration-time profiles (see Note 14) (Fig. 4).

Table 3
Stability of KR-62980 in rat plasma ($n=5$)

Condition tested	QCL (0.03 μg/mL)			QCM (0.5 μg/mL)			QCH (8 μg/mL)		
	Mean	RSD[a] (%)	RE[b] (%)	Mean	RSD (%)	RE (%)	Mean	RSD (%)	RE (%)
Short-term stability									
Control samples (freshly prepared)	–	7.9	–	–	5.2	–	–	4.3	–
Freeze-thaw (−80°C, 3 cycle)	0.033	0.9	11.4	0.429	5.3	−14.2	6.992	5.7	−12.6
Bench (room temperature, 1 day)	0.003	12.2	−90.2	0.029	9.6	−94.2	0.688	11.7	−91.4
Refrigerator (4°C, 1 day)	0.009	6.0	−70.2	0.160	7.9	−68.0	4.584	7.1	−42.7
Freezer (−20°C, 1 day)	0.031	1.2	2.9	0.512	5.8	2.4	8.152	4.0	1.9
Freezer (−80°C, 1 day)	0.031	4.3	3.1	0.477	14.2	−4.6	7.448	4.8	−6.9
post-preparative stability (4°C, 1 day)	0.031	0.8	4.3	0.491	7.8	−1.8	7.928	3.3	−0.9
post-preparative stability (4°C, 1 week)	0.032	8.4	5.7	0.482	1.2	−3.6	7.136	6.6	−10.8
post-preparative stability (room temperature, 1 day)	0.027	2.3	−9.2	0.523	7.6	4.6	7.312	2.0	−8.6
Long-term stability									
Freezer (−80°C, 30 days)	0.030	4.0	−0.2	0.484	4.9	−3.2	7.434	4.4	−7.1
Freezer (−20°C, 30 days)	0.023	9.4	−22.7	0.375	4.0	−25.0	5.432	3.0	−32.1

[a]RSD (%) = standard deviation of the concentration/mean concentration × 100
[b]RE (%) = (calculated concentration − theoretical concentration)/theoretical concentration × 100

4. Notes

1. For validation, blank plasma samples should be obtained from six different sources.
2. The sample is extracted with a suitable solvent to remove the sample matrix and extract the KR-62980. Ethyl acetate, a relatively polar solvent, extracts both target compounds and polar impurities, with a stable baseline and no interference peaks appearing in MRM mode.
3. To ensure correct identification and to prevent false positives, two or more different ions should be selected for each analyte, and the peak area ratio of two selected ions (quantitative ion and confirmative ion) should be compared with that of the standard compound. A positive identification yields a peak area ratio within 15–20% of that of the standard compound.
4. The peak shape of KR-62980 and IS is sharp, but the retention time tend to decline, by increasing the amount of organic solvent in the mobile phase (see Fig. 2).
5. The calibration curves are generated by a linear least squares regression analysis of the KR-62980/IS peak area ratio versus the amount of spiked KR-62980. The correlation coefficient (R^2) for KR-62980 is greater than 0.99, indicating excellent linearity.
6. Intra-day accuracy ranged from −7.1 to 13.2% (defined as RE), with RSD values ranging from 1.2 to 3.4%, and inter-day accuracy ranged from −5.4 to 1.1%, with RSD values ranging from 8.5 to 10.6%, indicating excellent accuracy (see Table 1).
7. The mean matrix effects at KR-62980 concentrations of 0.03, 0.5, and 8 μg/mL are 90.0, 94.1, and 96.5%, respectively, which are well within acceptable limits (<20%); the mean percentage recoveries at the three concentrations are 96.4, 88.6, and 85.4%, respectively. Process efficiency obtained by this method showed adequate efficiency for KR-62980 in plasma matrix (82.4–86.8%) (see Table 2).
8. The stock solution of KR-62980 (1 mg/mL) was investigated for 3 weeks at −20°C; KR-62980 was stable, ranging from 92.5 to 105.4%.
9. The concentration of KR-62980 (8 μg/mL) at room temperature and 4°C for 1 day decreased significantly, to 8.6 and 57.3%, respectively (see Table 3).

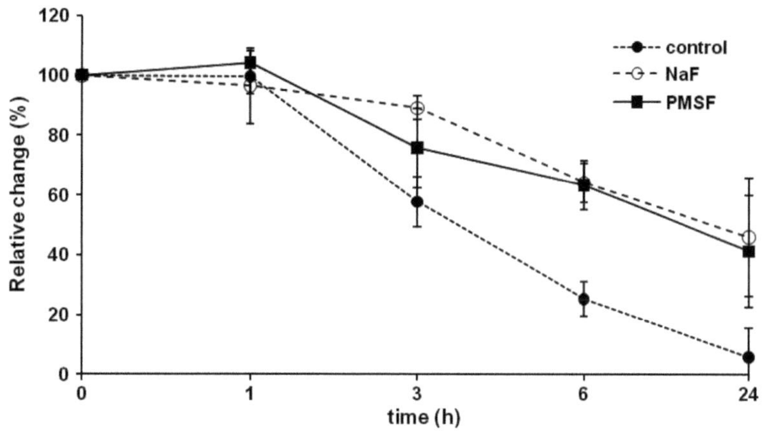

Fig. 3. Stability of KR-62980 in rat plasma samples (0.5 μg/mL level) at room temperature without (control), with NaF and PMSF (n=5).

10. The stability of a drug in plasma depends on the labile functional groups, such as esters, amide, carbamates, lactones, lactams, sulfates, sulfonamides, phosphates, peptides, and peptide mimetics (see ref. (11, 12)). Such functional groups are vulnerable to hydrolysis by plasma enzymes.
11. KR-62980 sample (0.5 μg/mL level in rat plasma) stored for 1 day at room temperature without (control) or with NaF (10 mg/mL) or PMSF (0.18 mg/mL) (see Fig. 3).
12. After 1 h, the concentration of KR-62980 appreciably decreased in control. However, the degradation rate of KR-62980 in rat plasma decreased with addition of NaF or PMSF, esterase inhibitors.
13. One of the main reasons for the poor stability of KR-62980 in rat plasma is believed to be the ester groups of this compound. Therefore, careful sample storage conditions (below -80°C) and sample preparation performed within 1 h are necessary.
14. The concentration of KR-62980 was readily measurable in plasma samples collected up to 24 h post-dose. The terminal half-life and AUC_{∞} values of KR-62980 were 2.33 ± 0.43 h and 3.59 ± 0.43 μg h/mL. The $AUC_{0\rightarrow last}/AUC_{0\rightarrow\infty}$ ratio was higher than 99% for all subjects (mean values, 99.8 ± 0.1%). The systemic clearance (CL), volume of distribution (Vss), and MRT was 2.82 ± 0.33 L/h/kg, 5.40 ± 0.90 L/kg, and 1.91 ± 0.54 h, respectively (see Fig. 4).

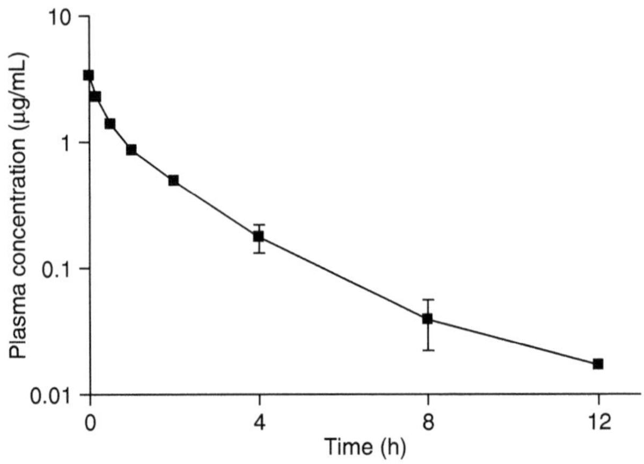

Fig. 4. Mean plasma concentration-time plot of KR-62980 after intravenous administration of KR-62980 at 10 mg/kg to rats (mean ± standard deviation ($n = 3$)).

Acknowledgments

This research was supported by the Center for Biological Modulators of the 21st Century Frontier R&D program, the Ministry of Science and Technology, Korea, and the Ministry of Knowledge Economy.

References

1. Calnek DS, Mazzella L, Roser S et al (2003) Peroxisome proliferator-activated receptor γ ligands increase release of nitric oxide from endothelial cells. Arterioscler Thromb Vasc Biol 23:52–57
2. Odegaard JI, Ricardo-Gonzalez R, Goforth M et al (2007) Macrophage-specific PPARc controls alternative activation and improves insulin resistance. Nature 447:1116–1121
3. Chui PC, Guan H, Lehrke M et al (2005) PPARγ regulates adipocyte cholesterol metabolism via oxidized LDL receptor. J Clin Invest 115:2244–2256
4. Rangwala SM, Rhoades B, Shapiro J et al (2003) Genetic modulation of PPARγ phosphorylation regulates insulin sensitivity. Dev Cell 5:657–663
5. Brown KK, Henke B, Blanchard J et al (1999) A novel n-aryl tyrosine activator of peroxisome proliferator-activated receptor-γ reverses the diabetic phenotype of the Zucker diabetic fatty rat. Diabetes 48:1415–1424
6. Adams M, Montague C, Prins J et al (1997) Activators of peroxisome proliferator–activated receptorγ have depot-specific effects on human preadipocyte differentiation. J Clin Invest 100:3149–3153
7. Ye J, Doyle P, Iglesias M et al (2001) Peroxisome proliferator–activated receptor (PPAR)-α activation lowers muscle lipids and improves insulin sensitivity in high fat–fed rats. Diabetes 50:411–417
8. Kim KR, Lee J, Kim S et al (2006) KR-62980: a novel peroxisome proliferator-activated receptor γ agonist with weak adipogenic effects. Biochem Pharmacol 72:446–454
9. Jung H, Lee J, Jang E et al (2009) Augmentation of PPARγ-TAZ interaction contributes to the anti-adipogenic activity of KR62980. Biochem Pharmacol 78:1323–1329
10. Kim MS, Song J, Roh H et al (2011) Determination of a peroxisome proliferator-activated receptor agonist, 1-(trans-methylimino-N-oxy)-6-(2-morpholinoethoxy-3-phenyl-1H-indene-2-carboxylic acid ethyl ester

(KR-62980) in rat plasma by liquid chromatography–tandem mass spectrometry. J Pharm Biomed Anal 54:121–126

11. Di L, Kerns E, Hong Y et al (2005) Development and application of high throughput plasma stability assay for drug discovery. Int J Pharm 297:110–119
12. Di L, Kerns EH (2008) Solution stability—plasma, gastrointestinal, bioassay. Curr Drug Metab 9:860–868

Chapter 24

Behavioral Paradigms to Evaluate PPAR Modulation in Animal Models of Brain Injury

Dana Greene-Schloesser, Caroline I. Schnegg, and Mike E. Robbins

Abstract

The use of behavioral testing has become an invaluable tool for assessing the efficacy of therapeutics for a variety of disorders of the central nervous system. This chapter will describe in detail several behavioral paradigms to evaluate the efficacy of PPAR agonists to modulate cognitive impairments in rodent models. When used together as a battery these procedures allow for a global assessment of cognition. These tests are explained in detail below, and include: (1) Novel Object Recognition (NOR), (2) Morris Water Maze (MWM), (3) Delay Match to Place (DMP), and (4) Cue Strategy.

Key words: PPAR agonists, Cognition, Novel object recognition, Morris water maze, Delay match to place, Cue strategy

1. Introduction

PPAR agonists appear to have therapeutic potential in the treatment of neurologic diseases including: Alzheimer's disease (1), multiple sclerosis (2, 3), Parkinson's disease (4–7), excitotoxic injury (8, 9), AIDS dementia (10, 11), brain tumors including glioma (12, 13), and radiation-induced brain injury (14, 15). The therapeutic implications for PPAR agonists necessitate the use of animal models and procedures to evaluate potential efficacy.

Several behavioral paradigms have been successfully used to evaluate the efficacy of PPAR agonists to modulate cognitive impairments. These involve procedures that allow for the interrogation of several different brain regions in order to globally assess cognition. These tests are explained in detail below, and include: (1) Novel Object Recognition (NOR), (2) Morris Water Maze (MWM), (3) Delay Match to Place (DMP), and (4) Cue Strategy.

Mostafa Z. Badr and Jihan Youssef (eds.), *Peroxisome Proliferator-Activated Receptors (PPARs): Methods and Protocols*, Methods in Molecular Biology, vol. 952, DOI 10.1007/978-1-62703-155-4_24, © Springer Science+Business Media New York 2013

2. Novel Object Recognition (NOR) Task

Rodents are attracted by novel objects and engage readily in exploration using their vibrissae, nose, and forepaws. The premise of the NOR is that rodents are able to discriminate a novel from a familiar object; they can also detect a mismatch between the past and present location of a familiar object (object in place recognition), and can discriminate different objects in terms of which object has been encountered earlier (temporal order memory) (16). The version of NOR task described below is mainly a perirhinal cortex (PRh)-dependent version (17).

2.1. Testing Apparatus

The arena should be sufficiently large enough to allow for ample exploration. For our rat studies we use an opaque white polypropylene container that is 24″ × 18″ × 20″. A video camera is mounted directly above the arena. The NOR can be scored using an automated system; however, care must be taken to ensure that only true exploration is counted. The standard practice is to record "object exploration" when the animal's nose is within 1–2 cm of the object and the vibrissae are moving. Object exploration is not recorded when the animal uses the object to prop itself up while looking around and sniffing, or when the rat is touching the object with some part of the body but headed in another direction. There are systems, such as the Noldus EthoVision® (Leesburg, Virginia, USA) that allow for precise tracking of the nose in order to collect object exploration data, as well as multiple parameters to evaluate other non-object directed behavior (e.g., rearing, grooming, etc.). Several other vendors market similar systems that would be suitable for acquisition and analysis of the NOR task (Stoelting Co., Wood Dale, IL, USA; TSE Systems Inc, Chesterfield, MO, USA). Alternatively, a trained observer, who is blind to group membership, can easily conduct the scoring from the previously recorded video. When setting up your testing paradigm, special consideration should be given to the color of the rats to be tested, as the tracking systems typically need contrasting backgrounds for successful video acquisition (e.g., we used brown rats in a white arena). The objects should be secured to the arena floor so the animal cannot displace them during exploration. We use Velcro patches for this purpose, which also ensures that the objects are always placed in the same location within the arena.

2.2. Methodology

The NOR task is comprised of three phases: familiarization, delay, and test phases. Before testing, each animal is habituated to the testing arena by allowing it to explore the empty arena for 5 min a day for 4 days. After habituation, the following sessions are divided into a sample phase and a test phase. During the sample phase (test 1), the rodent is allowed to explore and become familiar with

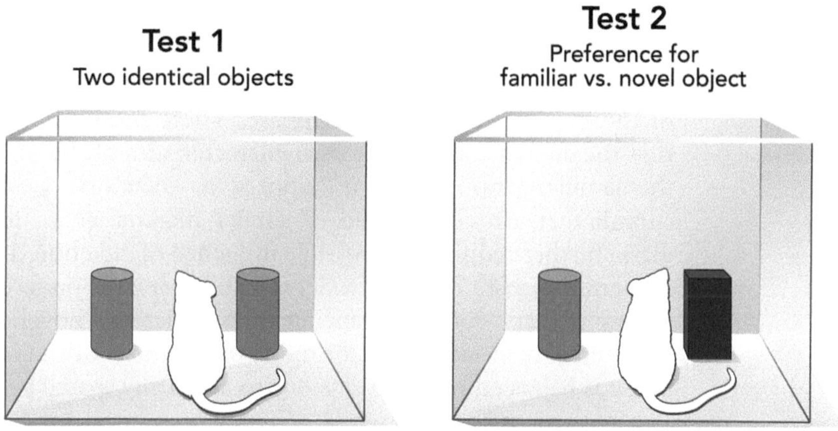

Fig. 1. The novel object test phases. Test 1 is the sample phase: the rodent is allowed to explore and become familiar with two identical 3-D objects located within the testing arena for 3 min. A delay is introduced between test phases and the animal is returned to its home cage. A test phase (test 2) follows in which one of the two objects from test 1 is paired with a novel object; the subject is left in the arena for a total of 3 min during which activity is recorded.

two identical 3-D objects located within the testing arena for 3 min (Fig. 1). Following the sample phase, a delay is introduced (we use a 5 min delay) during which the animal is removed from the arena and returned to its home cage. A test phase (test 2) follows in which one of the two objects from the sample phase is paired with a novel object (Fig. 1). The subject is left in the arena for a total of 3 min during the test phase and activity is recorded. The objects are secured to the arena floor so the subject cannot displace them during exploration. Velcro patches can be used for this purpose; these patches also serve as fiduciary marks that ensure the objects are always placed in the same location within the arena. Recognition memory is inferred from the preferential exploration (measured as time) of the novel object compared with that of the familiar object.

The basic measurement is the time the rat spent exploring an object, defined as placing his/her nose within ≤2 cm of the object and actively exploring it. The following parameters are generally obtained, E1: the total time spent exploring the identical objects A1 and A2 in the sample phase (3 min); E2: the total time spent exploring object A3 (identical to A1 and A2) and the novel object B1 in test phase (3 min); and D1: the index of discrimination defined as the difference in time spent exploring objects A3 and B1 in the test phase (i.e., B1—A3). Rats with an E1 <10 s are eliminated from further analysis due to insufficient exploratory behavior. The discrimination ratio (D1/E2), a measure of the rat's recognition memory, is then calculated, and the average discrimination ratio for each treatment group compared.

2.3. Special Considerations

Care must be taken to ensure that scent-marks and trails are removed between each phase and animal, which can be done with thorough cleaning of the arena as well as by using identical copies of the objects during the test phase. These measures will ensure that the animal is recalling visual characteristics of the object from the familiarization phase as opposed to olfactory cues. Pairing stimuli that are constructed of similar material (e.g., glass with glass) further reduces the possible influence of olfaction during the recognition test. Because rodents exhibit innate preferences for certain objects, counterbalancing the familiar and novel objects is required. The position (left or right) of the novel object in the test phase is balanced between sessions to avoid any spatial preference.

3. Morris Water Maze (MWM) Task

The MWM has been the most extensively used and accepted task to assess and compare spatial learning and memory in rodents (18). The MWM task is a challenging procedure for rodents that employs a variety of sophisticated mnemonic processes (19). In order to successfully navigate the maze and find the hidden platform, the animal must acquire and spatially localize relevant visual cues and then process, consolidate, retain, and then retrieve these cues (18, 19). The general processes used for "visuospatial navigation" in rats also appear to contribute considerably to human day-to-day cognitive processes and thus the MWM task has been used to investigate models relevant to the study of a variety of neurological illnesses where cognition is impaired (e.g., Alzheimer's disease, Parkinson's disease, schizophrenia, and radiation-induced brain injury) (6, 20–28).

3.1. Invisible Platform Tests

In the classic design, rats are repetitively released in a circular pool filled with opaque water, in which the only escape opportunity is provided by a small submerged platform. The platform is invisible and occupies a fixed position throughout training. To escape, rats must learn and remember the position of the platform relative to visible cues placed around the arena.

3.1.1. Testing Apparatus

MWM testing should be conducted in a large circular pool (rats, diameter: 68″, height: 24″; mice, diameter: 40″, height: 12″) made of plastic. The pool should be filled to a depth of ~30 in. of water (mice, ~10″) to cover a white 4.25″ diameter platform (mice, ~2″) and maintained at 25°C ± 1.0°C. The platform is submerged approximately 1.0 cm below the surface of the water and placed in the center of the fourth quadrant. The water is rendered opaque by the addition of a white food coloring agent. The pool should be

Fig. 2. The standard set-up for the Morris water maze (MWM). The subjects are repetitively released in a circular pool filled with opaque water, in which the only escape opportunity is provided by a small submerged platform. The platform is invisible and occupies a fixed position throughout training. To escape, subject must learn and remember the position of the platform relative to visible contrasting cues placed on a curtain that surrounds the arena. The curtain also allows for the experimenter to be hidden from the subject undergoing testing, reducing distraction during testing. A camera is mounted directly above the arena to allow for automated tracking and data collection.

located in an isolated room, with the pool surrounded by a curtain. Attached to the curtain should be a number of extramaze visual cues, including highly visible contrasting geometric images (squares, triangles, circles, etc.). The curtain also allows for the experimenter to be hidden from the rat undergoing testing, reducing distraction of the rat during testing (Fig. 2). Swimming activity of each rat is monitored via a video camera mounted overhead, which relays information including latency to the platform, total distance traveled, time and distance spent in each quadrant, etc., to a video tracking system. When setting up a testing paradigm, special consideration should be given to the color of the rats to be tested, as the tracking systems typically need contrasting backgrounds for successful video acquisition (e.g., we used brown rats in a white pool, with white opaque water). Our laboratory has used the Noldus EthoVision® system (Leesburg, VA, USA) with good results; however, there are several other vendors which market similar systems (Columbus Instruments, Columbus, OH, USA; TSE Systems Inc., Stoelting Co.).

3.1.2. Methodology

Rats are tested individually by being placed into various quadrants of the pool, and the time elapsed and/or the distance traversed to reach the hidden platform recorded. With each subsequent entry into the maze the rats progressively become more efficient at locating the platform, thus escaping the water by learning the location of the platform relative to the distal visual cues. The learning curves are then compared between groups. An illustration of a typical MWM set-up (as used in our laboratory) is shown in Fig. 2. The MWM consists of two testing phases; a standard phase followed by a reversal phase. In the standard phase spatial learning and reference memory are measured. The escape platform is placed in quadrant 4 throughout all training trials. At the beginning of each training trial, the rats are introduced into the quadrants in a systematically random pattern, and each rat's performance is recorded using an automated tracking system during the 90 seconds (s) trial. Two training trials are given each day for 2 days followed by the third day which includes one training trial followed by one probe trial (1 block; Table 1). During each training trial, the rat is allowed to search for 90 s to locate the platform. Once the rat has found the platform it is allowed a 30 s "rest" period, in order for the rat to obtain spatial visual cues after each trial before being removed. If the rat does not locate the hidden platform, then a 90 s time is recorded for that rat and it is led and placed on the platform for 30 s, in order to ensure all rats have the opportunity to obtain spatial cues equally. After the completion of each trial, the rat is dried with a towel to remove excess water and returned to their home cage. The measurements acquired during this phase include: (1) the latency to the platform, (2) total distance traveled, and (3) the time and distance spent in each quadrant. The latency and distance traveled should decrease over the sample blocks, indicating acquisition of the task (Fig. 3). During the probe trial, the platform is lowered beyond reach; the rat is then allowed to search for 30 s. The time elapsed and distance swam in the previous platform quadrant (quadrant 4) is recorded. In addition to measuring the quadrant, an annulus ring can be placed around the previous platform location (via computer program) to localize the area more precisely. The number of crossings through this region is typically recorded and compared between groups. This schedule is repeated for four blocks.

In the reversal phase of the MWM, the platform location is moved to the opposite quadrant (quadrant 2) with the visual cues

Table 1
The experimental protocol and timing for the standard Morris water maze. The experimental paradigm is 13 days which are divided between four blocks, each containing of 3 days. Each day consists of two trails a day, with the third and final day of each block consisting of one trial (trial 5) and a probe trial. The very last trial (Day 13) is the reversal, where the platform is moved to the opposite quadrant; this is also Day 1 of the delayed-match-to-place task

Standard Morris Water Maze

Block 1	Day 1	Trial 1
		Trial 2
	Day 2	Trial 3
		Trial 4
	Day 3	Trial 5
		Probe
Block 2	Day 4	Trial 1
		Trial 2
	Day 5	Trial 3
		Trial 4
	Day 6	Trial 5
		Probe
Block 3	Day 7	Trial 1
		Trial 2
	Day 8	Trial 3
		Trial 4
	Day 9	Trial 5
		Probe
Block 4	Day 10	Trial 1
		Trial 2
	Day 11	Trial 3
		Trial 4
	Day 12	Trial 5
		Probe
	Day 13	Reversal

remaining in the same positions. This task measures spatial bias for the previous platform location and requires the rats to inhibit their previously learned response and form a new strategy to find the escape platform in a novel location. This is accomplished by measuring the percentage of time spent (and distance swam) in the previous target quadrant as well as the number of crossings over the previous platform location. These assessments provide a second estimate of the strength and accuracy of the memory of the previous platform location.

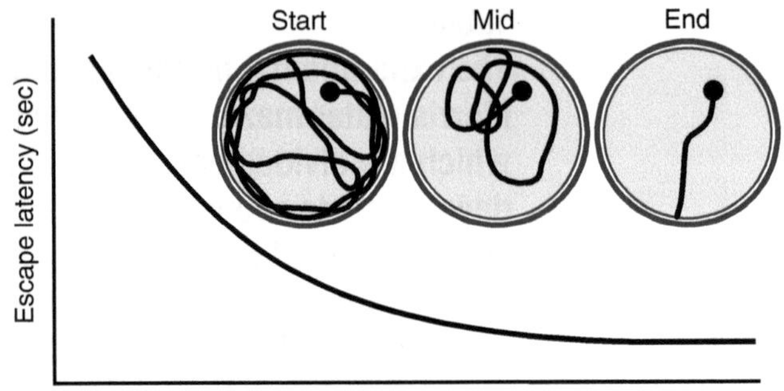

Fig. 3. A typical measurement of task acquisition of the MWM. The representative plot shows the expected trend for escape latency (time to locate the hidden platform) over the duration of the trials; escape latency decreases over time indicating task acquisition. The circular diagrams represent the MWM arena and the relative path lengths from the beginning, middle, and end of the training trials. Overall path lengths decrease and the path becomes more direct as the subject progresses through the trials.

3.2. Visible Platform Tests

A visible platform test is performed to determine if alterations in visual acuity confound the data analysis, since the successful completion of the task depends on the use of visual cues. However, care must be exercised to ensure that certain behaviors that might be interpreted as impaired visual acuity are taken into account (19). For example, the absence of search behaviors or thymgotaxis (swimming constantly along the perimeter of the pool) might be misinterpreted as visual deficits since the animal does not locate the platform in a reasonable period of time (19). Thus, animals must make attempts to cross the pool and then be impaired at locating the platform in order for an interpretation of visual deficits to be made (19).

Immediately following the final probe trial, the platform is placed in the pool in the quadrant located diametrically opposite the original position. A visible cue (we use an inflated glove) is attached to the platform which is raised slightly above the surface of the water. Each rat is allowed one trial in order to acclimate to the new set of conditions and locate the platform visually. This is accomplished by lowering the rat into the water into quadrant 4 and allowing the rat to locate the platform; no time limit is placed on this first trial. Once the platform is located, allow the rat 30 s on the platform. The rat should then immediately be given a second trial in the same manner and the latency to the platform measured as a comparison of visual acuity.

4. Delay Match to Place (DMP)

The DMP version of the MWM is sometimes referred to as a working memory protocol. In this DMP version, the platform is moved to a new location each day resulting in the animal not knowing

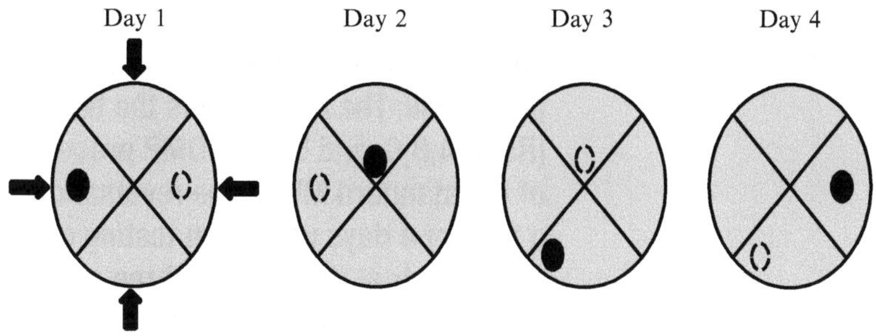

Fig. 4. The delay match to place (DMP) task set-up for MWM. Following the standard MWM, the DMP version can be performed. On Day 1 of the trial, the submerged platform is moved to the opposite quadrant (dark circle) from where it was in the previous standard MWM trial (dotted circle). The subject is introduced to the arena at various quadrants throughout the training and trials on Day 1 (*arrows*) and for all other days (Days 2–4). The platform location is changed each day (dark circle is new location; dotted circle is location from previous day).

where the platform is hidden on trial 1 of each day (Fig. 4). Once the animal finds the platform, it can generally encode this new location in one trial. This is shown by the animal finding the platform much faster on trial 2 and subsequent trials of that day (29).

In our studies, we use a DMP procedure that starts on the last day of the standard MWM (Table 1). The reversal day of the standard MWM is the first day of the DMP and lasts for a total of 4 days. Each DMP testing day consists of four trials, a sample trial (trial 1) and three match trials (trial 2–4; Table 2, Fig. 4). The hidden escape platform in the sample phase each day is located 2 cm under the water surface, in a unique location (Fig. 4), and remains in the same location for the remaining match trials for that day. The delays between trials are constant within days and are as follows across the 4 days of testing: 20 min, 15 s, and 15 s (Table 2).

5. Cue Strategy Task

A spatial/place strategy, which involves the flexible use of spatial cues, is dependent upon intact hippocampal circuitry. Conversely, a response/cue strategy, or the formation of associations between discreet cues and behavioral responses, is dependent upon intact striatal circuitry.

Strategy preference can be assessed 24 h after the last day of DMP training. A visible platform is placed in the quadrant opposite of the quadrant where the hidden platform was previously located. The rats are given two 60 s strategy probe trials.

Table 2
The experimental protocol and timing of the delayed match to place task. The table shows the timing of the DMP for four rats (Rats: A,B,C, and D). The DMP procedure starts on the last day of the standard MWM, the reversal day. DMP is performed over a total of 4 days with each testing day consisting of four trials: a sample trial (trial 1) and three match trials (trial 2–4). The hidden escape platform in the sample phase each day is located 2 cm under the water surface, in a unique location and remains in the same location for the remaining match trials for that day. The delays between trials are constant within the 4 days of testing and are as follows: 20 min, 15 s, and 15 s

Delayed match to place version of the Morris Water Maze — **Starts on reversal (Day 13) of MWM**

40 min Testing period			Trial
20 min Interval	0 --> 5	Rat A	1
	5 --> 10	Rat B	1
	10 --> 15	Rat C	1
	15 --> 20	Rat D	1
2 min			
15 s Interval	20 --> 25	Rat A	2
			3
			4
	25 --> 30	Rat B	2
			3
			4
	30 --> 35	Rat C	2
			3
			4
	35 --> 40	Rat D	2
			3
			4

Start locations are on either side of the tank, equidistant from the visible cue platform and the prior hidden platform location. A "place strategy" is recorded if the rat crossed the annulus of the prior hidden platform location before escaping to the visible platform. The annulus is defined as a 5 cm perimeter around the prior hidden escape platform location. A "cue strategy" is recorded if the rat did not cross the prior hidden platform annulus before swimming to the visible platform. Data are scored by a blinded observer using recorded swim paths (Ethovision, Noldus).

Acknowledgments

Supported by National Institutes of Health grants CA112593 and CA113267.

References

1. Watson GS, Cholerton BA, Reger MA, Baker LD, Plymate SR, Asthana S, Fishel MA, Kulstad JJ, Green PS, Cook DG, Kahn SE, Keeling ML, Craft S (2005) Preserved cognition in patients with early Alzheimer disease and amnestic mild cognitive impairment during treatment with rosiglitazone: a preliminary study. Am J Geriatr Psychiatry 13:950–958
2. Racke MK, Gocke AR, Muir M, Diab A, Drew PD, Lovett-Racke AE (2006) Nuclear receptors and autoimmune disease: the potential of PPAR agonists to treat multiple sclerosis. J Nutr 136:700–703
3. Schmidt S, Moric E, Schmidt M, Sastre M, Feinstein DL, Heneka MT (2004) Anti-inflammatory and antiproliferative actions of PPAR-+¦ agonists on T lymphocytes derived from MS patients. J Leukoc Biol 75:478–485
4. Carta AR, Pisanu A, Carboni E (2011) Do PPAR-gamma agonists have a future in Parkinson's disease therapy? Parkinsons Dis 2011:689181
5. Chaturvedi RK, Beal MF (2008) PPAR: a therapeutic target in Parkinson's disease. J Neurochem 106:506–518
6. Kumar P, Kaundal RK, More S, Sharma SS (2009) Beneficial effects of pioglitazone on cognitive impairment in MPTP model of Parkinson's disease. Behav Brain Res 197: 398–403
7. Schintu N, Frau L, Ibba M, Caboni P, Garau A, Carboni E, Carta AR (2009) PPAR-gamma-mediated neuroprotection in a chronic mouse model of Parkinson's disease. Eur J Neurosci 29:954–963
8. Luna-Medina R, Cortes-Canteli M, Sanchez-Galiano S, Morales-Garcia JA, Martinez A, Santos A, Perez-Castillo A (2007) NP031112, a thiadiazolidinone compound, prevents inflammation and neurodegeneration under excitotoxic conditions: potential therapeutic role in brain disorders. J Neurosci 27: 5766–5776
9. Glatz T, Stock I, Nguyen-Ngoc M, Gohlke P, Herdegen T, Culman J, Zhao Y (2010) Peroxisome-proliferator-activated receptors gamma and peroxisome-proliferator-activated receptors beta/delta and the regulation of interleukin 1 receptor antagonist expression by pioglitazone in ischaemic brain. J Hypertens 28:1488–1497
10. Huang W, Eum SY, Andras IE, Hennig B, Toborek M (2009) PPAR{alpha} and PPAR{gamma} attenuate HIV-induced dysregulation of tight junction proteins by modulations of matrix metalloproteinase and proteasome activities. FASEB J 23:1596–1606
11. Ramirez SH, Heilman D, Morsey B, Potula R, Haorah J, Persidsky Y (2008) Activation of peroxisome proliferator-activated receptor gamma (PPAR{gamma}) suppresses Rho GTPases in human brain microvascular endothelial cells and inhibits adhesion and transendothelial migration of HIV-1 infected monocytes. J Immunol 180:1854–1865
12. Drukala J, Urbanska K, Wilk A, Grabacka M, Wybieralska E, Del Valle L, Madeja Z, Reiss K (2010) ROS accumulation and IGF-IR inhibition contribute to fenofibrate/PPARalpha -mediated inhibition of glioma cell motility in vitro. Mol Cancer 9:159
13. Reiss K, Urbanska K, DelValle L, Mencel PJ (2008) PPAR{alpha} agonist fenofibrate inhibits IGF-I - mediated growth and DNA repair responses and sensitizes human glioblastoma cells to cisplatin. ASCO Meeting Abstracts 26:13020
14. Ramanan S, Kooshki M, Zhao W, Hsu FC, Riddle DR, Robbins ME (2009) The PPARalpha agonist fenofibrate preserves hippocampal neurogenesis and inhibits microglial activation after whole-brain irradiation. Int J Radiat Oncol Biol Phys 75:870–877
15. Ramanan S, Kooshki M, Zhao W, Hsu FC, Robbins ME (2008) PPARalpha ligands inhibit radiation-induced microglial inflammatory responses by negatively regulating NF-kappaB and AP-1 pathways. Free Radic Biol Med 45:1695–1704
16. Dere E, Huston JP, De Souza Silva MA (2007) The pharmacology, neuroanatomy and neurogenetics of one-trial object recognition in rodents. Neurosci Biobehav Rev 31: 673–704
17. Winters BD, Saksida LM, Bussey TJ (2008) Object recognition memory: neurobiological

mechanisms of encoding, consolidation and retrieval. Neurosci Biobehav Rev 32: 1055–1070

18. Morris R (1984) Developments of a water-maze procedure for studying spatial learning in the rat. J Neurosci Methods 11:47–60
19. Jr Terry AV (2009) Spatial navigation (water maze) tasks. In: Buccafusco JJ (ed) Methods of behavior analysis in neuroscience. CRC Press, Boca Raton
20. Da Cunha C, Wietzikoski S, Wietzikoski EC, Silva MH, Chandler J, Ferro MM, Andreatini R, Canteras NS (2007) Pre-training to find a hidden platform in the Morris water maze can compensate for a deficit to find a cued platform in a rat model of Parkinson's disease. Neurobiol Learn Mem 87:451–463
21. Janus C (2004) Search strategies used by APP transgenic mice during navigation in the Morris water maze. Learn Mem 11:337–346
22. Scearce-Levie K (2011) Monitoring spatial learning and memory in Alzheimer's disease mouse models using the Morris water maze. Methods Mol Biol 670:191–205
23. Raber J, Rola R, LeFevour A, Morhardt D, Curley J, Mizumatsu S, VandenBerg SR, Fike JR (2004) Radiation-induced cognitive impairments are associated with changes in indicators of hippocampal neurogenesis. Radiat Res 162:39–47
24. Hagan JJ, Jones DNC (2005) Predicting drug efficacy for cognitive deficits in schizophrenia. Schizophr Bull 31:830–853
25. Wisman LAB, Sahin G, Maingay M, Leanza G, Kirik D (2008) Functional convergence of dopaminergic and cholinergic input is critical for hippocampus-dependent working memory. J Neurosci 28:7797–7807
26. Rola R, Raber J, Rizk A, Otsuka S, VandenBerg SR, Morhardt DR, Fike JR (2004) Radiation-induced impairment of hippocampal neurogenesis is associated with cognitive deficits in young mice. Exp Neurol 188:316–330
27. Shi L, Adams MM, Long A, Carter CC, Bennett C, Sonntag WE, Nicolle MM, Robbins M, D'Agostino R, Brunso-Bechtold JK (2006) Spatial learning and memory deficits after whole-brain irradiation are associated with changes in NMDA receptor subunits in the hippocampus. Radiat Res 166:892–899
28. Shi L, Olson J, D'Agostino R, Linville C, Nicolle MM, Robbins ME, Wheeler KT, Brunso-Bechtold JK (2011) Aging masks detection of radiation-induced brain injury. Brain Res 1385:307–316
29. Steele RJ, Morris RG (1999) Delay-dependent impairment of a matching-to-place task with chronic and intrahippocampal infusion of the NMDA-antagonist D-AP5. Hippocampus 9: 118–136

ERRATUM

In Vivo Studies of PPAR-Chromatin Interactions: Chromatin Immunoprecipitation for Single-Locus and Genomewide Analyses

Till Adhikary and Rolf Müller

Mostafa Z. Badr and Jihan Youssef (eds.), *Peroxisome Proliferator-Activated Receptors (PPARs): Methods and Protocols*, Methods in Molecular Biology, vol. 952, DOI 10.1007/978-1-62703-155-4, pp. 175–185
© Springer Science+Business Media New York 2013

DOI 10.1007/978-1-62703-155-4_25

The publisher regrets the following typographical errors in Chapter 12:

Page 177: The reversion mix should be made from volumes of microliters, not milliliters (16 microliters instead of 16 milliliters, in two locations, and 8 microliters instead of 8 milliliters).

Page 177: The same buffer recipe contains an additional error. It should read ten micrograms of RNase A, and not ten milligrams.

Page 184: The dot in front of the square root should be omitted.

The online version of the original chapter can be found at
http://dx.doi.org/10.1007/978-1-62703-155-4_12

INDEX

Mostafa Z. Badr and Jihan Youssef (eds.), *Peroxisome Proliferator-Activated Receptors (PPARs): Methods and Protocols*, Methods in Molecular Biology, vol. 952, DOI 10.1007/978-1-62703-155-4, © Springer Science+Business Media New York 2013

D

E

F

G

H

I

Q

R

S

T

U

V

W

MIX
Papier aus verantwortungsvollen Quellen
Paper from responsible sources
FSC® C105338

If you have any concerns about our products,
you can contact us on
ProductSafety@springernature.com

In case Publisher is established outside the EU,
the EU authorized representative is:
Springer Nature Customer Service Center GmbH
Europaplatz 3, 69115 Heidelberg, Germany

Printed by Libri Plureos GmbH
in Hamburg, Germany